Solutions Manual to Accompany

Shriver & Atkins

Inorganic Chemistry

Fifth Edition

Michael E. Hagerman
Union College

R. Chris Schnabel
Eckerd College

Kandalam Ramanujachary
Rowan University

OXFORD
University Press

W. H. FREEMAN AND COMPANY
New York

Solutions Manual to Accompany Shriver & Atkins Inorganic Chemistry, Fifth Edition

Printed in the United States of America

First printing

Published, under license, in the United States and Canada by
W. H. Freeman and Company
41 Madison Avenue
New York, NY 10010

www.whfreeman.com

ISBN-13: 978-1-4292-5255-3
ISBN-10: 1-4292-5255-3

Published in the rest of the world by
Oxford University Press
Great Clarendon Street
Oxford, OX2 6DP
United Kingdom

www.oup.com

ISBN-13: 978-0-19-958595-3
ISBN-10: 0-19-958595-4

TABLE OF CONTENTS

COLLEGIUM DE
SOMERVILLE
DONEC
RURSUS
IMPLEAT
ORBEM

PREFACE

This book covers all of the self-tests and exercises in the textbook *Shriver & Atkins Inorganic Chemistry,* 5th edition, by Peter Atkins, Tina Overton, Jonathan Rourke, Mark Weller, and Fraser Armstrong. It is written to help you study the fascinating subject of inorganic chemistry. As you read each chapter in *Inorganic Chemistry* the *Solutions Manual* will help you advance your learning process by presenting complete solutions that permit a more detailed understanding by explicitly tying the exercises and solutions to the key points developed in the textbook. The solutions include nearly all of the figures and drawings posed in the exercises. They also include many other figures that will help you to visualize new concepts.

Learning about inorganic chemistry will require more than a quick reading of the text and a glance at the solutions. There is no substitute for reading and re-reading the relevant material. One way to study is to carefully reflect on the key points in the textbook and to re-copy your lecture notes within a few days of each lecture emphasizing these key points. Use the text boxes to help you to better understand the context of the key points described and be sure to work through the self-tests while you are reading to enhance your understanding. These strategies will alert you to concepts you may not yet understand fully. When studying for an exam or quiz, you may want to work through the self-tests and exercises for a second time. If your solution is not as complete as the one in the book, you should review relevant sections of the textbook. We have directed you to these relevant sections throughout the solutions guides. Keep in mind that a good way to be sure that you understand something is to be able to explain it to someone else.

We hope you enjoy applying your *Foundations* and *Descriptive* knowledge to the *Frontiers* section of the textbook, which focuses on emerging and applicable inorganic chemistry that bridges the exciting interdisciplinary areas of materials science, nanotechnology, catalysis, and bioinorganic and environmental chemistry. We hope that you find the guide useful as a catalyst for learning inorganic chemistry and recognizing how interconnected inorganic chemistry is to other important fields of study.

Michael E. Hagerman

R. Chris Schnabel

Kandalam Ramanujachary

ACKNOWLEDGMENTS

We would like to thank Duward F. Shriver and Peter Atkins for their insightful comments and valuable assistance during the preparation of the fifth edition of the *Solutions Manual*. We would also like to thank the many faculty members who used the prior editions and wrote to us with their suggestions and comments and the many students in our classes that have made important suggestions to improve this edition. Finally, we would like to thank the following people for their contributions: Brady Clapsaddle, Jon Rourke, Mark Weller, Tina Overton, Fraser Armstrong, Paul Salvador, Edward Stiefel, Thomas Spiro, and Mark Benvenuto.

Chapter 1 Atomic Structure

Chemists frequently use energy level diagrams like this one. You will find similar ones throughout the text (Figures 1.19 is an example). The vertical axis is an energy axis, with increasing energy from bottom to top. The short horizontal lines represent the energies of orbitals. Frequently, small arrows are used to represent electrons in those orbitals. In this diagram, the electron configuration of a ground-state boron atom is shown.

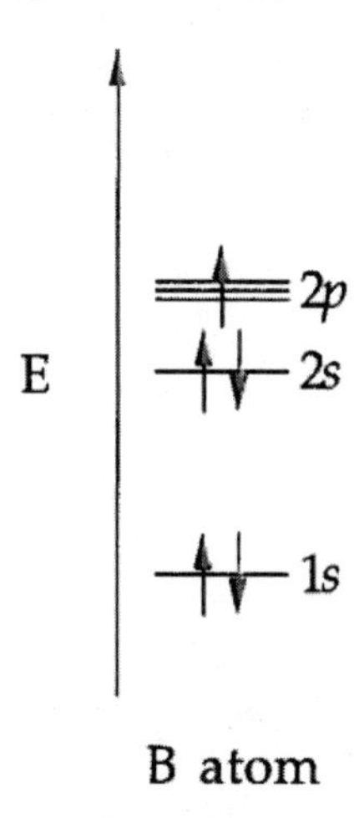

B atom

S1.1 **Neutron capture by $^{80}_{35}$Br?** You should use the same reasoning as in the example. Neutron capture by an atom involves an increase in mass number, in this case from 80 to 81. The atomic number remains the same, therefore, the atom in question is a bromine atom before *and* after the neutron capture. Any excess energy will appear as a photon in the γ-ray region of the electromagnetic spectrum. The balanced nuclear reaction is:

$$^{80}_{35}\text{Br} + {}^{1}_{0}\text{n} \longrightarrow {}^{81}_{35}\text{Br} + \gamma$$

S1.2 **Which set of orbitals are defined by $n = 3$, $l = 2$?** This is the third shell, and the subshell where $l = 2$ consists of the *d* orbitals. When $l = 2$, $m_l = -2, -1, 0, 1, 2$, which indicates there are five orbitals in the subshell. This set of quantum numbers indicates the start of the transition metals, beginning with Sc, which has its last electron in a 3*d* orbital in the ground state. Figure 1.15 shows the electron density maps for d orbitals.

S1.3 **How many radial nodes does the 5s orbital have?** The number of radial nodes is give by the expression: n-l-1. For the 5s orbital, $n = 5$ and $l = 0$. Therefore: 5-0-1 = 4. So there are 4 radial nodes in a 5s orbital. Remember, the first occurrence of a radial node for an s orbital is the 2s orbital, which has 1 radial node, the 3s has 2, the 4s has 3, and finally the 5s has 4. If you forget the expression for determining radial nodes, just count by a unit of one form the first occurrence of a radial node for that particular "shape" of orbital. Figure 1.9 shows the radial wavefuntions of 1s, 2s, and 3s hydrogenic orbitals. Where the lines intercept 0 on the y-axis is where there is a radial node for that particular orbital.

S1.4 **Probability of being close to the nucleus?** There is no figure showing the radial distribution functions for 3*p* and 3*d* orbitals, so you must reason by analogy. In the example, you saw that an electron in a *p* orbital has a smaller probability than in an *s* orbital of close approach to the nucleus, because an electron in a *p* orbital has a greater angular momentum than in an *s* orbital. Visually, Figure 1.12 shows this. The area under the graph represents where the electron has the highest probability of being found. The origin of the graph is the nucleus, so one can see that the 2s orbital, on average, spends more time closer to the nucleus than a 2p orbital. Similarly, an electron in a *d* orbital has a greater angular momentum than in a *p* orbital. In other words, $l(d) > l(p) > l(s)$. Therefore, an electron in a *p* orbital has a greater probability than in a *d* orbital of close approach to the nucleus.

S1.5 **Relative changes in Z_{eff}?** The configuration of the valence electrons, called the valence configuration, is as follows for the four atoms in question:

Li: $2s^1$ B: $2s^2 2p^1$

Be: $2s^2$ C: $2s^2 2p^2$

When an electron is added to the $2s$ orbital on going from Li to Be, Z_{eff} increases by 0.63, but when an electron is added to an empty p orbital on going from B to C, Z_{eff} increases by 0.72. The s electron already present in Li repels the incoming electron more strongly than the p electron already present in B repels the incoming p electron, because the incoming p electron goes into a new orbital. Therefore, Z_{eff} increases by a smaller amount on going from Li to Be than from B to C. However, extreme caution must be exercised with arguments like this, because the effects of electron–electron repulsions are very subtle. This is illustrated in period 3, where the effect is opposite to that just described for period 2.

S1.6 **Electron configuration of Ni and Ni^{2+}?** Following the example, for an atom of Ni with $Z = 28$ the electron configuration is:

$$\text{Ni}: \ 1s^2 2s^2 2p^6 3s^2 3p^6 3d^8 4s^2 \ \text{or} \ [\text{Ar}]3d^8 4s^2$$

Once again, the $4s$ electrons are listed last since the energy of the $4s$ orbital is higher than the energy of the $3d$ orbitals. Despite this ordering of the individual $3d$ and $4s$ energy levels for elements past Ca (see Figure 1.19), interelectronic repulsions prevent the configuration of an Ni atom from being $[Ar]3d^{10}$. For an Ni^{2+} ion, with two fewer electrons than an Ni atom but with the same Z as an Ni atom, interelectronic repulsions are less important. Because of the higher energy $4s$ electrons as well as smaller Z_{eff} than the $3d$ electrons, the 4s electrons are removed from Ni to form Ni^{2+}, and the electron configuration of the ion is:

$$\text{Ni}^{2+}: \ 1s^2 2s^2 2p^6 3s^2 3p^6 3d^8 \ \text{or} \ [\text{Ar}]3d^8$$

S1.7 **What element has the electron configuration of $1s^2 2s^2 2p^6 3s^2 3p^6 4s^2$?** The valence electrons are in the $n = 4$ shell. Therefore the element is in period 4 of the periodic table. It has two valence electrons that are in a 4s orbital, indicating that it is an alkaline earth metal. Therefore the element is Calcium, Ca.

S1.8 **$I_1(Cl) < I_1(F)$?** When considering questions like these, it is always best to begin by writing down the electron configurations of the atoms or ions in question. If you do this routinely, a confusing comparison may become more understandable. In this case the relevant configurations are:

$$\text{F}: \ 1s^2 2s^2 2p^5 \ \text{or} \ [\text{He}]2s^2 2p^5$$

$$\text{Cl}: \ 1s^2 2s^2 2p^6 3s^2 3p^5 \ \text{or} \ [\text{Ne}]3s^2 3p^5$$

The electron removed during the ionization process is a $2p$ electron for F and a $3p$ electron for Cl. The principal quantum number, n, is lower for the electron removed from F ($n = 2$ for a $2p$ electron), so this electron is bound more strongly by the F nucleus than a $3p$ electron in Cl is bound by its nucleus.
A general trend: within a group, *larger* atoms have *lower* ionization energies. There are only a few exceptions to this trend, and they are found in groups 13 (IIIA) and 14 (IVA).

S1.9 **Which group does the element belong to?** When considering questions like these, look for the highest jump in energies. This occurs for the 5th ionization energy of this element. $I_4 = 6229$ kJ mol^{-1}, while $I_5 = 37838$ kJ mol^{-1}, indicating breaking into a complete subshell. Therefore the element is in the chalcogens (O, S, Se, etc.).

S1.10 **$A_e(C) > A_e(N)$?** The electron configurations of these two atoms are:

$$\text{C}: \ [\text{He}]2s^2 2p^2 \qquad \text{N}: \ [\text{He}]2s^2 2p^3$$

An additional electron can be added to the empty $2p$ orbital of C, and this is a favorable process ($A_e = 122$ kJ/mol). However, all of the $2p$ orbitals of N are already half occupied, so an additional electron added to N would experience sufficiently strong interelectronic repulsions. Therefore, the electron-gain process for N is unfavorable (A_e = –8 kJ/mol). This is despite the fact that the 2*p* Z_{eff} for N is larger than the 2*p* Z_{eff} for C (see Table 1.2). This tells you that attraction to the nucleus is not the only force that determines electron affinities (or, for that matter, ionization energies). Interelectronic repulsions are also important.

S1.11 **Which would be more polarizing?** Cs^+ has a larger ionic radius than Na^+. Cs^+ is isoelectronic to Xe and Na^+ is isoelectronic to Ne. Xe has a larger atomic radius than Ne. According to Fajan's rules, small, highly charged cations have polarizing ability. Both cations have the same charge, but Na^+ is smaller than Cs^+; thus Na^+ is more polarizing.

1.1 **Nuclear reactions? (a) $^{14}_{7}N$ + $^{4}_{2}He$?** You can tackle these questions easily if you do an accounting of protons and neutrons right away. The nuclear reactants ^{14}N and ^{4}He together contain 9 protons (7 + 2) and 9 neutrons(7 + 2). You are told that one of the products is ^{17}O, which contains 8 protons and 9 neutrons. (Remember, you can tell how many protons an atom contains by consulting a periodic table and noting its atomic number.) Therefore, since one proton is not yet accounted for on the products side of the equation, we add a proton, the balanced equation is:

$$^{14}_{7}N + ^{4}_{2}He \rightarrow ^{17}_{8}O + ^{1}_{1}p + \gamma$$

(b) $^{12}_{6}C$ + p? The atomic number of carbon is 6. If you add one proton to a carbon nucleus, the product must be the nucleus of the element with an atomic number of 6 + 1 = 7, which of course is a nitrogen nucleus. Remember, the mass number will also increase whenever a proton, or a neutron, is added to a nucleus and no subsequent fission occurs. Therefore, the product is a ^{13}N nucleus and the balanced equation is:

$$^{12}_{6}C + ^{1}_{1}p \rightarrow ^{13}_{7}N + \gamma$$

(c) $^{14}_{7}N$ + $^{1}_{0}n$? Here you are told that the products are $^{3}_{1}H$ and $^{12}_{6}C$. You should confirm that the equation

$$^{14}_{7}N + ^{1}_{0}n \rightarrow ^{3}_{1}H + ^{12}_{6}C$$

is indeed balanced. The reactants together contain 7 protons and 8 neutrons (7 + 1). The products also contain 7 protons (1 + 6) and 8 neutrons (2 + 6). Therefore, the equation is balanced.

1.2 **Balance the equation:** Use the accounting procedure described in the answer to Exercise 1.1 above. Curium has atomic number 96 (therefore 96 protons) and carbon has atomic number 6 (6 protons) therefore 96 + 6 equals 112. So the new element has 112 protons! Element number 112 was discovered by scientists at the Centre for Heavy Ion Research in Darmstadt, Germany. It is the largest superheavy element to be officially recognized and it has the proposed name of copernicium, Cp. The name should be official by January 2010.

$$^{246}_{96}Cm + ^{12}_{6}C \rightarrow ^{257}_{112}Uub + ^{1}_{0}n$$

1.3 ***I_2* of some period 4 elements?** The second ionization energies of the elements calcium through manganese increase from left to right in the periodic table with the exception that I_2(Cr) > I_2(Mn). The electron configurations of the elements are:

Ca	Sc	Ti	V	Cr	Mn
[Ar]$4s^2$	[Ar]$3d^1 4s^2$	[Ar]$3d^2 4s^2$	[Ar]$3d^3 4s^2$	[Ar]$3d^5 4s^1$	[Ar]$3d^5 4s^2$

Both the first and the second ionization processes remove electrons from the 4*s* orbital of these atoms, with the exception of Cr. In general, the 4*s* electrons are poorly shielded by the 3*d* electrons, so $Z_{eff}(4s)$ increases from left to right and I_2 also increases from left to right. While the I_1 process removes the sole 4*s* electron for Cr, the I_2 process must remove a 3*d* electron. The higher value of I_2 for Cr relative to Mn is a consequence of the special stability of half-filled subshell configurations and the higher Z_{eff} of a 3*d* electron verses a 4*s* electron.

1.4 **^{22}Ne + α?** Use the accounting procedure described in the answer to Exercise 1.1 above. Neon has atomic number 10 (therefore 10 protons) and magnesium has atomic number 12 (12 protons). An α particle is the same thing as a helium-4 nucleus, ^{4}He. Therefore, the nuclear reactants together contain 12 protons (10 +2) and 14 neutrons (12 + 2). One of the products is a ^{25}Mg nucleus, which contains 12 protons and 13 neutrons. Since one neutron is not accounted for, it must appear on the product side for mass balance. The equation is:

$$^{22}_{10}\mathrm{Ne}+^{4}_{2}\mathrm{He}\rightarrow^{25}_{12}\mathrm{Mg}+^{1}_{0}\mathrm{n}$$

1.5 **^{9}Be undergoes α decay to produce ^{12}C and neutrons?** Alpha decay means that α particles (^{4}He) are produced. Remember, you must have mass balance on both sides of the reaction. Since Be has only has 4 protons, and it is the only reactant, you need two Be nuclei, a total of 8 protons, to produce a C nucleus which has 6 protons and a ^{4}He nucleus which has 2 protons. Now the protons are balance on both sides, then all you have to do is balance the neutrons.

$$^{9}_{4}\mathrm{Be}+^{9}_{4}\mathrm{Be}\rightarrow^{12}_{6}\mathrm{C}+^{4}_{2}\mathrm{He}+2^{1}_{0}\mathrm{n}$$

1.6 **The natural abundances of adjacent elements?** The cosmic abundance of elements with an even number of protons (even atomic number, Z) is greater than that of adjacent elements with an odd Z. The reason for this is connected to the nucleosynthesis of the light elements (read section 1.1). Hydrogen is the most abundant element in the universe, when it fuses it produced helium nuclei, which are the basic building blocks for the synthesis of the rest of the light elements. Since helium has 2 protons, when you fuse two Helium-4 nuclei together, you will obtain an even Z, in this case Berilium-8.

$$^{4}_{2}\mathrm{He}+^{4}_{2}\mathrm{He}\rightarrow^{8}_{4}\mathrm{Be}$$

Therefore, since helium-4 is the basic building block, most additional fusion processes will produce nuclei with even atomic numbers.

1.7 **Does a nuclear reaction release energy?** Energy in the form of high-energy photons may be released during the course of a nuclear reaction. This can be calculated by referring to a table of very accurate particle rest masses. In exercise 1.2, the reactants are ^{246}Cm and ^{12}C, and the products are ^{257}Uub and a neutron. Take the summation of the rest masses of all the nuclei of the products minus the masses of the nuclei of the reactants. If you get a negative number, energy will be released. But what you have calculated is the mass difference, which in the case of a nuclear reaction is converted to energy. Using Einstein's Equation, $E = mc^2$, you can determine the amount of energy released, where m is the mass difference. Of course, if the mass difference is a positive number, the reaction requires energy to proceed to products.

1.8 **$E(\mathrm{He}^+)/E(\mathrm{Be}^{3+})$?** The ground-state energy of a hydrogenic ion, like He^+ or Be^{3+}, is defined as the orbital energy of its single electron, which is given by Equations 1.3 and 1.4:

$$\mathrm{E} = Z^2 m_e e^4/32\pi^2(\varepsilon_0)^2(h/2\pi)^2 n^2$$

For the ratio $E(\mathrm{He}^+)/E(\mathrm{Be}^{3+})$, the constants can be ignored, and:

$$E(\mathrm{He}^+)/E(\mathrm{Be}^{3+}) = Z(\mathrm{He}^+)^2/Z(\mathrm{Be}^{3+})^2 = 2^2/4^2 = 0.25$$

1.9 **$E(\mathrm{H}, n = 1) - E(\mathrm{H}, n = 6)$?** The expression for E given in Equations 1.3 and 1.4 (see above) can be used for a hydrogen atom as well as for hydrogenic ions. The ratio $E(\mathrm{H}, n = 1)/E(\mathrm{H}, n = 6)$ can be determined as follows:

$$E(\mathrm{H}, n = 1)/E(\mathrm{H}, n = 6) = (1/1^2)/(1/6^2) = 36$$

Therefore, $E(\mathrm{H}, n = 6) = (E(\mathrm{H}, n = 1))/36 = -0.378$ eV, and the difference is:

$$E(\mathrm{H}, n = 1) - E(\mathrm{H}, n = 6) = -13.2 \text{ eV}$$

1.10 What is the wavenumber/wavelength of the first transition in the visible region for H? The visible region starts when $n_1 = 2$. The next transition is where n_2 equals 3. This can be determined using the Balmer-Rydberg equation (Equation 1.1).

$$\frac{1}{\lambda} = R\left(\frac{1}{2^2} - \frac{1}{3^2}\right) = 1.524 \times 10^{-3}\,\text{nm}^{-1}$$

$$\lambda = 656.3\ \text{nm}$$

1.11 Prediction of the Lyman series? The version of the Rydberg formula which generated the Lyman series is:

$$\frac{1}{\lambda} = R\left(\frac{1}{1^2} - \frac{1}{n^2}\right) \qquad R = 1.097 \times 10^7\ \text{m}^{-1}$$

Where n is a natural number greater than or equal to n = 2 (i.e. $n = 2,3,4,...\infty$). There are infinitely many spectral lines, but they become very dense as they approach ∞, so only some of the first lines and the last one appear. The wavelengths (nm) in the Lyman series are all ultraviolet. If we let n = ∞, we get an approximation for the first line:

$$\frac{1}{\lambda} = R\left(\frac{1}{1^2} - \frac{1}{\infty^2}\right) = 1.0974 \times 10^7\, m^{-1}$$

$$\lambda = 91.124\,nm$$

The next line is for n = 4:

$$\frac{1}{\lambda} = R\left(\frac{1}{1^2} - \frac{1}{4^2}\right) = 1.0288 \times 10^7\, m^{-1}$$

$$\lambda = 97.199\,nm$$

The next line is for n = 3:

$$\frac{1}{\lambda} = R\left(\frac{1}{1^2} - \frac{1}{3^2}\right) = 9.7547 \times 10^6\, m^{-1}$$

$$\lambda = 102.52\,nm$$

The final line is for n = 2:

$$\frac{1}{\lambda} = R\left(\frac{1}{1^2} - \frac{1}{2^2}\right) = 8.2305 \times 10^6\, m^{-1}$$

$$\lambda = 121.499\,nm$$

So all these numbers predict the Lyman series within a few significant figures.

1.12 Principal quantum number and its relation to *l*? The principal quantum number n labels one of the shells of an atom. For a hydrogen atom or a hydrogenic ion, n alone determines the energy of all of the orbitals contained in a given shell (since there are n^2 orbitals in a shell, these would be n^2-fold degenerate). For a given value of n, the angular momentum quantum number l can assume all integer values from 0 to $n - 1$.

1.13 How many orbitals for a given value of *n*? For the first shell ($n = 1$), there is only one orbital, the 1*s* orbital. For the second shell ($n = 2$), there are four orbitals, the 2*s* orbital and the three 2*p* orbitals. For $n = 3$, there are 9 orbitals, the 3*s* orbital, three 3*p* orbitals, and five 3*d* orbitals. The progression of the number of orbitals so far is 1, 4, 9, which is the same as n^2 (for example, $n^2 = 1$ for $n = 1$, $n^2 = 4$ for $n = 2$, etc.). As a further verification, consider the fourth shell ($n = 4$), which, according to the analysis so far, should contain $4^2 = 16$ orbitals. Does it?

Yes; the fourth shell contains the 4*s* orbital, three 4*p* orbitals, five 4*d* orbitals, and seven 4*f* orbitals, and 1 + 3 + 5 + 7 = 16.

1.14 Complete the table.

n	l	m_l	Orbital designation	Number of orbitals
2	1	+1, 0, −1	2*p*	3
3	2	+2, +1, …, −2	3*d*	5
4	0	0	4*s*	1
4	3	+3, +2, …, −3	4*f*	7

1.15 What are the values for the quantum numbers for the set of 5*f* orbitals? When n = 5, l = 3 (for the f orbitals) and m_l = -3,-2,-1,0,1,2,3, which represent the 7 orbitals that complete the 5*f* subshell. Figure 1.16 shows a picture of the seven different 5*f* orbitals. The 5f orbitals represent the start of the Actinides, starting with Th and ending with Lr.

1.16 Comment on the values of the screening (shielding) constant going across period 2. The values for the shielding constant are calculated below. The formula for effective nuclear charge is $Z_{eff} = Z - \sigma$, where σ is the shielding or screening constant.

Li: $\sigma = Z - Z_{eff}$; σ = 3-1.28 = 1.72
Be: $\sigma = Z - Z_{eff}$; σ = 4-1.19 = 2.09
B: $\sigma = Z - Z_{eff}$; σ = 5-2.42 = 2.58
C: $\sigma = Z - Z_{eff}$; σ = 6-3.14 = 2.86
N: $\sigma = Z - Z_{eff}$; σ = 7-3.83 = 3.17
O: $\sigma = Z - Z_{eff}$; σ = 8-4.45 = 3.55
F: $\sigma = Z - Z_{eff}$; σ = 9-5.10 = 3.90

The trend is as expected, the shielding constant increases going across a period as a product of electron electron repulsion as you fill the 2s and 2p sub-shells. This, off course, has consequences on the atomic radii, ionization energies, and electron affinities of the period 2 elements. The most obvious being the atomic radii; as you go from left to right across period 2, the atomic radii decreases because the outer shell electrons "see" more positive charge while filling the same shell, n = 2.

1.17 Consider the process of shielding in the Be atom. In an atom with many electrons like Beryllium, the outer electrons (the 2s electrons in this case) are simultaneously attracted to the positive nucleus (the protons in the nucleus) and repelled by the negatively charged electrons in a given atomic orbital (in this case, the 2s orbital). The two electrons in the 1s orbital on average are statically closer to the nucleus than the 2s electrons, thus the 1s electrons "see" more positive charge than the 2s electrons. The 1s electrons also shield that positive charge form the 2s electrons, which are further out from the nucleus than the 1s electrons. Consequently, the 2s electrons "see" less positive charge than the 1s electrons for Beryllium.

1.18 Radial wavefunctions vs. radial distribution functions vs. angular wavefunctions? The plots of ψ vs. *r* shown in Figures 1.11 and 1.12 are plots of the radial parts of the total wavefunctions for the indicated orbitals. Notice that the plot of ψ(2*s*) vs. *r* takes on both positive (small *r*) and negative (larger *r*) values, requiring that for some value of *r* the wavefunction ψ(2*s*) = 0 (i.e., the wavefunction has a node at this value of *r*; for a hydrogen atom or a hydrogenic ion, ψ(2*s*) = 0 when $r = 2a_0/Z$). Notice also that the plot of ψ(2*p*) vs. *r* is positive for all values of *r*. Although a 2*p* orbital does have a node, it is not due to the radial wavefunction (the radial part of the total wavefunction). The plot of $4\pi r^2\psi^2$ vs. *r* for a 1*s* orbital in Figure 1.13 is a radial distribution function. For comparison, plots of $r^2\psi^2$ vs. *r* for 1*s* and 2*s* orbitals are shown below:

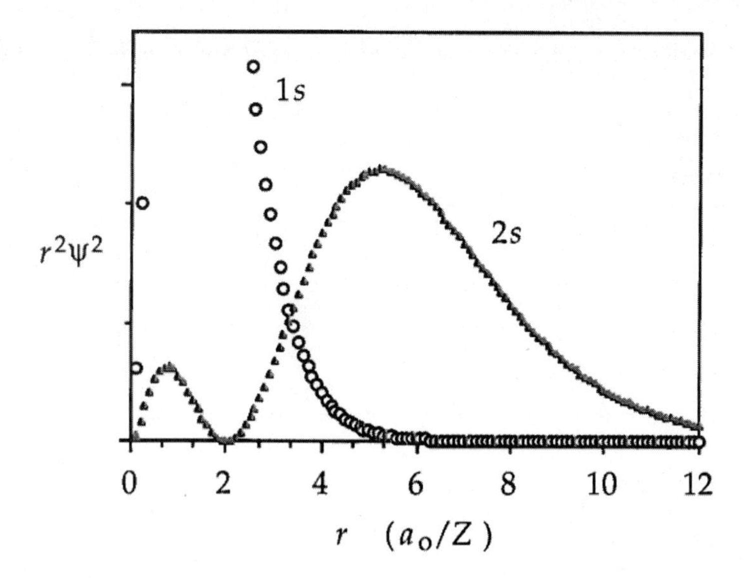

Whereas the radial distribution function for a 1*s* orbital has a single maximum, that for a 2*s* orbital has two maxima and a minimum (at $r = 2a_o/Z$ for hydrogenic 2*s* orbitals). The presence of the node at $r = 2a_o/Z$ for $\psi(2s)$ requires the presence of the two maxima and the minimum in the 2*s* radial distribution function. Using the same reasoning, the absence of a radial node for $\psi(2p)$ requires that the 2*p* radial distribution function have only a single maximum, as shown in Figure 1.12 and below:

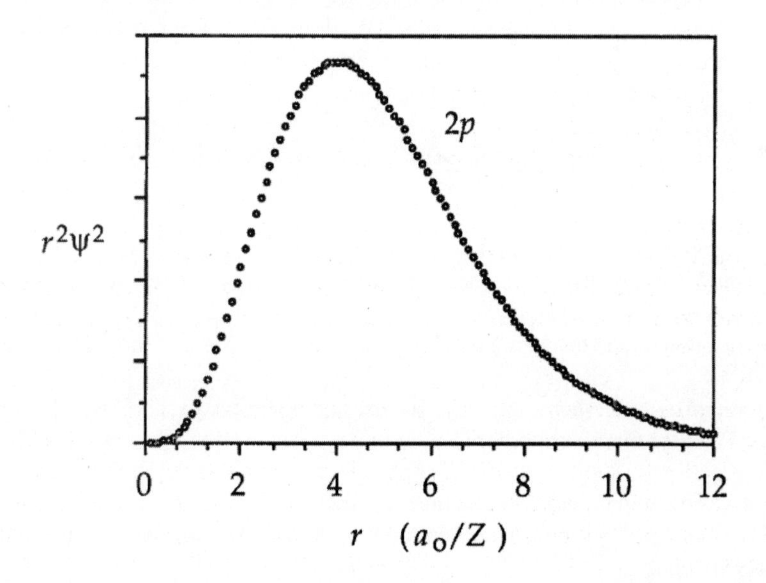

The angular wavefunctions are the familiar pictures that chemists draw to represent *s*, *p*, *d*, and *f* orbitals, such as the ones in Figures 1.13–1.16. The familiar nodal plane for a 2*p* orbital is a property the orbital possesses because of the mathematical form of its angular wavefunction, not because of the mathematical form of its radial wavefunction.

1.19 ***I*(Ca) vs. *I*(Zn)?** The first ionization energies of calcium and zinc are 6.11 and 9.39 eV. Both of these atoms have an electron configuration that ends with $4s^2$: Ca is [Ar]$4s^2$ and Zn is [Ar]$3d^{10}4s^2$. An atom of zinc has 30 protons in its nucleus and an atom of calcium has 20, so clearly zinc has a higher nuclear charge than calcium. Remember, though, that it is *effective* nuclear charge (Z_{eff}) that directly affects the ionization energy of an atom. Since $I(Zn) > I(Ca)$, it would seem that $Z_{eff}(Zn) > Z_{eff}(Ca)$. How can you demonstrate that this is as it should be?

The actual nuclear charge can always be readily determined by looking at the periodic table and noting the atomic number of an atom. The effective nuclear charge cannot be directly determined, i.e., it requires some interpretation on your part. Read Section 1.6, *Penetration and shielding*, again. Study the trend for the period 2 *p*-block elements in Table 1.2. The pattern that emerges is that not only Z but also Z_{eff} rises from boron to neon. Each successive element has one additional proton in its nucleus and one additional electron to balance the charge. However, the additional electron never completely shields the other electrons in the atom. Therefore, Z_{eff} rises from B to Ne. Similarly, Z_{eff} rises through the *d* block from Sc to Zn, and that is why Z_{eff}(Zn) > Z_{eff}(Ca).

1.20 **I(Sr) vs. I(Ba) vs. I(Ra)?** The first ionization energies of strontium, barium, and radium are 5.69, 5.21, and 5.28 eV. Normally, atomic radius increases and ionization energy decreases down a group in the periodic table. However, in this case I(Ba) < I(Ra). Study the periodic table, especially the alkaline earths. Notice that Ba is eighteen elements past Sr, but Ra is thirty-two elements past Ba. The difference between the two corresponds to the fourteen 4*f* elements (the lanthanides) between Ba and Lu. Therefore, even though radium would be expected to have a larger radius than barium, it has a higher first ionization energy because it has such a large Z_{eff}, due to the insertion of the lanthanides.

1.21 **I_2 of some period 4 elements?** The second ionization energies of the elements calcium through manganese increase from left to right in the periodic table with the exception that I_2(Cr) > I_2(Mn). The electron configurations of the elements are:

Ca	Sc	Ti	V	Cr	Mn
[Ar]$4s^2$	[Ar]$3d^1 4s^2$	[Ar]$3d^2 4s^2$	[Ar]$3d^3 4s^2$	[Ar]$3d^5 4s^1$	[Ar]$3d^5 4s^2$

Both the first and the second ionization processes remove electrons from the 4*s* orbital of these atoms, with the exception of Cr. In general, the 4*s* electrons are poorly shielded by the 3*d* electrons, so $Z_{eff}(4s)$ increases from left to right and I_2 also increases from left to right. While the I_1 process removes the sole 4*s* electron for Cr, the I_2 process must remove a 3*d* electron. The higher value of I_2 for Cr relative to Mn is a consequence of the special stability of half-filled subshell configurations.

1.22 **Ground-state electron configurations? (a) C?** Four elements past He. Helium ends period 1; therefore carbon is [He]$2s^2 2p^2$.
(b) F? Seven elements past He; therefore [He]$2s^2 2p^5$.
(c) Ca? Two elements past Ar, which ends period 3, leaving the 3*d* subshell empty; therefore [Ar]$4s^2$.
(d) Ga^{3+}? Thirteen elements, but only ten electrons, past Ar, and since it is a cation, there is no doubt that $E(3d) < E(4s)$; therefore [Ar]$3d^{10}$.
(e) Bi? Twenty-nine elements past Xe, which ends period 5, leaving the 5*d* and the 4*f* subshells empty; therefore [Xe]$4f^{14}5d^{10}6s^2 6p^3$.
(f) Pb^{2+}? Twenty-eight elements, but only twenty-six electrons, past Xe, which ends period 5, leaving the 5*d* and the 4*f* subshells empty; therefore [Xe]$4f^{14}5d^{10}6s^2$.

1.23 **More ground-state electron configurations? (a) Sc?** Three elements past Ar; therefore [Ar]$3d^1 4s^2$.
(b) V^{3+}? Five elements, but only two electrons, past Ar, and since it is a cation, there is no doubt that $E(3d) < E(4s)$; therefore [Ar]$3d^2$.
(c) Mn^{2+}? Seven elements, but only five electrons, past Ar; therefore [Ar]$3d^5$.
(d) Cr^{2+}? Six elements, but only four electrons, past Ar; therefore [Ar]$3d^4$.
(e) Co^{3+}? Nine elements, but only six electrons, past Ar; therefore [Ar]$3d^6$.
(f) Cr^{6+}? Six elements past Ar, but with a +6 charge it has the *same* electron configuration as Ar, which is written as [Ar]. Sometimes inorganic chemists will write the electron configuration as [Ar]$3d^0$ to emphasize that there are no *d* electrons for this *d*-block metal ion in its highest oxidation state.
The relative sizes of Cr^{6+}, [Ar]$3d^0$, and Cr^{2+}, [Ar]$3d^4$:

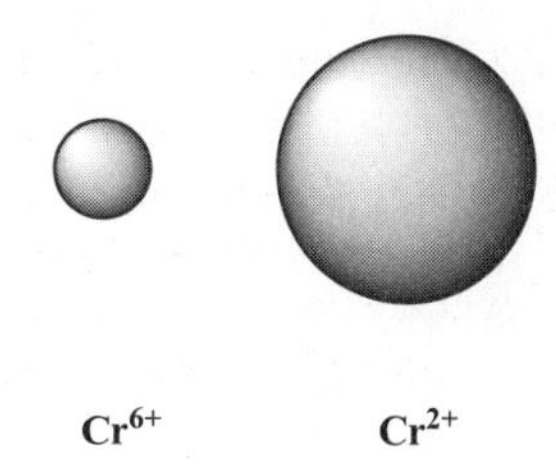

(g) Cu? Eleven elements past Ar, but its electron configuration is not [Ar]$3d^94s^2$. The special stability experienced by completely filled subshells causes the actual electron configuration of Cu to be [Ar]$3d^{10}4s^1$.
(h) Gd^{3+}? Ten elements, but only seven electrons, past Xe, which ends period 5 leaving the 5*d* and the 4*f* subshells empty; therefore [Xe]$4f^7$.

1.24 More ground-state electron configurations? (a) W? Twenty elements past Xe, fourteen of which are the 4*f* elements. If you assumed that the configuration would resemble that of chromium, you would write [Xe]$4f^{14}5d^56s^1$. It turns out that the actual configuration is [Xe]$4f^{14}5d^46s^2$. The configurations of the heavier *d*- and *f*-block elements show some exceptions to the trends for the lighter *d*-block elements.
(b) Rh^{3+}? Nine elements, but only six electrons, past Kr; therefore [Kr]$4d^6$.
(c) Eu^{3+}? Nine elements, but only six electrons, past Xe, which ends period 5, leaving the 5*d* and the 4*f* subshells empty; therefore [Xe]$4f^6$.
(d) Eu^{2+}? This will have one more electron than Eu^{3+}. Therefore, the ground-state electron configuration of Eu^{2+} is [Xe]$4f^7$.
(e) V^{5+}? Five elements past Ar, but with a 5+ charge it has the *same* electron configuration as Ar, which is written as [Ar] or [Ar]$3d^0$.
(f) Mo^{4+}? Six elements, but only two electrons, past Kr; therefore [Kr]$4d^2$.

1.25 Identify the elements?
(a) S
(b) Sr
(c) V
(d) Tc
(e) In
(f) Sm

1.26 Draw the periodic table? See Figure 1.4 and the inside front cover of this book. You should start learning the names and positions of elements that you do not know. Start with the alkali metals and the alkaline earths. Then learn the elements in the *p* block. A blank periodic table can be found on the inside back cover of this book. You should make several photocopies of it and should test yourself from time to time, especially after studying each chapter.

1.27 I_1, A_e, and χ for period 3? The following values were taken from Tables 1.5, 1.6, and 1.7:

Element	Electron configuration	I_1(eV)	A_e (eV)	χ
Na	[Ne]$3s^1$	5.14	0.548	0.93
Mg	[Ne]$3s^2$	7.64	−0.4	1.31
Al	[Ne]$3s^23p^1$	5.98	0.441	1.61
Si	[Ne]$3s^23p^2$	8.15	1.385	1.90
P	[Ne]$3s^23p^3$	11.0	0.747	2.19
S	[Ne]$3s^23p^4$	10.36	2.077	2.58
Cl	[Ne]$3s^23p^5$	13.10	3.617	3.16
Ar	[Ne]$3s^23p^6$	15.76	−1.0	

In general, I_1, A_e, and χ all increase from left to right across period 3 (or from top to bottom in the table above). All three quantities reflect how tightly an atom holds on to its electrons, or how tightly it holds on to additional electrons. The cause of the general increase across the period is the gradual increase in Z_{eff}, which itself is caused by the incomplete shielding of electrons of a given value of n by electrons with the same n. The exceptions are explained as follows: I_1 (Mg) > I_1(Al) and A_e(Na) >A_e(Al)—both of these are due to the greater stability of 3s electrons relative to 3p electrons; A_e(Mg) and A_e(Ar) < 0—filled subshells impart a special stability to an atom or ion (in these two cases the additional electron must be added to a higher energy subshell (for Mg) or shell (for Ar)); I_1(P) > I_1(S) and A_e(Si) > A_e(P)—the loss of an electron from S and the gain of an additional electron by Si both result in an ion with a half-filled p subshell, which, like filled subshells, imparts a special stability to an atom or ion.

1.28 **Metallic radii of Nb and Ta?** If you look at the elements just before these two in Table 1.3, you will see that this is a general trend. Normally, the period 6 elements would be expected to have larger metallic radii than their period 5 vertical neighbors; only Cs and Ba follow this trend: Cs is larger than Rb and Ba is larger than Sr. Lutetium, Lu, is significantly smaller than yttrium, Y, and Hf is just barely the same size as Zr. After Nb and Ta, the "normal" expectation is observed. There are no intervening elements between Sr and Y, but there are fourteen intervening elements, the lanthanides, between Ba and Lu. A contraction of the radii of the elements starting with Lu is due to incomplete shielding by the 4f electrons.

1.29 **Frontier orbitals of Be?** Recall from Section 1.9(c), *Electron affinity*, that the frontier orbitals are the highest occupied and the lowest unoccupied orbitals of a chemical species (atom, molecule, or ion). Since the ground-state electron configuration of a beryllium atom is $1s^22s^2$, the frontier orbitals are the 2s orbital (highest occupied) and the 2p orbitals (lowest unoccupied). Note that there can be more than two frontier orbitals if either the highest occupied and/or lowest unoccupied energy levels are degenerate.

1.30 **Electronegativities across period 2?** Plots of electronegativity across period 2 and ionization energies across period 2 are superimposed on the figure below. The general trend is the same in both plots; both χ and I_1 increase from left to right across a period, and this is because the effective nuclear charge increases for the n = 2 orbitals across period 2. The two deviations in the upper plot result from different phenomena. For boron, the outermost electron occupies a 2p orbital, which has a higher energy than a beryllium atom's 2s orbital. The higher energy of the 2p electron offsets a boron atom's greater nuclear charge. For oxygen, two electrons are paired in one of the 2p orbitals, and the mutual repulsion they experience offsets an oxygen atom's greater nuclear charge relative to a nitrogen atom. Even though this exercise did not require the use of electron affinity values (Table 1.6), it is useful to think about the "connections" between the various atomic properties. Note that the electron affinities of beryllium and nitrogen are negative, and the explanation for these apparent anomalies is the same as the explanation given above for the departure of I_1(B) and I_1(O) from the general upward trend.

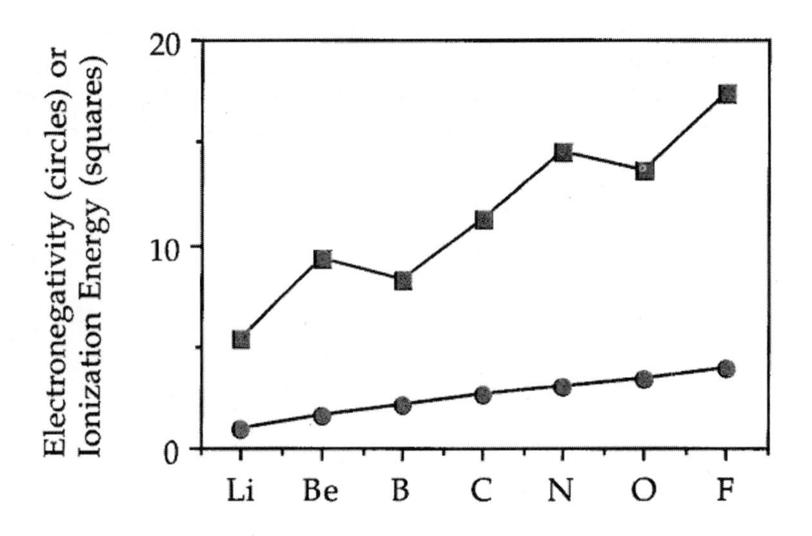

Chapter 2 Molecular Structure and Bonding

One of the most powerful concepts in chemistry is that molecular orbitals result from the overlap of atomic orbitals. In this figure, two hydrogen 1*s* atomic orbitals overlap to form a σ molecular orbital.

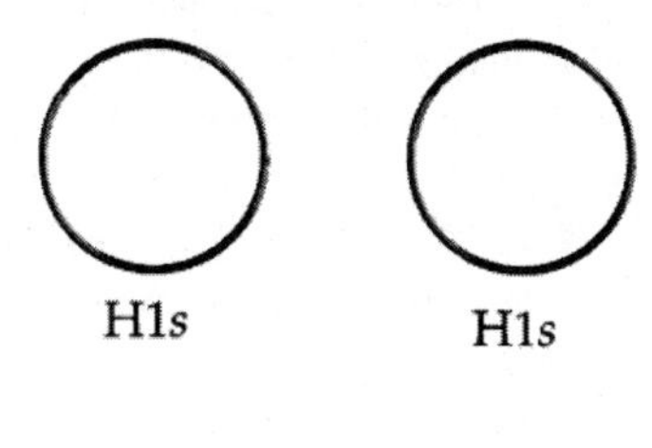

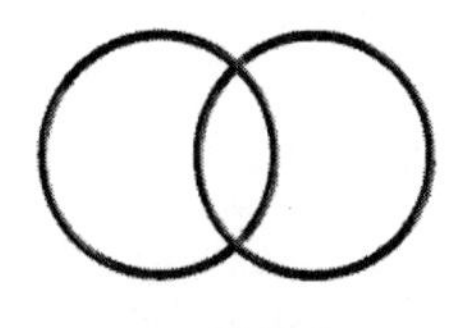

σ orbital

S2.1 **Lewis structure for PCl_3?** The four atoms supply 5 + (3 × 7) = 26 valence electrons. Since P is less electronegative than Cl, it is likely to be the central atom, so the 13 pairs of electrons are distributed as shown at the right. In this case, each atom obeys the octet rule. Whenever it is possible to follow the octet rule without violating other electron counting rules, you should do so.

:C̈l:P̈:C̈l:
:C̈l:

:C̈l—P̈—C̈l:
|
:C̈l:

S2.2 **Predict the shape of H_2S and XeO_4?** The Lewis structures and the shapes of H_2S and XeO_4 are shown below. According to the VSEPR model, electrons in bonds and in lone pairs can be thought of as charge clouds that repel one another and stay as far apart as possible. First, write a Lewis dot structure for the molecule, and then arrange the lone pairs and atoms around the central atom, such that the lone pairs are as far away from each other as possible.

Lewis Structure	Geometry
H–S̈–H	H–S–H
	angular (bent) **H-S-H angle about 90°**

Xe with four O (Lewis structure) — **square planar** — **or tetrahedral**

All atoms in both structures have formal charges of zero. Note that H_2O is bent with an angle of about 109°, while H_2S is angular with an angle of about 90°, indicating more *p* character in bonding. The actual structure of XeO_4 is tetrahedral; it is a highly unstable colorless gas.

S2.3 **Predict the shape of XeF_2?** The Lewis structure is shown below, accommodating an octet for the 4 F atoms and an expanded valence shell of 10 electrons for the Xe atom, with the 8 + (2 × 7) = 22 valence electrons provided by the three atoms. The five electron pairs around the central Xe atom will arrange themselves at the corners of a trigonal bipyramid (as in PF_5). The three lone pairs will be in the equatorial plane, to minimize lone pair–lone pair repulsions. The resulting shape of the molecule, shown at the right, is linear (i.e., the F–Xe–F bond angle is 180°).

$$:\ddot{\underset{\cdot\cdot}{Xe}}\langle \begin{matrix} :\ddot{\underset{\cdot\cdot}{F}}: \\ :\ddot{\underset{\cdot\cdot}{F}}: \end{matrix}$$

F — Xe — F

S2.4 **Electron configurations for S_2^{2-} and Cl_2^-?** The first of these two anions has the same Lewis structure as peroxide, O_2^{2-}. It also has a similar electron configuration to that of peroxide, except for the use of sulfur atom valence 3*s* and 3*p* atomic orbitals instead of oxygen atom 2*s* and 2*p* orbitals. There is no need to use sulfur atom 3*d* atomic orbitals, which are higher in energy than the 3*s* and 3*p* orbitals, since the 2(6) + 2 = 14 valence electrons of S_2^{2-} will not completely fill the stack of molecular orbitals constructed from sulfur atom 3*s* and 3*p* atomic orbitals. Thus, the electron configuration of S_2^{2-} is $1\sigma_g^2 2\sigma_u^2 3\sigma_g^2 1\pi_u^4 2\pi_g^4$. The Cl_2^- anion contains one more electron than S_2^{2-}, so its electron configuration is $1\sigma_g^2 2\sigma_u^2 3\sigma_g^2 1\pi_u^4 2\pi_g^4 4\sigma_u^1$.

S2.5 **Electron configuration of ClO^-?** ClO^- is isoelectronic (same number of electrons) with ICl. The orbitals to be used are Chlorine's 3s and 3p valence shell orbitals and the Oxygen's 2s and 2p valence shell orbitals. The bonding orbitals will be predominantly O in character being that O is more electronegative. So even the MO diagram of ClO^- will be similar to ICl. We have a total of 7 + 6 + 1(for charge) valence electrons giving us 14. Therefore the ground-state electron configuration is $1\sigma_g^2 2\sigma_u^2 3\sigma_g^2 1\pi_u^4 2\pi_g^4$, same as ICl.

S2.6 **Predict the bond order of the carbide anion C_2^{2-}?** The number of valence electrons for C_2^{2-} is equal to 10 (4 + 4 + 2 (for charge)). Thus C_2^{2-} is isoelectronic with N_2(which has 10 valence electrons as well). The configuration of C_2^{2-} would be $1\sigma_g^2 1\sigma_u^2 1\pi_u^4 2\sigma_u^2$. The bond order would be ½[2-2+4+2] = 3. So C_2^{2-} has a triple bond.

S2.7 **Predict the order of bond strength and bond length for C–N, C=N, and C≡N?** In general, the more bonds you have between two atoms, the shorter the bond length and the stronger the bond. Therefore, the ordering for bond length going from shortest to longest is C≡N, C=N, and C–N. For bond strength, going from strongest to weakest, the order is C≡N > C=N > C–N.

S2.8 **Is any XH_2 molecule linear?** According to Figure 2.37, a XH_2 molecule is expected to be linear if it contains four or fewer electrons. This is because the bottom two orbitals, which can contain up to four electrons, are lowest in energy when the H–X–H bond angle is 180°. Based on this analysis, both NaH_2 and MgH_2 should be linear, because they contain three and four valence electrons, respectively. The molecule AlH_2 contains five electrons and so is not expected to be linear.

S2.9 **Estimate $\Delta_f H$ for H_2S?** You can "form" this molecule by considering the following reaction:

$$^{1}/_{8}S_8 + H_2 \rightarrow H_2S$$

On the left side, you must break one H–H bond and also produce one sulfur atom from cyclic S_8. Since there are eight S–S bonds holding eight S atoms together, you must supply the mean S–S bond enthalpy *per S atom*. On the right side, you form two H–S bonds. From the values given in Table 2.8, you can estimate:

$$\Delta_f H = (436 \text{ kJ/mol}) + (264 \text{ kJ/mol}) - 2(338 \text{ kJ/mol}) = 24 \text{ kJ/mol}$$

This estimate indicates a slightly endothermic enthalpy of formation, but the experimental value, –21 kJ mol^{-1}, is slightly exothermic.

S2.10 **Oxidation numbers? (a) O_2^+?** The charge on the oxygenyl ion is +1, and that charge is shared by two oxygen atoms. Therefore, O.N.(O) = +1/2. This is an unusual oxidation number for oxygen.
(b) Phosphorus in PO_4^{3-}? The charge on the phosphate ion is –3, so O.N.(P) + 4 × O.N.(O) = –3. Oxygen is normally given an oxidation number of –2. Therefore, O.N.(P) = –3 – (4)(–2) = +5. The central phosphorus atom in the phosphate ion has the maximum oxidation number for group V/15.

2.1 **What shape would you expect for (a) H_2S?** The Lewis structure for hydrogen sulfide is shown below. The shape would be expected to be bent with the H–S–H angle less than 109°. However, the angle is actually close to 90°, indictative of considerable *p* character in the bonding between S and H.
(b) BF_4^-? The Lewis structure is shown below. The shape is tetrahedral with all angles 109.5°.
(c) NH_4^+? The Lewis structure of the ammonium ion is shown below. Again, the shape is tetrahedral with all angles 109.5°.

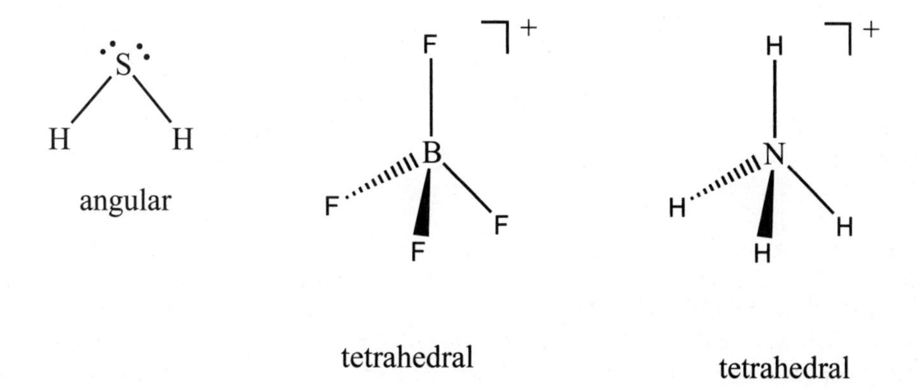

2.2 **What shape would you expect for (a) SO_3?** The Lewis structure of sulfur trioxide is shown below. With three σ bonds and no lone pairs, you should expect a trigonal-planar geometry (like BF_3). The shape of SO_3 is also shown below.
(b) SO_3^{2-}? The Lewis structure of sulfite ion is shown below. With three σ bonds and one lone pair, you should expect a trigonal-pyramidal geometry such as NH_3. The shape of SO_3^{2-} is also shown below.
(c) IF_5? The Lewis structure of iodine pentafluoride is shown below. With five σ bonds and one lone pair, you should expect a square-pyramidal geometry. The shape of IF_5 is also shown below.

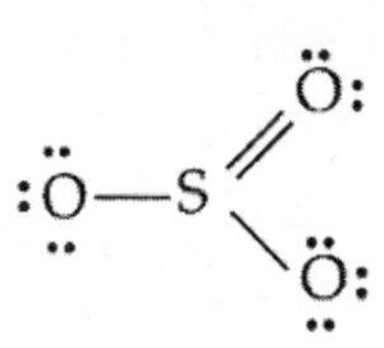

SO_3 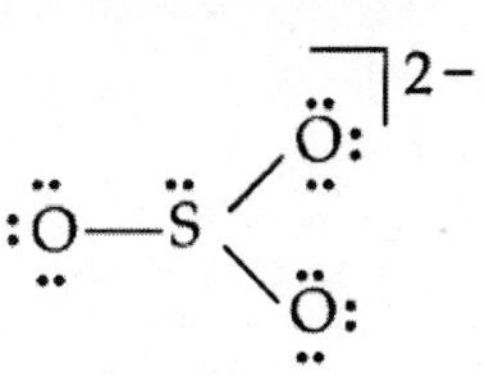 SO_3^{2-} 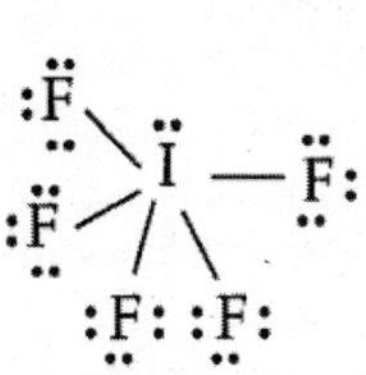 IF_5

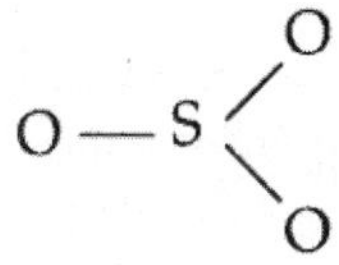

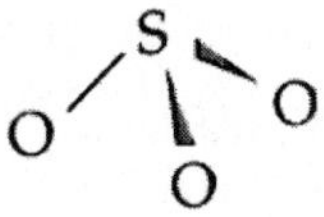

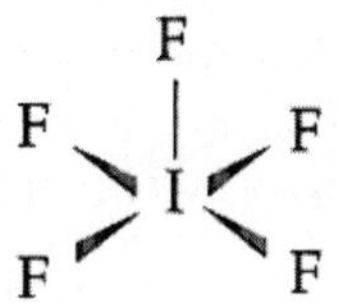

2.3 **The shapes of ClF_3, ICl_4^-, and I_3^-?** **(a)** The Lewis structure of ClF_3 is shown below. The chlorine atom in ClF_3 is bonded to the three fluorine atoms through sigma bonds and has two nonbonding electron lone pairs. Both lone pairs occupy equatorial positions (the largest angles in a trigonal bipyramid), resulting in a T shape for the molecule.

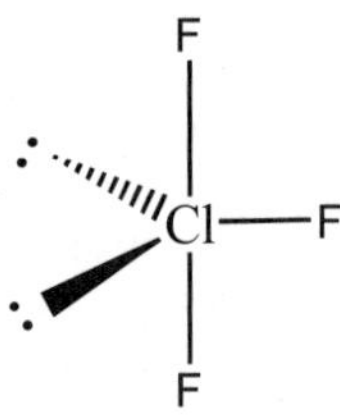

(b) The Lewis structure for ICl_4^- is shown below. The iodide atom is bonded to the four chlorine atoms through sigma bonds and has two sets of lone pairs. The lone pairs are opposite each other, occupying the axial sites of an octahedron. The overall shape of the molecule is square planar.

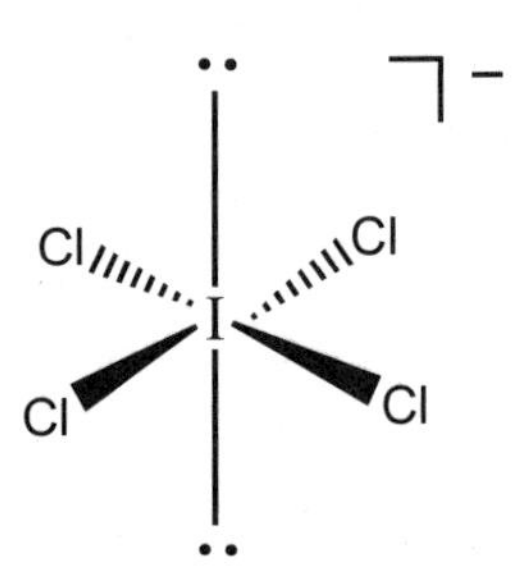

(c) The Lewis structure for I_3^- is shown below. The iodide atom is bonded to two other iodide atoms through sigma bonds and has three sets of lone pairs. The lone pairs occupy the equatorial sites of a trigonal bipyramid. The overall shape of the molecule is linear.

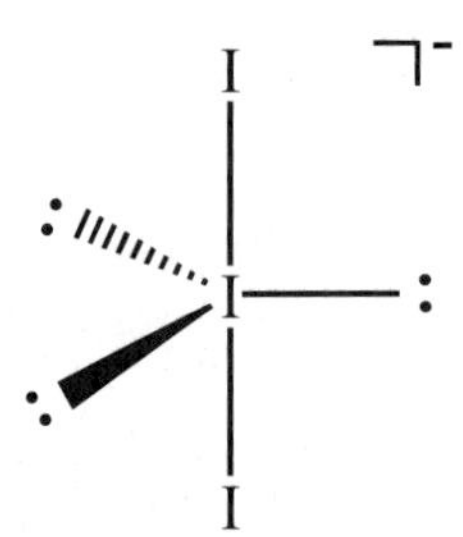

2.4 **In which of the species ICl_4^- or SF_4 is the bond angle closest to that predicted by VSEPR?** The structure based on VSEPR theory of ICl_4^- is shown below. The actual structure is very close, the lone pairs are on opposite sides of the Iodine atom, repelling each other as much as possible can only result in a square planar geometry.

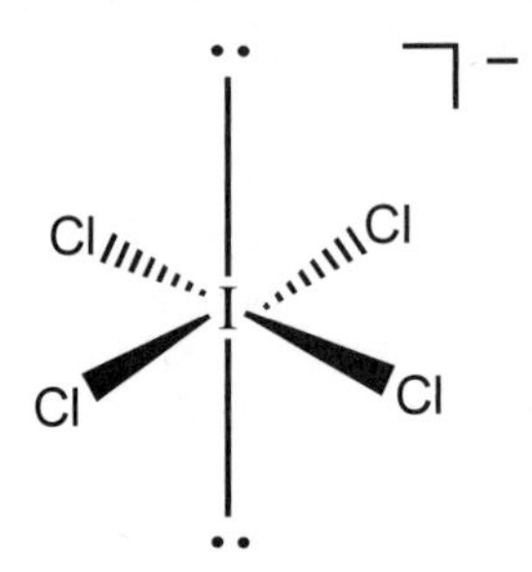

The VSEPR model and actual structure of SF_4 are shown below. Remember, lone pairs repel bonding regions, read section 2.3 (b). The VSEPR theory predicts a see-saw structure with a bond angle of 120° between the S and equatorial F's and a bond angle of 180° between the S and the axial F's. The bond angle is actually 102° for the S and equatorial F's and 173° for the S and the axial F's. This is due to the equatorial lone pair repelling the bonding S-F atoms.

F F 173° F F 120° S 180° F F F S 102° F F

VESPR Model Actual Structure

2.5 The shapes of PCl_4^+ and PCl_6^-? The Lewis structures of these two ions are shown below. With four σ bonds and no lone pairs for PCl_4^+, and six σ bonds and no lone pairs for PCl_6^-, the expected shapes are tetrahedral (like CCl_4) and octahedral (like SF_6), respectively. In the tetrahedral PCl_4^+ ion, all P–Cl bonds are the same length and all Cl–P–Cl bond angles are 109.5°. In the octahedral PCl_6^- ion, all P–Cl bonds are the same length and all Cl–P–Cl bond angles are either 90° or 180°. The P–Cl bond distances in the two ions would not necessarily be the same length.

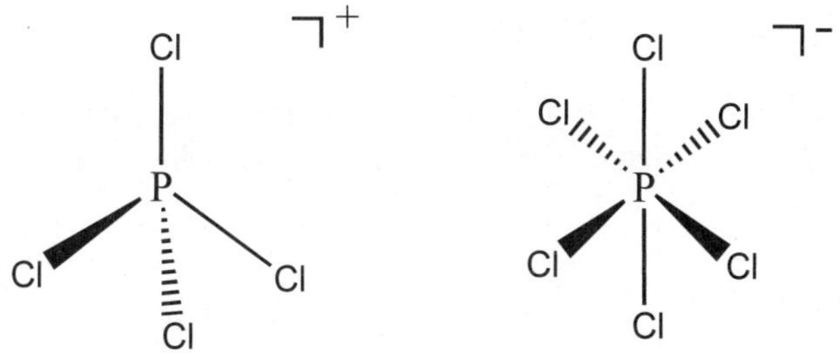

2.6 Calculate bond lengths. (a) CCl_4 (observed value = 1.77 Å)? From the covalent radii values given in Table 2.7, 0.77 Å for C and 0.99 Å for Cl, the C–Cl bond length in CCl_4 is predicted to be 0.77 Å + 0.99 Å = 1.76 Å. The agreement with the experimentally observed value is excellent.

(b) $SiCl_4$ (observed value = 2.01 Å)? The covalent radius for Si is 1.18 Å. Therefore, the Si–Cl bond length in $SiCl_4$ is predicted to be 1.18 Å + 0.99 Å = 2.17 Å. This is 8% longer than the observed bond length, so the agreement is not as good in this case.

(c) $GeCl_4$ (observed value = 2.10 Å)? The covalent radius for Ge is 1.22 Å. Therefore, the Ge–Cl bond length in $GeCl_4$ is predicted to be 2.21 Å. This is 5% longer than the observed bond length.

2.7 Si=O or Si–O in silicon-oxygen compounds? You need to consider the enthalpy difference between one mole of Si=O double bonds and two moles of Si–O single bonds. The difference is:

$$2(\text{Si-O}) - (\text{Si=O}) = 2(466 \text{ kJ}) - (640 \text{ kJ}) = 292 \text{ kJ}$$

Therefore, the two single bonds will always be better enthalpically than one double bond. If silicon atoms only have single bonds to oxygen atoms in silicon-oxygen compounds, the structure around each silicon atom will be tetrahedral: each silicon will have four single bonds to four different oxygen atoms.

2.8 Why is elemental nitrogen N_2 and elemental phosphorus P_4? Diatomic nitrogen has a triple bond holding the atoms together, whereas six P–P single bonds hold together a molecule of P_4. If N_2 were to exist as N_4 molecules with the P_4 structure, then two N≡N triple bonds would be traded for six N–N single bonds, which are intrinsically weak. The net enthalpy change can be estimated from the data in Table 2.8 to be 2(945 kJ) – 6(163 kJ) = 912 kJ, which indicates that the tetramerization of nitrogen is *very* unfavorable. On the other hand, multiple bonds between period 3 and larger atoms are not as strong as two times the analogous single bond, so P_2 molecules, each with a P≡P triple bond, would not be as stable as P_4 molecules, containing only P–P single bonds. In this case, the net enthalpy change for $2P_2 \rightarrow P_4$ can be estimated to be 2(481 kJ) – 6(201 kJ) = –244 kJ.

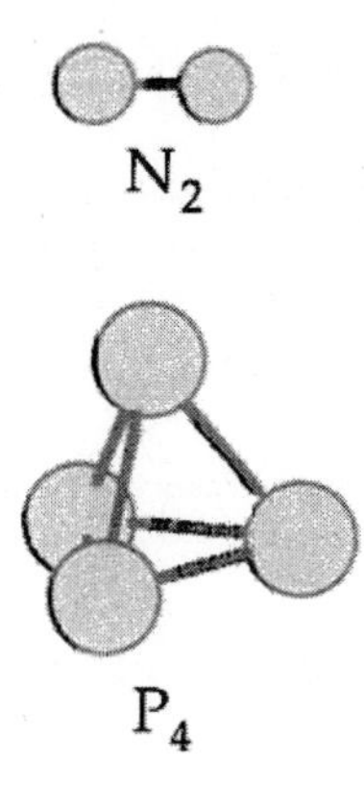

2.9 **Calculate ΔH from mean bond enthalpies?** For the reaction:

$$2H_2(g) + O_2(g) \rightarrow 2H_2O(g)$$

Since you must break two moles of H–H bonds and one mole of O=O bonds on the left-hand side of the equation and form four moles of O–H bonds on the right-hand side, the enthalpy change for the reaction can be estimated as:

$$\Delta H = 2(436\text{ kJ}) + 497\text{ kJ} - 4(463\text{ kJ}) = -483\text{ kJ}$$

The experimental value is –484 kJ, which is in closer agreement with the estimated value than ordinarily expected. Since Table 2.8 contains average bond enthalpies, there is frequently a small error when comparing estimates to a specific reaction.

2.10 **Predict standard enthalpies?** Consider the first reaction:

$$S_2^{2-}(g) + {}^1/_4 S_8(g) \rightarrow S_4^{2-}(g)$$

Hypothetically, two S–S single bonds (of S_8) are broken to produce two S atoms, which combine with S_2^{2-} to form two new S–S single bonds in the product S_4^{2-}. Since two S–S single bonds are broken and two are made, the net enthalpy change is zero. Now consider the second reaction:

$$O_2^{2-}(g) + O_2(g) \rightarrow O_4^{2-}(g)$$

Here there is a difference. A mole of O=O double bond of O_2 is broken, and two moles of O–O single bonds are made. The overall enthalpy change, based on the mean bond enthalpies in Table 2.8, is:

$$O{=}O - 2(O{-}O) = 497\text{ kJ} - 2(146\text{ kJ}) = 205\text{ kJ}$$

The large positive value indicates that this is not a favorable process.

2.11 **Place the compounds AB, AD, BD, and AC in order of increasing covalent character?** Difference in electronegativities are AB 0.5, AD 2.5, BD 2.0, and AC 1.0. The increasing covalent character is AD < BD < AC < AB.

2.12 **What type of bonding for BCl_3, KCl, and BeO?** **(a)** Using the electronegativity values in Table 1.7 and the Ketelaar triangle in Figure 2.38, the $\Delta\chi$ for $BCl_3 = 3.16 - 2.04 = 1.12$ and $\chi_{mean} = 2.60$. This value places BCl_3 in the covalent region of the triangle.
(b) $\Delta\chi$ for KCl $= 3.16 - 0.82 = 2.34$ and $\chi_{mean} = 1.99$. This value places KCl in the ionic region of the triangle.
(c) $\Delta\chi$ for BeO $= 3.44 - 1.57 = 1.87$ and $\chi_{mean} = 2.51$. This value places BeO in the ionic region of the triangle.

2.13 **Predict the hybridization of orbitals? (a)** BCl_3 has a trigonal planar geometry, according to Table 2.4, the most likely hybridization would be sp^2.
(b) NH_4^+ has a tetrahedral geometry, so the most likely hybridization would be sp^3.
(c) SF_4 has distorted see-saw geometry, with the lone pair occupying one of the equatorial sites, see 2.4 above. Therefore it would be sp^3d or spd^3.
(d) XeF_4 has a square planar geometry, so it's hybridization is p^2d^2 or sp^2d.

2.14 **How many unpaired electrons? (a) O_2^-?** You must write the electron configurations for each species, using Figure 2.17, and then apply the Pauli exclusion principle to determine the situation for incompletely filled degenerate orbitals. In this case the electron configuration is $1\sigma_g^2 2\sigma_u^2 3\sigma_g^2 1\pi_u^4 2\pi_g^3$. With three electrons in the pair of $2\pi_g$ molecular orbitals, one electron must be unpaired. Thus, the superoxide anion has a single unpaired electron.
(b) O_2^+? The configuration is $1\sigma_g^2 2\sigma_u^2 3\sigma_g^2 1\pi_u^4 2\pi_g^1$, so the oxygenyl cation also has a single unpaired electron.
(c) BN? You can assume that the energy of the $3\sigma_g$ molecular orbital is *higher* than the energy of the $1\pi_u$ orbitals, since that is the case for CO (see Figure 2.23). Therefore, the configuration is $1\sigma_g^2 2\sigma_u^2 1\pi_u^4$, and, as observed, this diatomic molecule has no unpaired electrons. If the configuration were $1\sigma_g^2 2\sigma_u^2 3\sigma_g^2 1\pi_u^2$, the molecule would have two unpaired electrons since each of the $1\pi_u$ orbitals would contain an unpaired electron, in accordance with the Pauli exclusion principle.
(d) NO^-? The exact ordering of the $3\sigma_g$ and $1\pi_u$ energy levels is not clear in this case, but it is not relevant either as far as the number of unpaired electrons is concerned. The configuration is either $1\sigma_g^2 2\sigma_u^2 1\pi_u^4 3\sigma_g^2 2\pi_g^2$ or it is $1\sigma_g^2 2\sigma_u^2 3\sigma_g^2 1\pi_u^4 2\pi_g^2$. In either case, this anion has two unpaired electrons, and these electrons occupy the set of antibonding $2\pi_g$ molecular orbitals.

2.15 **Writing electron configurations using Figure 2.17? (a) Be_2?** Having only four valence electrons for two Be atoms gives the electron configuration $1\sigma_g^2 2\sigma_u^2$. The HOMO for Be_2 is a σ antibonding orbital, shown below.

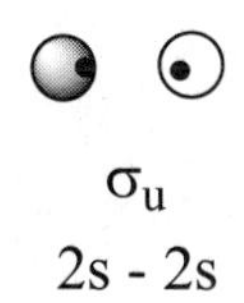

(b) B_2? The electron configuration is $1\sigma_g^2 2\sigma_u^2 1\pi_u^2$. The HOMO for B_2 is a π bonding MO, shown below.

(c) C_2^-? The electron configuration is $1\sigma_g^2 2\sigma_u^2 1\pi_u^4 3\sigma_g^1$. The HOMO for C_2^+ is a σ bonding MO formed from mixing two 2*p* atomic orbitals, shown below.

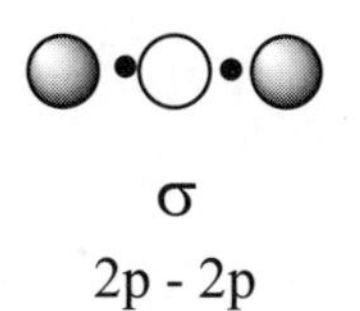

(d) F_2^+? The electron configuration is $1\sigma_g^2 2\sigma_u^2 3\sigma_g^2 1\pi_u^4 2\pi_g^3$. The HOMO for F_2^+ is a π antibonding MO, shown below.

2.16 **Describe the bonding in copper (I) acetylide?** The configuration of C_2^{2-} would be $1\sigma_g^2 1\sigma_u^2 1\pi_u^4 2\sigma_u^2$. The bond order would be ½[2-2+4+2] = 3. So C_2^{2-} has a triple bond, as discussed in the self test **S2.6,** above. The HOMO (highest occupied molecular orbital for C_2^{2-}) is in a sigma orbital. Those two electrons on each carbon atom can donate in a sigma fashion to an empty orbital on each of the copper(I) atoms shown below.

$$\overset{\oplus}{\text{Cu}}-\overset{\ominus}{\text{C}}\equiv\overset{\ominus}{\text{C}}-\overset{\oplus}{\text{Cu}}$$

The configuration for the neutral C_2 would be $1\sigma_g^2 1\sigma_u^2 1\pi_u^4$(see figure 2.17). The bond order would be ½[2-2+4] = 2.

2.17 **Describe the bonding in IBr? (a)** The orbital that would be used to construct the M.O. diagram are the 5p and 5s of I, and the 4p and 4s of Br. IBr is isoelectronic to ICl, so the ground state configuration is the same, $1\sigma_g^2 2\sigma_u^2 3\sigma_g^2 1\pi_u^4 2\pi_g^4$.
(b) The bond order would be ½[2+2-2+4-4] = 1, there would be single bond between I and Br.
(c) IBr^- would have the ground state configuration of $1\sigma_g^2 2\sigma_u^2 3\sigma_g^2 1\pi_u^4 2\pi_g^4 4\sigma_u^1$, with a bond order of ½[2+2-2+4-4-1] = ½, it would not be very stable at all. While IBr^{2-} would have the ground state configuration of $1\sigma_g^2 2\sigma_u^2 3\sigma_g^2 1\pi_u^4 2\pi_g^4 4\sigma_u^2$, with a bond order of ½[2+2-2+4-4-2] = 0, it would not exist, there is no bond between the two atoms.

2.18 **Determining bond orders?** The Lewis structures for the three species are shown below:

$$:\ddot{\text{S}}=\ddot{\text{S}}: \qquad :\ddot{\text{Cl}}-\ddot{\text{Cl}}: \qquad \left[:\ddot{\text{N}}=\ddot{\text{O}}:\right]^-$$

(a) S_2? The electron configuration of this diatomic molecule is $1\sigma_g^2 2\sigma_u^2 3\sigma_g^2 1\pi_u^4 2\pi_g^2$. The bonding molecular orbitals are $1\sigma_g$, $1\pi_u$, and $3\sigma_g$, while the antibonding molecular orbitals are $2\sigma_u$, and $2\pi_g$. Therefore, the bond order is (1/2)((2 + 4 + 2) – (2 + 2)) = 2, which is consistent with the double bond between the S atoms suggested by the Lewis structure.
(b) Cl_2? The electron configuration is $1\sigma_g^2 2\sigma_u^2 3\sigma_g^2 1\pi_u^4 2\pi_g^4$. The bonding and antibonding orbitals are the same as for S_2, above. Therefore, the bond order is (1/2)((2 + 4 + 2) – (2 + 4)) = 1, which is in harmony with the single bond between the Cl atoms indicated by the Lewis structure.
(c) NO^-? The electron configuration of NO^-, $1\sigma_g^2 2\sigma_u^2 1\pi_u^4 3\sigma_g^2 2\pi_g^2$, is the same as the configuration for S_2, shown above. Thus, the bond order for NO^- is 2, as for S_2, once again in harmony with the conclusion based on the Lewis structure.

2.19 **Changes in bond order and bond distance? (a) $O_2 \rightarrow O_2^+ + e^-$?** The molecular orbital electron configuration of O_2 is $1\sigma_g^2 2\sigma_u^2 3\sigma_g^2 1\pi_u^4 2\pi_g^2$. The two $2\pi_g$ orbitals are π antibonding orbitals, so when one of the $2\pi_g$ electrons is removed, the oxygen-oxygen bond order increases from 2 to 2.5. Since the bond in O_2^+ becomes stronger, it should become shorter as well.
(b) $N_2 + e^- \rightarrow N_2^-$? The molecular orbital electron configuration of N_2 is $1\sigma_g^2 2\sigma_u^2 1\pi_u^4 3\sigma_g^2 2\pi_g^4$. The next electron must go into the $4\sigma_u$ orbital, which is σ antibonding (refer to Figures 2.13 and 2.14). This will decrease the nitrogen-nitrogen bond order from 3 to 2.5. Therefore, N_2^- has a weaker and longer bond than N_2.
(c) $NO \rightarrow NO^+ + e^-$? The configuration of the NO molecule is either $1\sigma_g^2 2\sigma_u^2 1\pi_u^4 3\sigma_g^2 2\pi_g^4 4\sigma_u^1$ or $1\sigma_g^2 2\sigma_u^2 3\sigma_g^2 1\pi_u^4 2\pi_g^4 4\sigma_u^1$. Removal of the $4\sigma_u$ antibonding electron will increase the bond order from 2.5 to 3. Therefore, NO^+ has a stronger and shorter bond than NO. Notice that NO^+ and N_2 are isoelectronic.

2.20 **Linear H_4 MOs?** Four atomic orbitals can yield four independent linear combinations. The four relevant ones in this case, for a hypothetical linear H_4 molecule, are shown at the right in order of increasing energy. The most stable orbital has the fewest nodes (i.e., the electrons in this orbital are not excluded from the internuclear regions), the next orbital in energy has only one node, and so on to the fourth and highest energy orbital, with three nodes (a node between each of the four H atoms).

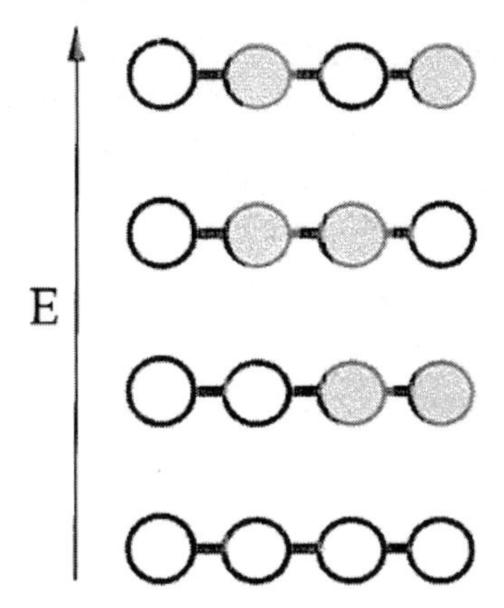

2.21 Molecular orbitals of linear $[HHeH]^{2+}$? By analogy to linear H_3, the three atoms of $[HHeH]^{2+}$ will form a set of three molecular orbitals; one bonding, one nonbonding, and one antibonding. They are shown below. The forms of the wavefunctions are also shown, without normalizing coefficients. You should conclude that He is more electronegative than H because the ionization energy of He is nearly twice that of H. Therefore, the bonding MO has a larger coefficient (larger sphere) for He than for H, and the antibonding MO has a larger coefficient for H than for He. The bonding MO is shown at lowest energy, since it has no nodes. The nonbonding and antibonding orbitals follow at higher energies, since they have one and two nodes, respectively. Since $[HHeH]^{2+}$ has four electrons, only the bonding and nonbonding orbitals are filled. However, the species is probably not stable in isolation because of +/+ repulsions. In solution it would be unstable with respect to proton transfer to another chemical species that can act as a base, such as the solvent or counterion. *Any* substance is more basic than helium.

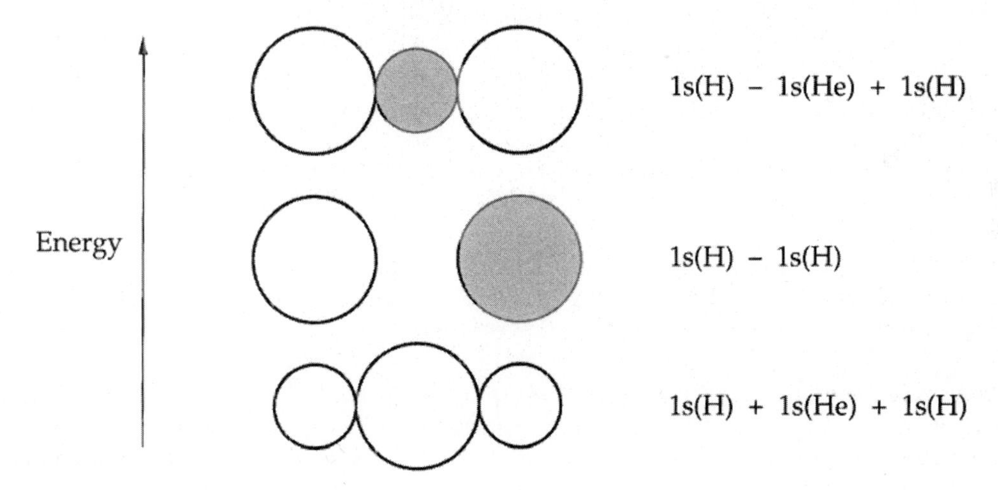

2.22 Average bond order in NH_3? The molecular orbital energy diagram for ammonia is shown in Figure 2.30. The interpretation given in the text was that the $2a_1$ molecular orbital is almost nonbonding, so the electron configuration $1a_1^2 1e^4 2a_1^2$ results in only three bonds ((2 + 4)/2 = 3). Since there are three N–H "links," the average N–H bond order is 1 (3/3 = 1).

2.23 Describe the character of the HOMOs and LUMOs of SF_6? The molecular orbital energy diagram for sulfur hexafluoride is shown in Figure 2.31. The nonbonding *e* HOMOs are pure F atom symmetry-adapted orbitals, and they do not have any S atom character whatsoever. They could only have S atom character if they were bonding or antibonding orbitals composed of atomic orbitals of both types of atoms in the molecule. On the other hand, the antibonding *t* orbitals have both a sulfur and a fluorine character. Since sulfur is less electronegative than fluorine, its valence orbitals lie at higher energy than the valence orbitals of fluorine (from which the *t* symmetry-adapted combinations were formed). Thus, the *t* bonding orbitals lie closer in energy to the F atom *t* combinations and hence they contain more F character; the *t* antibonding orbitals, the LUMOs, lie closer in energy to the S atom $3p$ orbitals and hence they contain more S character.

2.24 **Electron precise or electron deficient? (a) Square H_4^{2+}?** The drawing below shows a square array of four hydrogen atoms. Clearly, each line connecting any two of the atoms is not a (2c,2e) bond, because this molecular ion has only two electrons. Instead, this is a hypothetical example of (4c,2e) bonding. We cannot write a Lewis structure for this species. It is not likely to exist; it should be unstable with respect to two separate H_2^+ diatomic species with (2c,1e) bonds.

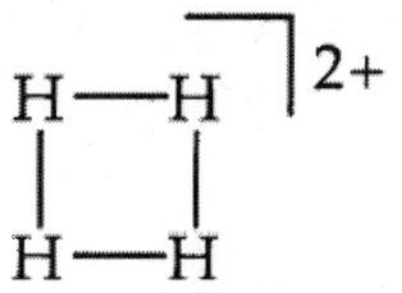

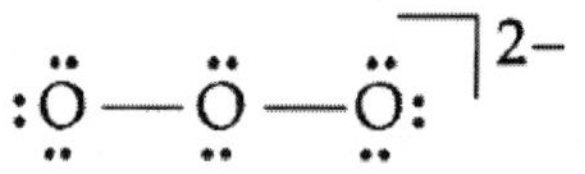

H—H 2+
| |
H—H

:Ö—Ö—Ö: 2−

(b) Bent O_3^{2-}? A proper Lewis structure for this 20-electron ion is shown above. Therefore, it is electron precise. It could very well exist.

Chapter 3 The Structures of Simple Solids

The rock-salt structure of NaCl is an example of an ionic solid built up from a close-packed array of anions. In the drawing below, the small open circles represent Na^+ ions. The larger circles represent Cl^- ions, which are stacked together in a cubic close-packed (i.e., ABCABC ...) array. The direction perpendicular to the plane of close-packed Cl^- ions is along a body diagonal of the unit cell cube: the black Cl^- ions are in A positions, the lightly shaded ones are in B positions, and the heavily shaded ones are in C positions. The Na^+ ions are in *all* of the octahedral holes formed by the Cl^- ion array.

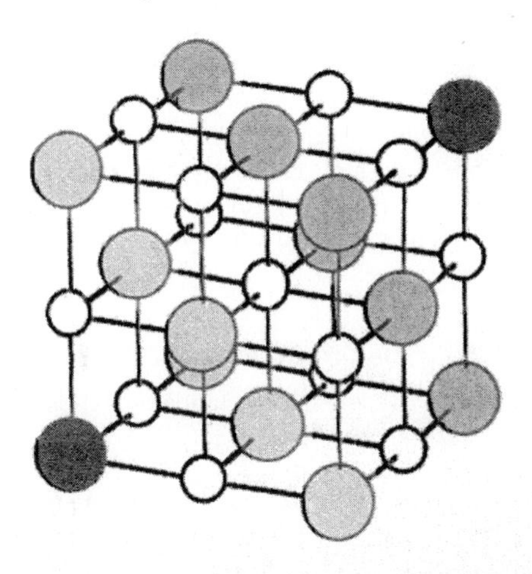

S3.1 **Lattice type of CsCl?** By examining Fig. 3.7 and Fig. 3.32, we note that the caesium cations sit on a primitive cubic unit cell (lattice type P) with chloride anion occupying the cubic hole in the body center. Alternatively, one can view the structure as P type lattice of chloride anions with caesium cation in cubic hole.

S3.2 **Convert projection diagram of SiS_2 into 3D representation?** Silicon sulfide has the wurtzite structure shown in Figure 3.35.

S3.3 **Fraction of space occupied in primitive cubic unit cell?** See Figure 3.3. Each unit cell contains 1 sphere (equivalent to 8 × 1/8 spheres on the vertices in contact along the edges). With a radius r the volume within the sphere is $4/3\ \pi\ r^3$ and the volume of the unit cell is $(2r)^3$, so the fraction filled is $4/3\ \pi\ r^3/\ 8r^3 = 0.52$. Thus, 52% of the free volume in the primitive cubic unit cell is filled.

S3.4 **The size of a tetrahedral hole?** See Figure 3.19b. Note that line S–T = $r + r_h$, by definition. Therefore, you must express S–T in terms of r. Note also that S–T is the hypoteneuse of the right triangle STM, with sides S ... M and M–T. Point M is at the midpoint of line S ... S, and since S ... S = $2r$, S ... M = r. The angle θ is 54.74°, one-half of the tetrahedral angle S–T–S (109.48°). Therefore, sin 54.74° = $r/(r + r_h)$, and $r_h = 0.225r$. This is the same as $r_h = ((3/2)^{1/2} - 1)r$.

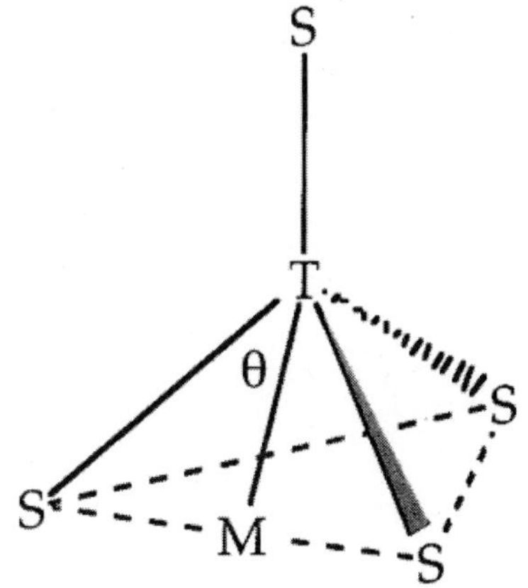

S3.4 **Maximum size of sphere in face of bcc unit cell?** Considering Figure 3.22 and the answer to Example 3.4, the distorted octahedral site has a minimum vertex to vertex distance of $\sqrt{2}a - 2r$ where a is the bcc unit cell dimension and r is the sphere radius. Thus the maximum ion radius for this site is $\frac{1}{2}(\sqrt{2}(4r/\sqrt{3}) - 2r) = 0.63r$.

S3.5 **Lattice parameter of Ag?** The fcc unit cell contains 4 atoms that weigh together (4 mol Ag × 107.87 g/mol)/(6.022 × 10^{23} Ag/mol) g or 7.165 × 10^{-22} g. The volume of the unit cell is a^3, where a is the length of the edge of the unit cell. The density equals mass divided by volume or:

$$10.5\ \text{g cm}^{-3} = 7.165 \times 10^{-22}\ \text{g} / a^3 \text{ with a in cm}$$
$$a = 4.09 \times 10^{-8}\ \text{cm or } 409\ \text{pm}$$

S3.6 **Lattice parameter for Po?** For the bcc structure, $a = 4r/\sqrt{3}$. (This relation can be derived simply by considering the right triangle formed from the body diagonal, face diagonal and edge of the bcc unit cell. Using the Pythagorean theorem, we have $(4r)^2 = a^2 + (a\sqrt{2})^2$. Solving for a in terms of r, we have $a = 4r/\sqrt{3}$.) From Example 3.6, we know that the metallic radius of Po is 174 pm. Therefore, using the derived relation we can calculate a = 401 pm.

S3.7 **Stoichiometry and lattice type of the iron/chromium alloy?** Again the translation symmetry is that of primitive cubic and the stoichiometry yields 1.5 Cr [4(1/8) corners + 1(1) body] atoms and 0.5 Fe [4(1/8) corners] atoms per unit cell. The formula for the alloy is therefore $FeCr_3$.

S3.8 **Stoichiometry of an hcp array with two-thirds of the octahedral sites occupied?** For close packed structures, n cp ions leads to n octahedral sites. Therefore, is we assume 3 close packed anions (A), then there are two cations (X) in the octahedral sites and the stoichiometry for the solid is X_2A_3.

S3.9 **The coordination number of Ti in rutile?** The rutile structure is shown in Figure 3.39. There is a Ti^{4+} ion in the center of the structure, and it is connected to six O^{2-} ions to form an octahedral TiO_6 coordination unit. Therefore, the coordination number of Ti in rutile is 6.

S3.10 **Perovskite composition with La(III)?** La(III) would have to go on the larger A site, and to maintain charge balance it would be paired with In(III) and thus the composition would be $LaInO_3$.

S3.11 **Calculate ΔH_L for $MgBr_2$?** You should proceed as in the example, calculating the total enthalpy change for the Born-Haber cycle and setting it equal to $\Delta_L H$. In this case, it is important to recognize that two Br^- ions are required, so the enthalpy changes for (i) vaporization of $Br_2(l)$ and (ii) breaking the Br–Br bond in $Br_2(g)$ are used without dividing by 2, as was done for KCl. Furthermore, the first *and* second ionization enthalpies for Mg(g) must be added together for the process $Mg(g) \rightarrow Mg^{2+}(g) + 2e^-$. The Born-Haber cycle for $MgBr_2$ is shown below, with all of the enthalpy changes given in kJ mol^{-1}. These enthalpy changes are not to scale. The lattice enthalpy is equal to 2421 kJ mol^{-1}. Note that $MgBr_2$ is a stable compound despite the enormous enthalpy of ionization of magnesium. This is because the very large lattice enthalpy more than compensates for this positive enthalpy term. Note the standard convention used; lattice enthalpies are positive enthalpy changes.

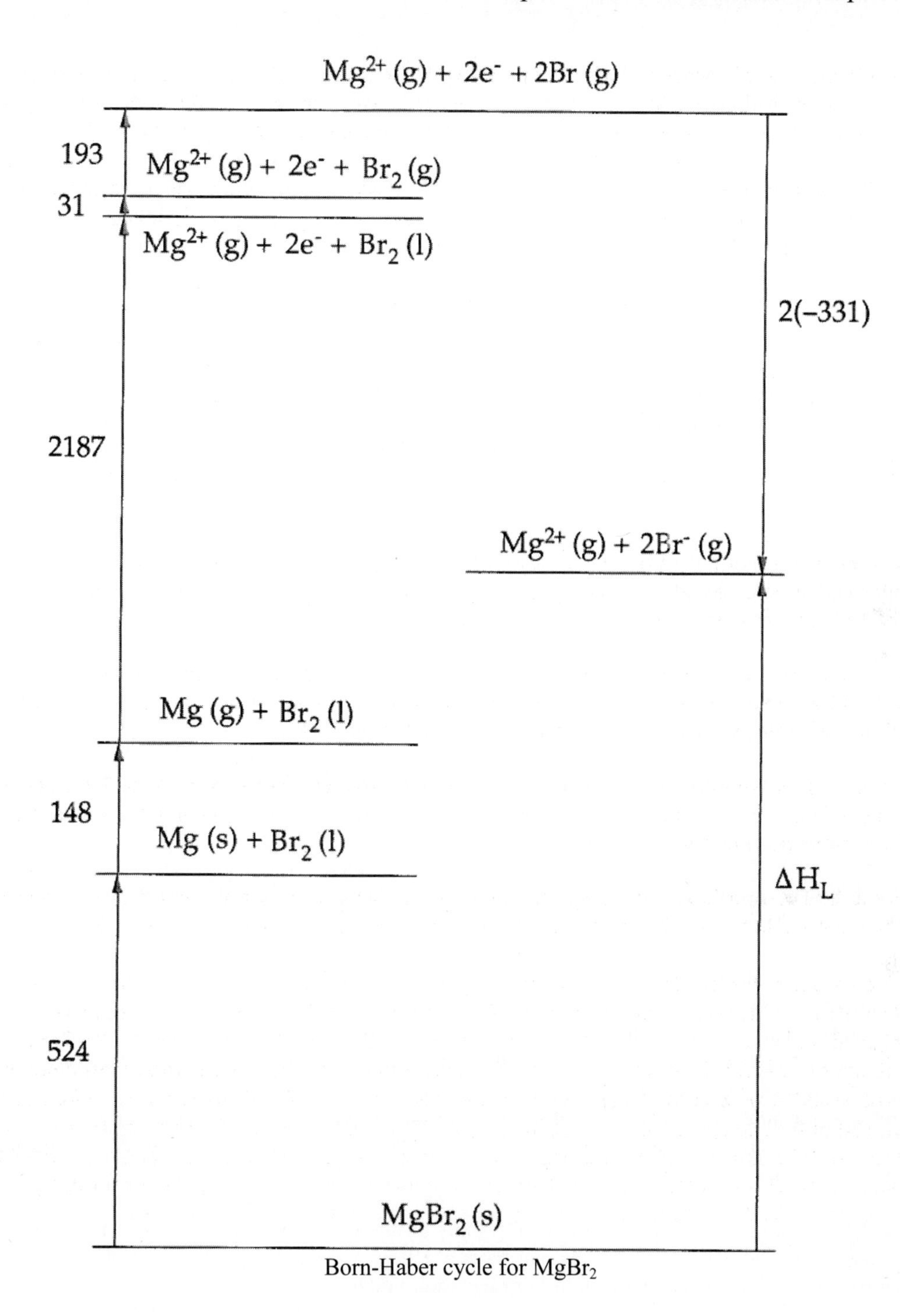

Born-Haber cycle for $MgBr_2$

S3.12 **Is $CsCl_2$ with fluorite structure likely to exist?** This compound is unlikely to exist owing to a large positive value for the heat of formation for $CsCl_2$ that is mainly from the large second ionization energy for Cs. The compound is predicted to be unstable with respect to its elements mainly because the large ionization enthalpy to form Cs^{2+} is not compensated by the lattice enthalpy.

S3.13 Order of decomposition temperatures for alkaline earth sulfates? The enthalpy change for the reaction

$$MSO_4(s) \rightarrow MO(s) + SO_3(g)$$

includes several terms, including the lattice enthalpy for MSO_4, the lattice enthalpy for MO, and the enthalpy change for removing O^{2-} from SO_4^{2-}. The last of these is constant as you change M^{2+} from Mg^{2+} to Ba^{2+}, but the lattice enthalpies change considerably. The lattice enthalpies for $MgSO_4$ and MgO are both larger than those for $BaSO_4$ and BaO, simply because Mg^{2+} is a smaller cation than Ba^{2+}. However, the *difference* between the lattice enthalpies for $MgSO_4$ and $BaSO_4$ is a smaller number than the difference between the lattice enthalpies for MgO and BaO (the larger the anion, the less changing the size of the cation affects ΔH_L). Thus going from $MgSO_4$ to MgO is thermodynamically more favorable than going from $BaSO_4$ to BaO, because the *change* in ΔH_L is greater for the former than for the latter. Therefore, magnesium sulfate will have the lowest decomposition temperature and barium sulfate the highest and the order will be $MgSO_4 < CaSO_4 < SrSO_4 < BaSO_4$.

S3.14 Which is more soluble in water, $NaClO_4$ or $KClO_4$? You should study Section 3.15(c), *Solubility*. The most important concept to remember is the general rule that compounds that contain ions with widely different radii are more soluble in water than compounds containing ions with similar radii. The six-coordinate radii of Na^+ and K^+ are 1.02 and 1.38 Å, respectively (see Table 1.4), while the thermochemical radius of the perchlorate ion is 2.36 Å (see Table 3.10). Therefore, since the radii of Na^+ and ClO_4^- differ more than the radii of K^+ and ClO_4^-, the salt $NaClO_4$ should be more soluble in water than $KClO_4$.

S3.15 Intrinsic defects for CsF? CsF has the rock-salt structure and ionic bonding. This type of compound generally forms Schottky defects.

S3.16 What elements other than As might be used to form extrinsic defects in silicon? We need to identify ions of similar charge (+4) and size (r = 40 pm) to silicon. Ionic radii are listed in Resource Section 1. Two obvious choices are phosphorus (r = 31 pm) and aluminium (r = 53).

S3.17 Which *d*-orbitals overlap in metal with primitive structure? The d_{x2-y2} and d_{z2} have lobes pointing along the cell edges to the nearest neighbor metals. See Figure 1.15 for review of the shape of *d*-orbitals.

S3.18 *p*- or *n*-type semiconductors? (a) V_2O_5? *n*-type is expected when a metal is in a high oxidation state such as vanadium(V) and is likely to undergo reduction. **(b) CoO?** *p*-type is expected when a metal is in a lower oxidation state and is likely to undergo oxidation. Recall that upon oxidation, holes are created in the conduction band of the metal and the charge carriers are now positive, leading the classification.

3.1 Unit cell parameters in orthorhombic crystal system? By consulting Table 3.1 for the orthorhombic crystal system we have $a \neq b \neq c$ and $\alpha = 90°$, $\beta = 90°$, $\gamma = 90°$. See Figure 3.2 for a three-dimensional structure of this type of unit cell.

3.2 Fractional coordinates for fcc unit cell? Points on the cell corners at (0,0,0), (1,0,0), (0,1,0), (0,0,1), (1,1,0), (1,0,1), (0,1,1), and (1,1,1) and in the cell faces at (½,½,0), (½,1,½), (0,½,½) (½,½,1), (½,1,½), and (1,½,½). See Figure 3.9 for insight on the projection representation of the fcc unit cell.

3.3 Which of the following are close-packed? (a) ABCABC ...? Any ordering scheme of planes is close-packed if no two adjacent planes have the same position (i.e., if no two planes are *in register*). When two planes are in register, the packing looks like the figure below and to the right, whereas the packing in a close-packed structure allows the atoms of one plane to fit more efficiently into the spaces between the atoms in an adjacent plane, like the figure below and to the left. Notice that the empty spaces between the atoms in the figure to the left are much smaller than in the figure to the right. The efficient packing exhibited by close-packed structures is why, for a given type of atom, close-packed structures are denser than any other possible structure. In the case of an ABCABC ... structure, no two adjacent planes are in register, so the ordering scheme is close-packed.

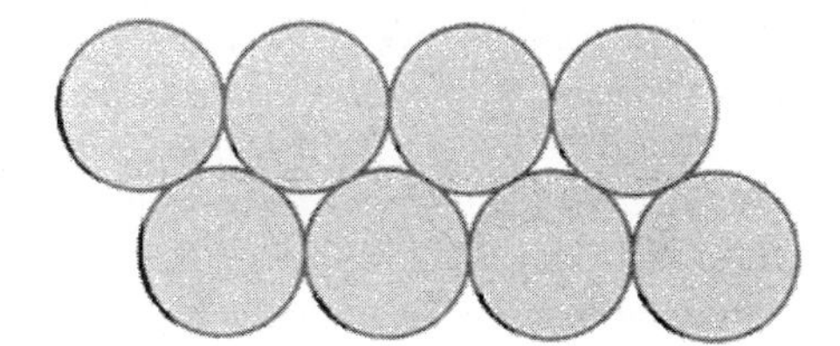
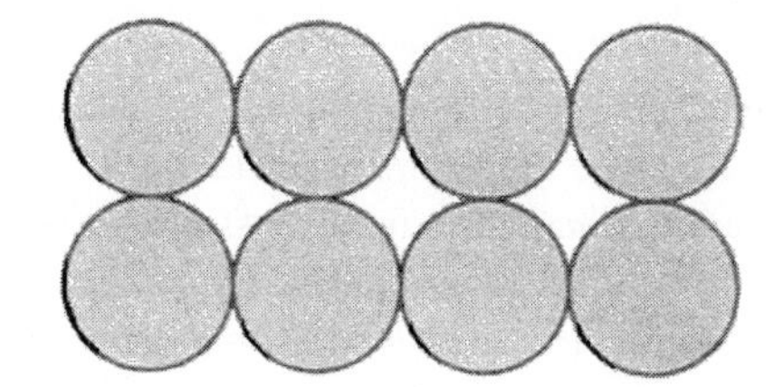

(b) ABAC …? Once again, no two adjacent planes are in register, so the ordering scheme is close-packed.
(c) ABBA …? The packing of planes using this sequence will put two B planes next to each other as well as two A planes next to each other, so the ordering scheme is not close-packed.
(d) ABCBC …? No two adjacent planes are in register, so the ordering scheme is close-packed.
(e) ABABC …? No two adjacent planes are in register, so the ordering scheme is close-packed.
(f) ABCCB …? The packing of planes using this sequence will put two C planes next to each other, so the ordering scheme is not close-packed.

3.4 **Formula?** In any close packed structure with n closed packed ions there are n octahedral holes and 2n tetrahedral holes. If we assume four close packed anions in the unit cell, then we have eight tetrahedral holes available. If one quarter of those are filled, then we have two X cations. The ratio of cations to anions is one to two and the formula is XA_2.

3.5 **Potassium fulleride stoichiometry?** Within the fcc lattice of fullerides shown in Figure 3.16, there are the equivalent of four close packed molecules (1/8(8) corners + ½(6) faces). This lattice contains the equivalent of four octahedral holes and eight tetrahedral holes as shown in Figure 3.18. If the potassium cations occupy all of these holes then we have twelve K^+ ions for every four C_{60} anions and the formula is K_3C_{60}. See Figure 24.69 for further visualization of the structure.

3.5 **Atomic radius for 12 coordinate Cs?** By consulting Table 3.3 we note that the relative radii of 12 and 8 coordination are 1 : 0.97 in the bcc lattice. When Cs is in 8-fold coordination, the radius is 272 pm. Multiply 272 by 1/0.97 to obtain the new 12-coordinate atomic radius equal to 281 pm.

3.6 **Length of the edge of the unit cell for metallic sodium (density of 970 kg m^{-3})?** The bcc unit cell contains 2 atoms that weigh together (2 mol Na × 22.99 g/mol)/(6.022 × 10^{23} Na/mol) g or 7.635 × 10^{-26} kg. The volume of the unit cell is a^3, where a is the length of the edge of the unit cell. The density equals mass divided by volume or:

$$970 \text{ kg m}^{-3} = 7.635 \times 10^{-26} \text{ kg} / a^3 \text{ with a in m}$$

$$a = 4.29 \times 10^{-10} \text{ m or } 429 \text{ pm}$$

3.7 **Copper and gold alloy?** The composition can be determined by counting atoms in the unit cell shown in Figure 3.75. Six face copper atoms times one-half gives 3 Cu atoms per unit cell and 8 corner gold atoms times one-eighth gives one Au per unit cell with an overall composition for the alloy of Cu_3Au. The unit cell considering just the Au atoms is primitive cubic. The mass % of Au in this alloy is approximately 50 %. Pure gold is 24 carat (100 % gold). This alloy that contains 50% by mass gold would therefore be 12 carat.

3.9 **Ketelaar's triangle and Sr_2Ga?** The electronegative difference is 0.86 for Sr and Ga and the mean is 1.38. Using Ketelaar's triangle and these values for the *y* and *x* axes, respectively, we find that the compound is an alloy (not in the Zintl phase region).

3.10 **Rubidium chloride? (a) Coordination numbers?** The rock-salt polymorph of RbCl is based on a ccp array of Cl^- ions in which the Rb^+ ions occupy all of the octahedral holes. An octahedron has six vertices, so the Rb^+ ions are six-coordinate. Since RbCl is a 1:1 salt, the Cl^- ions must be six-coordinate as well. The cesium-chloride polymorph is based on a cubic array of Cl^- ions with Rb^+ ions at the unit cell centers. A cube has eight vertices, so the Rb^+ ions are eight-coordinate, and therefore the Cl^- ions are also eight-coordinate.

(b) Larger Rb^+ radius? If more anions are packed around a given cation, the hole that the cation sits in will be larger. You saw an example of this when the radii of tetrahedral ($0.225r$) and octahedral holes ($0.414r$) were compared (r is the radius of the anion). Therefore, the cubic hole in the cesium-chloride structure is larger than the octahedral hole in the rock-salt structure. A larger hole means a longer distance between the cation and anion, and hence a larger apparent radius of the cation. Therefore, the apparent radius of rubidium is larger when RbCl has the cesium-chloride structure and smaller when RbCl has the rock-salt structure.

3.11 **Caesium-chloride structure and Cs^+ second-nearest neighbors?** The unit cell for this structure is shown in Figure 3.30. Each unit cell is surrounded by six equivalent unit cells; each of these unit cells shares a face with its six neighbors. Since each of these unit cells contains a Cs^+ ion at its center, each Cs^+ ion has six second-nearest neighbors that are Cs^+ ions, one in the center of each of the six neighboring unit cells.

3.11 **Coordination around the anions?** In the perovskite structure, there are 2B type cations and 4A type cations arranged in a distorted octahedral arrangement around each anion. See Figure 3.42 for more information on the perovskite structure.

3.13 **Structures using radius ratios?** Consult Table 3.6 and Resource Section 1. **(a) PuO_2?** $\rho = 0.78$, so fluorite; **(b) FrI?** $\rho = 0.94$, so CsCl; **(c) BeO?** $\rho = 0.19$, so ZnS; **(d) InN?** $\rho = 0.46$, so NaCl.

3.14 **Structure of FrBr?** Consulting Table 3.6 and Resource Section 1 we have $\rho = 0.93$, so the structure is CsCl.

3.15 **Significant terms in the Born-Haber cycle for Ca_3N_2?** The most important terms will involve the lattice enthalpies for the di- and the trivalent ions. Also, the bond energy and third electron gain enthalpy for nitrogen will be large. See Section 3.11 for more details.

3.16 **Lattice enthalpy for MgO and AlN?** Since the charges on both the anion and the cation are doubled in the Born-Mayer equation, the lattice enthalpy for MgO will be equal to four times the NaCl value or 3144 $kJmol^{-1}$. For AlN the charges are tripled and the lattice enthalpy will be equal to nine times the NaCl value or 7074 $kJmol^{-1}$.

3.17 **Lattice enthalpies?** You will need to consult Table 3.10 and Resource Section 1 to solve this problem.
(a) BkO_2? Using $r(Bk^+) = 97$ pm and $r(O^{2-}) = 128$ pm and equation 3.4 we have:

$$\Delta H_L = [(3 \times 4 \times 2)/(97 + 128)][(1 - 34.5)/(97 + 128)]\ (1.21\ \text{MJ mol}^{-1})$$

$$= 0.0179 \times 0.845 \times 1.21 \times 10^5\ \text{kJ mol}^{-1} = 10906\ \text{kJmol}^{-1}$$

(b) K_2SiF_6? Using $r(K^+) = 152$ pm and $r([SiF_6]^{2-}) = 194$ pm and equation 3.4 we have:

$$\Delta H_L = [(3 \times 2)/(152 + 194)] \times 0.9 \times 1.21 \times 10^5\ \text{kJ mol}^{-1} = 1888\ \text{kJ mol}^{-1}$$

(c) $LiClO_4$? Using $r(Li^+) = 90$ pm and $r([ClO_4]^-) = 236$ pm and equation 3.4 we have:

$$\Delta H_L = [(2 \times 1)/(236 + 90)] \times 0.895 \times 1.21 \times 10^5\ \text{kJ mol}^{-1} = 664\ \text{kJ mol}^{-1}$$

3.18 **Which is more soluble? (a) $MgSO_4$ or $SrSO_4$?** In general, difference in size of the ions favours solubility in water. The thermochemical radius of sulfate ion is 2.30 Å, while the six-coordinate radii of Mg^{2+} and Sr^{2+} are 0.72 and 1.16 Å, respectively. Since the difference in size between Mg^{2+} and SO_4^{2-} is greater than the difference in size between Sr^{2+} and SO_4^{2-}, $MgSO_4$ is predicted to be more soluble in water than $SrSO_4$, and this is in fact the case: $K_{sp}(MgSO_4) > K_{sp}(SrSO_4)$.
(b) NaF or $NaBF_4$? This exercise can be answered without referring to tables in the text. The ions Na^+ and F^- are isoelectronic, so clearly Na^+ is smaller than F^-. It should also be obvious that the radius of BF_4^- is larger than the radius of F^-. Therefore, the difference in size between Na^+ and BF_4^- is greater than the difference in size between Na^+ and F^-; $NaBF_4$ is more soluble in water than NaF.

3.19 **Order of increasing lattice enthalpies?** CsI < RbCl < LiF < CaO < NiO < AlN

The values of ΔH_L are proportional to the product of the charges on ions divided by the sum of their ionic radii. So the 3+/3– system will get a ×9 multiplier, while the 2+/2– will have a ×4 on the 1+/1– system. Within compounds that have the same charges, the bigger ions lead to larger sums on the radii and smaller lattice energies. Considering both of these factors, one can substantiate the trend shown above for increasing lattice enthalpies.

3.20 **Precipitation of carbonate ion in water?** Ba^{2+} is a good choice; recall that the solubilities decrease with increasing radius of the cation (see Example 3.14).

3.21 **Intrinsic defects? (a)** Ca_3N_2 has ionic bonding with anti-bixbyite structure common to manganese oxide with ion positions reversed. This compound is likely to exhibit Schottky defects owing to ites structure type and bonding. **(b)** HgS assumes hexagonal close packed structure with low coordination numbers and partial covalent bonding character. This compound is likely to exhibit Frenkel defects.

3.22 **Defects in a solid increase with increased temperature?** The formation of defects is normally endothermic because as the lattice is disrupted the enthalpy of the solid rises. However, the term –TS becomes more negative as defects are formed because they introduce more disorder into the lattice and the entropy rises. As long as we are not at absolute zero, the Gibbs energy will have a minimum at a nonzero concentration of defects and their formation will be spontaneous. As temperature is raised, this minimum in G shifts to higher defect concentrations as shown in Figure 3.52b, so solids have a greater number of defects as temperatures approaches their melting points.

3.23 **Dopant ions blue form of beryl known as aquamarine?** The likely dopants from Table B3.1 are Fe^{2+} and Ti^{4+} replacing Al^{3+} in adjacent octahedral sites. The origin of the blue color involves electron transfer from these cationic centres.

3.24 **Nonstoichiometry for magnesium oxide, vanadium carbide, manganese oxide?** As noted in section 3.17(a), nonstochiometry is common in the solid-state compounds of d-, f- and later p-block elements. We would therefore expect vanandium carbide and manganese oxide to exhibit nonstoichiometry but not magnesium oxide.

3.25 **Would VO or NiO be expected to show metallic properties?** Low oxidation number d-metal oxides can lose electrons through a process equivalent to the oxidation of the metal atoms, with the result that holes appear in the predominately metal band. The positive charge carriers result in their p-type semiconductor classification. NiO is an example of this p-type semiconduction. Early transitional metal oxides with low oxidation number such as TiO and VO have metallic properties owing to the extended overlap of the d orbitals of the cations. See Section 24.6b for more details.

3.26 **Semiconductor or semimetal?** A semiconductor is a substance with an electrical conductivity that decreases with increasing temperature. It has a small, measurable band gap. A semimetal is a solid whose band structure has a zero density of states and no measurable band gap as shown in Figure 3.70. Graphite is a classis example of a semimetal with conduction in the plane parallel to the sheets of carbon atoms.

3.27 **n- or p-type semiconductivity?** Ag_2S and CuBr (low oxidation number metal chalcogenide and halide) would be p-type and VO_2 (high oxidation number transition metal oxide) would be n-type.

Chapter 4 Acids and Bases

A simple Lewis acid–Lewis base complex, formed by mixing the Lewis acid BF_3 with the Lewis base CH_3CN. In this complex, the structure of the Lewis acid has changed from planar to pyramidal, while the structure of the Lewis base has hardly changed at all.

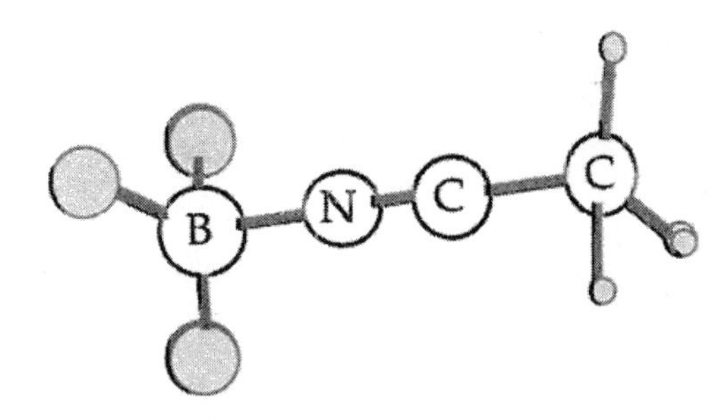

S4.1 **Identifying acids and bases? (a) $HNO_3 + H_2O \rightarrow H_3O^+ + NO_3^-$?** The compound HNO_3 transfers a proton *to* water, so it is an acid. The nitrate ion is its conjugate base. In this reaction, H_2O accepts a proton, so it is a base. The hydronium ion, H_3O^+, is its conjugate acid.

(b) $CO_3^{2-} + H_2O \rightarrow HCO_3^- + OH^-$? A carbonate ion accepts a proton from water, so it is a base. The hydrogen carbonate, or bicarbonate, ion is its conjugate acid. In this reaction, H_2O donates a proton, so it is an acid. A hydroxide ion is its conjugate base.

(c) $NH_3 + H_2S \rightarrow NH_4^+ + HS^-$? Ammonia accepts a proton from hydrogen sulfide, so it is a base. The ammonium ion, NH_4^+, is its conjugate acid. Since hydrogen sulfide donated a proton, it is an acid, while HS^- is its conjugate base.

S4.2 **What is the pH of a 0.10 M HF solution?** Weak acids only partially ionize in water, so the equation is:

$$HF + H_2O \rightleftharpoons H_3O^+ + F^-$$

Applying what you have learned in general chemistry, this type of problem can best be worked using the ICE technique, initial, change, and final concentrations according to the diagram below.

	$HF + H_2O$	$\rightleftharpoons$	H_3O^+	$+ F^-$
Initial (M)	0.10		0.0	0.0
Change (M)	-x		+x	+x
Equilibrium (M)	0.10 - x		x	x

Then plug the equilibrium values into the equilibrium expression.

$$K_a = \frac{[H_3O^+][F^-]}{[HF]} = 3.5 \times 10^{-4} = \frac{x^2}{0.10 - x}$$

Now, we can assume the $0.10 - x \approx 0.10$ then solve for x, which will be our $[H_3O^+]$.

$$3.5 \times 10^{-4} = \frac{x^2}{0.10}$$

$$x = 5.9 \times 10^{-3} = [H_3O^+]$$

To test the approximation, divide the calculated concentration of $[H_3O^+]$ by the initial concentration of the acid, and multiply be 100. This is a percent ionization. If this is below 5%, our approximation is valid.

$$\frac{5.9 \times 10^{-3}}{0.10} \times 100 = 5.9\%$$

Since this is greater than 5%, our approximation is not valid and we must solve the quadratic equation as follows:

$$x^2 + 3.5 \times 10^{-4}x - 3.5 \times 10^{-5} = 0$$

$$x = \frac{3.5 \times 10^{-4} \pm \sqrt{(3.5 \times 10^{-4})^2 - 4(1)(-3.5 \times 10^{-5})}}{2(1)}$$

The two roots you get are x = 5.7×10^{-3} M and -6.1×10^{-3} M, obviously the negative root is physically impossible, so our $[H_3O^+] = 5.7 \times 10^{-3}$ M. Taking the negative log gives the pH of the solution.

$$pH = -\log[H_3O^+] = 2.24$$

S4.3 **Calculate the pH of a 0.20 M tartaric acid solution?** This is a diprotic weak acid and must be solved in two equilibrium steps, similar to **S4.2.** The first equilibrium, and ICE table is shown below:

	$H_2C_4O_6$ + H_2O ⇌	H_3O^+	+ $HC_4O_6^-$
Initial (M)	0.20	0.0	0.0
Change (M)	-x	+x	+x
Equilibrium (M)	0.20 - x	x	x

Since the value of the K_a is near the magnitude of the concentration, an assumption will not work for the first deprotonation reaction, so we must solve the quadratic equation as follows:

$$x^2 + 1.0 \times 10^{-3}x - 2.0 \times 10^{-4} = 0$$

$$x = \frac{1.0 \times 10^{-3} \pm \sqrt{(1.0 \times 10^{-3})^2 - 4(1)(-2.0 \times 10^{-4})}}{2(1)}$$

$$x = 0.014 \text{ M} \quad \text{or } -0.015 \text{ M}$$

So, this means our $[H_3O^+] = 0.014$ M which is also equal to the $[HC_4O_6^-]$. Now we do the same process starting with $[HC_4O_6^-]$ as shown below.

	$HC_4O_6^-$ + H_2O ⇌	H_3O^+	+ $C_4O_6^{2-}$
Initial (M)	0.014	0.014	0.0
Change (M)	-x	+x	+x
Equilibrium (M)	0.014 - x	0.014 + x	x

Then plug the equilibrium values into the equilibrium expression:

$$4.6 \times 10^{-5} = \frac{(0.014 + x)(x)}{(0.014 - x)}$$

Now, we can assume the $0.014 \pm x \approx 0.014$ then x is simply equal to 4.6×10^{-5} M. We can now test if the assumption is valid.

$$\frac{4.6 \times 10^{-5}}{0.014} \times 100 = 0.33\%$$

This value is less than 5%, so our assumption is valid. This means our total $[H_3O^+] = 0.014 \text{ M} + 4.6 \times 10^{-5}$ which is essentially 0.014 M. Meaning the second deprotonation does not effect the pH in this solution. Our pH is:

$$pH = -\log[H_3O^+] = 1.85$$

S4.4 **Which solvent?** The two solvents in Figure 4.4 for which the window covers the range 25 to 27 are dimethylsulfoxide (DMSO) and ammonia.

S4.5 **Is aKBrF$_4$ an acid or a base in BrF$_3$?** We need to identify the autoionization products of the solvent and then decide whether the solute increases the concentration of the cation (an acid) or the anion (a base). The autoinization products of the the solvent are:

$$2BrF_3 \rightleftharpoons BrF_2^+ + BrF_4^-$$

The solute would ionize to give K^+ and BrF_4^-, since BrF_4^- will increase the amount of the anion, $KBrF_4$ is a base when dissolved in BrF_3.

S4.6 **Arrange in order of increasing acidity? $[Na(H_2O)_6]^+$, $[Sc(H_2O)_6]^{3+}$, $[Mn(H_2O)_6]^{2+}$, and $[Ni(H_2O)_6]^{2+}$?** Since the acid strength of aqua acids increases as the electrostatic parameter, $\xi = z^2/(r + d)$, increases, the strongest acid will have the highest charge, at least for a group of acids for which $r + d$ does not vary too much. Thus $[Na(H_2O)_6]^+$, with the lowest charge, will be the weakest of the four aqua acids, and $[Sc(H_2O)_6]^{3+}$, with the highest charge, will be the strongest. The remaining two aqua acids have the same charge, and so the one with the smaller ionic radius, r, will have the smaller value of $r + d$ and hence the greater acidity. Since Ni^{2+} has a greater Z_{eff} than Mn^{2+}, it has a smaller radius, and so $[Ni(H_2O)_6]^{2+}$ is more acidic than $[Mn(H_2O_6]^{2+}$. The order of increasing acidity is $[Na(H_2O)_6]^+ < [Mn(H_2O)_6]^{2+} < [Ni(H_2O)_6]^{2+} < [Sc(H_2O)_6]^{3+}$.

S4.7 **Predict pK_a values? (a) H$_3$PO$_4$?** Pauling's first rule for predicting the pK_a of a mononuclear oxoacid is p$K_a \approx 8 - 5p$ (where p is the number of oxo groups attached to the central element).
Since $p = 1$, the predicted value of pK_a for H_3PO_4 is $8 - (5 \times 1) = 3$. The actual value, given in Table 4.1, is 2.1.

O
P
HO OH
OH

(b) $H_2PO_4^-$? Pauling's second rule for predicting the pK_a of a mononuclear oxoacid is that successive pK_a values for polyprotic acids increase by five units for each successive proton transfer. Since pK_a(1) for H_3PO_4 was predicted to be 3 (see above), the predicted value of pK_a for $H_2PO_4^-$, which is pK_a(2) for H_3PO_4, is $3 + 5 = 8$. The actual value, given in Table 4.1, is 7.4.
(c) HPO_4^{2-}? The pK_a for HPO_4^{2-} is the same as pK_a(3) for H_3PO_4, so the predicted value is $3 + (2 \times 5) = 13$. The actual value, given in Table 4.1, is 12.7.

S4.8 **What happens to Ti(IV) in aqueous solution as the pH is raised?** According to Figure 4.5, Ti(IV) is amphoteric. Treatment of an aqueous solution containing Ti(IV) ions with ammonia causes the precipitation of TiO_2, but further treatment with NaOH causes the TiO_2 to redissolve.

S4.9 **Identify the acids and bases?** A general rule that works in many, but not all, instances is that negatively charged ions are Lewis bases and positively charged ions are Lewis acids. This rule certainly works in the three parts of this exercise. However, be alert for the possibility that a charged species may be neither acidic nor basic, just as an electrically neutral species may be neither acidic nor basic.
(a) $FeCl_3 + Cl^- \rightarrow [FeCl_4]^-$? The acid $FeCl_3$ forms a complex, $[FeCl_4]^-$, with the base Cl^-.
(b) $I^- + I_2 \rightarrow I_3^-$? The acid I_2 forms a complex, I_3^-, with the base I^-.

S4.10 **The difference in structure between $(H_3Si)_3N$ and $(H_3C)_3N$?** If the N atom lone pair of $(H_3Si)_3N$ is delocalized onto the three Si atoms, it cannot exert its normal steric influence as predicted by VSEPR rules. Therefore, the N atom of $(H_3Si)_3N$ is trigonal planar, whereas the N atom of $(H_3C)_3N$ is trigonal pyramidal.
The structures of $(H_3Si)_3N$ and $(H_3C)_3$ N, excluding the hydrogen atoms:

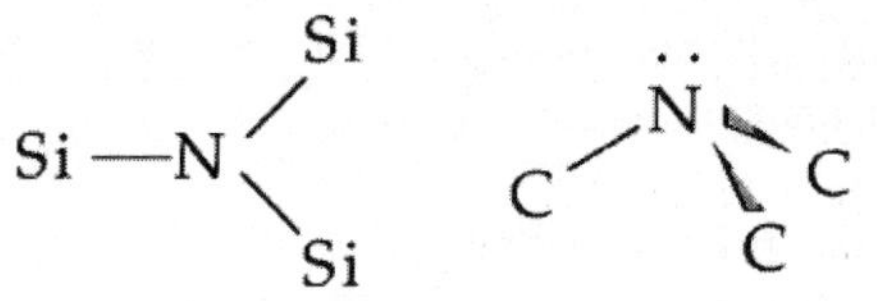

S4.11 Draw the structure of $BF_3 \cdot OEt_2$? The ether oxygen atom will form a dative bond with the boron atom of BF_3. The structure around the boron atom will go from trigonal-planar in BF_3 to tetrahedral in $F_3B–OEt_2$. The structure of the complex is shown below.

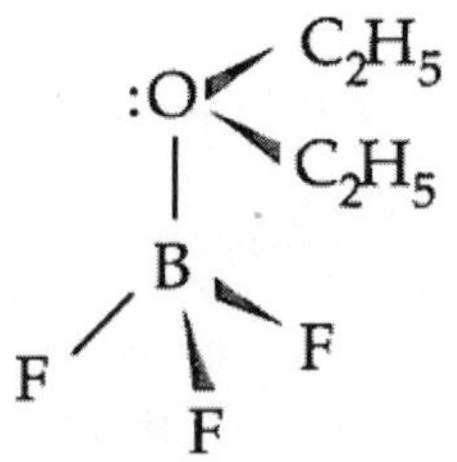

4.1 Sketch an outline of the *s* and *p* blocks of the periodic table, showing the elements that form acidic, basic, and amphoteric oxides? See the diagram below. If you cannot write out the *s* and *p* blocks from memory, you should spend some time learning that part of the periodic table. This knowledge will permit you to integrate many chemical facts into a logical pattern of trends.

The elements that form basic oxides are in plain type, those forming acidic oxides are in outline type, and those forming amphoteric oxides are in boldface type. Note the diagonal region from upper left to lower right that includes the elements forming amphoteric oxides. The elements Ge, Sn, Pb, As, Sb, and Bi form amphoteric oxides only in their lower oxidation states (II for Ge, Sn, and Pb; III for As, Sb, and Bi). They form acidic oxides in their higher oxidation state (IV for Ge, Sn, and Pb; V for As, Sb, and Bi).

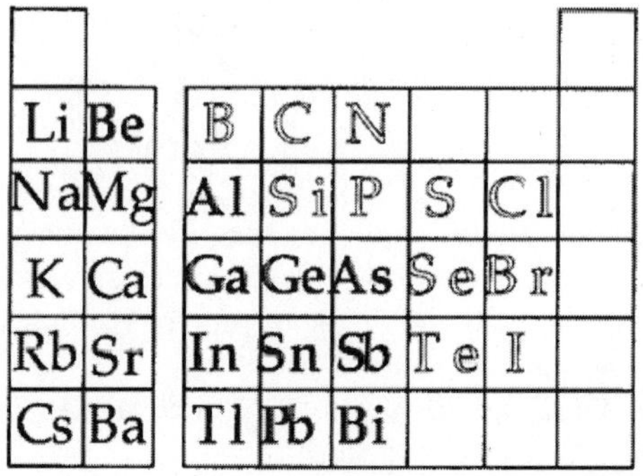

Li	**Be**	B	C	N			
Na	Mg	**Al**	Si	P	S	Cl	
K	Ca	**Ga**	**Ge**	**As**	Se	Br	
Rb	Sr	**In**	**Sn**	**Sb**	Te	I	
Cs	Ba	Tl	**Pb**	**Bi**			

4.2 Identify the conjugate bases of the following acids? (a) $[Co(NH_3)_5(OH_2)]^{3+}$? A conjugate base is a species with one fewer proton than the parent acid. Therefore, the conjugate base in this case is $[Co(NH_3)_5(OH)]^{2+}$, shown below (L = NH_3).

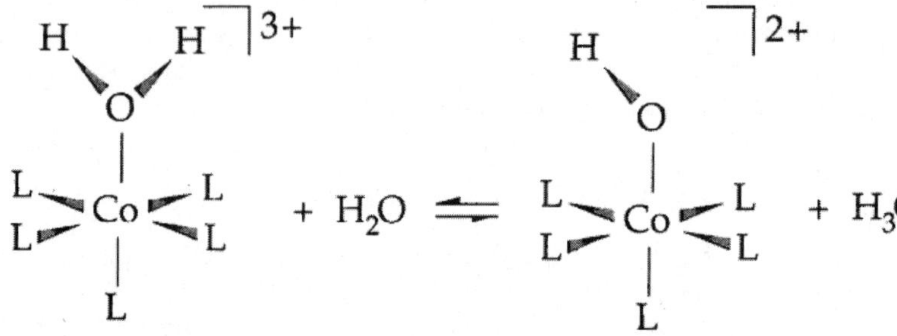

$$[Co(NH_3)_5(OH_2)]^{3+} + H_2O \rightleftharpoons [Co(NH_3)_5(OH)]^{2+} + H_3O^+$$

(b) HSO_4^-? The conjugate base is SO_4^-.
(c) CH_3OH? The conjugate base is CH_3O^-.
(d) $H_2PO_4^-$? The conjugate base is HPO_4^{2-}.
(e) $Si(OH)_4$? The conjugate base is $SiO(OH)_3^-$.
(f) HS^-? The conjugate base is S^{2-}.

4.3 **Identify the conjugate acids of the following bases? (a) C_5H_5N (pyridine)?** A conjugate acid is a species with one more proton than the parent base. Therefore, the conjugate acid in this case is the pyridinium ion, $C_5H_6N^+$, shown below.

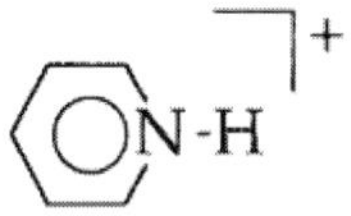

$C_5H_6N^+$ 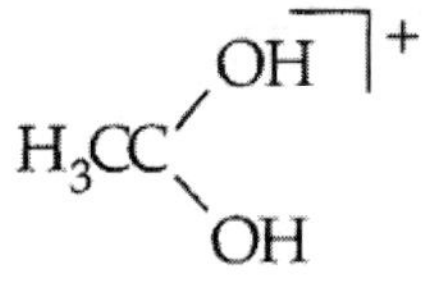 $CH_3C(OH)_2^+$

(b) HPO_4^{2-}? The conjugate acid is $H_2PO_4^-$.
(c) O^{2-}? The conjugate acid is OH^-.
(d) CH_3COOH? The conjugate acid is $CH_3C(OH)_2^+$, shown above.
(e) $[Co(CO)_4]^-$? The conjugate acid is $HCo(CO)_4$, shown below.
A drawing of the $HCo(CO)_4$ molecule, the conjugate acid of the tetrahedral $Co(CO)_4^-$ anion. The C atoms of the CO ligands are bound to the Co atom. The O atoms are unshaded.

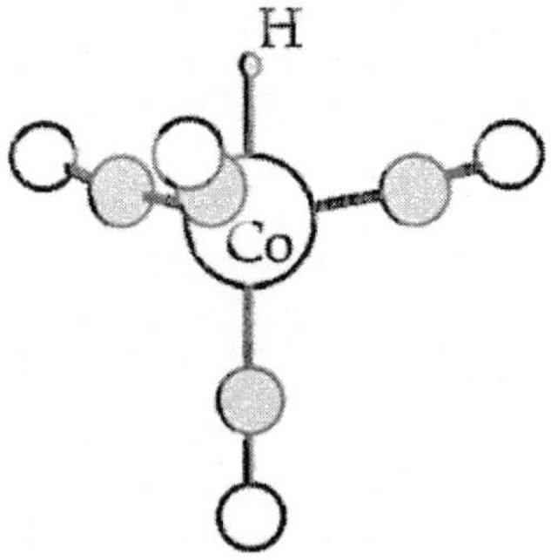

(f) CN^-? The conjugate acid is HCN.

4.4 **Calculate the $[H_3O^+]$ and pH of a 0.10 M butanoic acid solution?** Work this problem the same way we did in **S4.2**. The equilibrium, and ICE table is shown below:

	Bu-H + H_2O	$\rightleftharpoons$	H_3O^+	+ Bu^-
Initial (M)	0.10		0.0	0.0
Change (M)	-x		+x	+x
Equilibrium (M)	0.10 - x		x	x

Then plug the equilibrium values into the equilibrium expression:

$$1.86 \text{ X } 10^{-5} = \frac{x^2}{(0.10 - x)}$$

Now, we can assume the $0.10 - x \approx 0.10$ then solve for x, which will be our $[H_3O^+]$.

$$1.86 \text{ X } 10^{-5} = \frac{x^2}{0.10}$$

$$x = 1.4 \text{ X } 10^{-3} = [H_3O^+]$$

We can now test if the assumption is valid.

$$\frac{1.4 \text{ X } 10^{-3}}{0.10} \text{ X } 100 = 1.4\%$$

This value is less than 5%, so our assumption is valid. This means our total $[H_3O^+] = 1.4 \text{ X } 10^{-3}$. The pH of the solution is:

$$\text{pH} = -\log[H_3O^+] = 2.85$$

4.5 **What is the K_b of ethanoic acid?** To find the K_b, use Equation 4.4:

$$K_a \times K_b = K_w$$
$$K_b = K_w/K_a = 1.0 \times 10^{-14}/1.8 \times 10^{-5} = 5.6 \times 10^{-10}$$

4.6 **What is the K_a for $C_5H_5NH^+$?** The nitrogen on pyridine can be protonated with a Brønsted acid to give the pyridinium salt.

To find the K_a, and more usefully the pK_a, use Equations 4.4 and 4.5:

$$K_a \times K_b = K_w$$
$$K_a = K_w/K_b = 1.0 \times 10^{-14}/1.8 \times 10^{-9} = 5.6 \times 10^{-6}$$
$$\text{p}K_a = -\log K_a = -\log(5.6 \times 10^{-6}) = 5.26$$

4.7 **Predict if F^- will behave as an acid or a base in water?** The easiest way to work this problem is to convert the proton affinity into the proton-gain enthalpy, which is simply the negative of the proton affinity. In this case it would be -1150 kJmol^{-1}. When this value is large and negative, it means a proton attachment is favourable, indicating basic character. So, F^- will behave as a base in water.

4.8 **What are the structures and the pK_a values of chloric ($HClO_3$) and chlorous ($HClO_2$) acid?** The structures for chloric acid and chlorous acid are shown below. Chloric acid is π bond to two oxygen atoms and σ bond to another oxygen atom, and has one lone pair, giving it a geometry of trigonal pyramidal. Chlorous acid is π bond to one oxygen atom and σ bond to one oxygen atom, and has two sets of lone pairs, thus the geometry is bent.

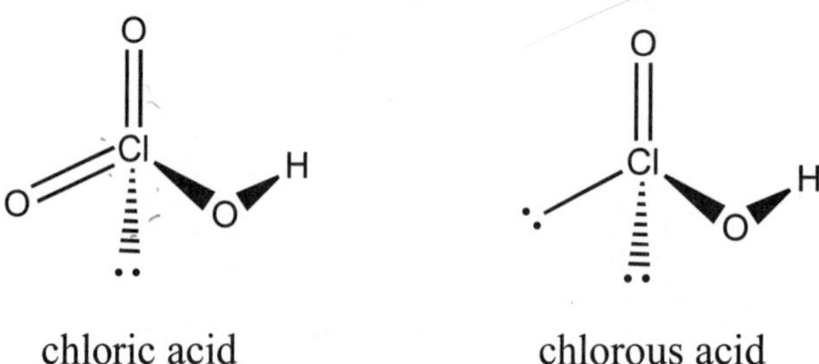

chloric acid chlorous acid

Pauling's first rule for predicting the pK_a of a mononuclear oxoacid is p$K_a \approx 8 - 5p$ (where p is the number of oxo groups attached to the central element). Since $p = 2$ for chloric acid, the predicted pK_a for $HClO_3$ is $8 - (5 \times 2) = -2$. The actual value, given in Table 4.1, is –1. For chlorous acid, $p = 1$, therefore the pK_a for $HClO_2$ is $8 - (5 \times 1) = 3$. The actual value, given in Table 4.1, is 2.

4.9 **Which bases are too strong or too weak to be studied experimentally? (a) CO_3^{2-} O^{2-}, ClO_4^-, and NO_3^- in water?** You can interpret the term "studied experimentally" to mean that the base in question exists in water (i.e., it is not completely protonated to its conjugate acid) *and* that the base in question can be partially protonated (i.e., it is not so weak that the strongest acid possible in water, H_3O^+, will fail to produce a measurable amount of the conjugate acid). Using these criteria, the base CO_3^{2-} is of directly measurable base strength, since the equilibrium $CO_3^{2-} + H_2O \rightarrow HCO_3^- + OH^-$ produces measurable amounts of reactants and products. The base O^{2-}, on the other hand, is completely protonated in water to produce OH^-, so the oxide ion is too strong to be studied experimentally in water. The bases ClO_4^{2-} and NO_3^- are conjugate bases of very strong acids, which are completely deprotonated in water. Therefore, since it is not possible to protonate either perchlorate or nitrate ion in water, they are too weak to be studied experimentally.

(b) HSO_4^-, NO_3^-, and ClO_4^-, in H_2SO_4? The hydrogen sulfate ion, HSO_4^-, is the strongest base possible in liquid sulfuric acid. However, since acids can protonate it, it is not too strong to be studied experimentally. Nitrate ion is a weaker base than HSO_4^-, a consequence of the fact that its conjugate acid, HNO_3, is a stronger acid than H_2SO_4. However, nitrate is not so weak that it cannot be protonated in sulfuric acid, so NO_3^- is of directly measurable base strength in liquid H_2SO_4. On the other hand, ClO_4^-, the conjugate base of one of the strongest known acids, is so weak that it cannot be protonated in sulfuric acid, and hence cannot be studied in sulfuric acid.

4.10 **Is the –CN group electron donating or withdrawing?** A comparison of the aqueous pK_a values is necessary to answer this question. These are:

HOCN, 4	H_2NCN, 10.5	CH_3CN, 20
H_2O, 14	NH_3, very large	CH_4, very large

You know that the values of pK_a are very large for ammonia and methane because these compounds are not normally thought of as acids (this implies that they are extremely weak acids). Now, in all three cases, the cyano-containing compound has a lower pK_a (a *higher* acidity) than the parent compound. In the case of H_2O and HOCN, the latter compound is 10 orders of magnitude more acidic than water. The deprotonation equilibrium involves the formation of an anion, the conjugate base of the acid in question. For example:

$$\text{HONC} + H_2O \rightleftharpoons \text{ONC}^- + H_3O^+$$
$$H_2O + H_2O \rightleftharpoons OH^- + H_3O^+$$

Since a lower pK_a means a larger K_a, this suggests that the anion OCN^- is better stabilized than OH^-. This occurs because the –CN group is more electron withdrawing than the –H substituent.

4.11 **Is the pK_a for $HAsO_4^{2-}$ consistent with Pauling's rules?** Pauling's first rule for predicting the pK_a of a mononuclear oxoacid is p$K_a \approx 8 - 5p$ (where p represents the number of oxo groups attached to the central element).

Since $p = 1$, the predicted value of p$K_a(1)$ for H_3AsO_4 is $8 - (5 \times 1) = 3$.

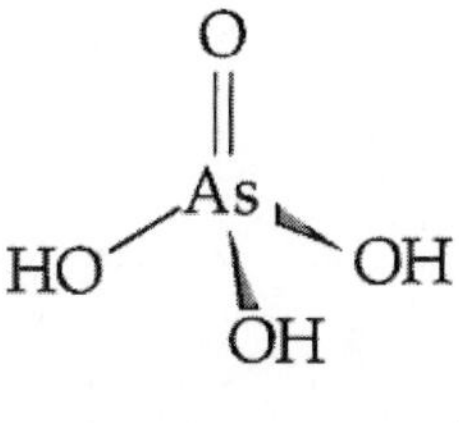

O
As
HO OH
OH

Pauling's second rule for predicting the pK_a of a mononuclear oxoacid is that successive pK_a values for polyprotic acids increase by five units for each successive proton transfer. Since p$K_a(1)$ for H_3AsO_4 was predicted to be 3, the predicted value of pK_a for $HAsO_4^{2-}$, which is p$K_a(3)$ for H_3AsO_4, is $3 + (2 \times 5) = 13$. The actual value, which differs by 1.5 pK_a units, is 11.5. This illustrates that Pauling's rules are only approximate.

4.12 **What is the order of increasing acid strength for HNO_2, H_2SO_4, $HBrO_3$, and $HClO_4$?** According to Pauling's first rule for predicting the pK_a of a mononuclear oxoacid, p$K_a \approx 8 - 5p$ (where p is the number of oxo groups attached to the central element).

pK_a for $HNO_2 = 8 - (5 \times 1) = 3$
pK_a for $H_2SO_4 = 8 - (5 \times 2) = -2$
pK_a for $HBrO_3 = 8 - (5 \times 2) = -2$
pK_a for $HClO_4 = 8 - (5 \times 3) = -7$

The lowest pK_a is for $HClO_4$, so this is the strongest acid. Next we have H_2SO_4 and $HBrO_3$, which have the same pK_a of –2 according to Pauling's rules for predicting the pK_a of mononuclear oxoacids. Bromine is more electronegative than sulfur; inductively, $HBrO_3$ is a stronger acid than H_2SO_4. HNO_2 has the highest pK_a and is the weakest acid. Therefore the order is $HClO_4 > HBrO_3 > H_2SO_4 > HNO_2$.

4.13 **Account for the trends in the pK_a values of the conjugate acids of SiO_4^{4-}, PO_4^{3-}, SO_4^{2-}, and ClO_4^-?** The structures of these four anions, which can be determined to be tetrahedral using VSEPR, are shown below. As can be seen, the charge on the anions decreases from –4 for the silicon-containing species to –1 for the chlorine–containing species. The charge differences alone would make SiO_4^{4-} the most basic species. Hence $HSiO_4^{3-}$ is the least acidic conjugate acid. The acidity of the four conjugate acids increases in the order $HSiO_4^{3-} < HPO_4^{2-} < HSO_4^- < HClO_4$.

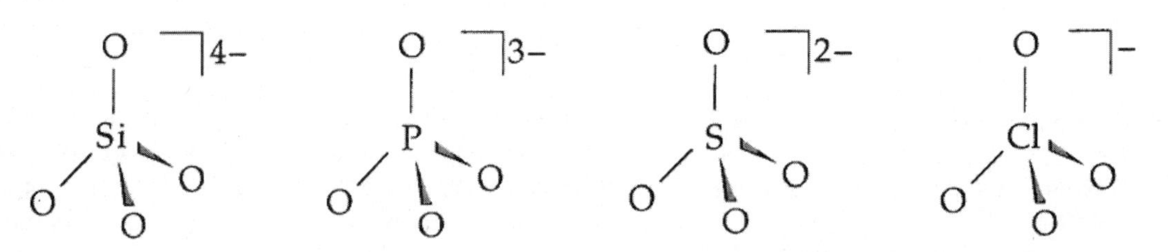

4.14 **Which of the following is the stronger acid? (a) $[Fe(OH_2)_6]^{3+}$ or $[Fe(OH_2)_6]^{2+}$?** The Fe(III) complex, $[Fe(OH_2)_6]^{3+}$, is the stronger acid by virtue of the higher charge. The electrostatic parameter, $\xi = z^2/(r + d)$, will be considerably higher for $z = 3$ than for $z = 2$. The minor decrease in $r + d$ on going from the Fe(II) to the Fe(III) species will enhance the differences in ξ for the two species.

(b) $[Al(OH_2)_6]^{3+}$ or $[Ga(OH_2)_6]^{3+}$? In this case, z is the same but $r + d$ is different. Since the ionic radius, r, is smaller for period 3 Al^{3+} than for period 4 Ga^{3+}, $r + d$ for $[Al(OH_2)_6]^{3+}$ is smaller than $r + d$ for $[Ga(OH_2)_6]^{3+}$ and the aluminum-containing species is more acidic.

(c) $Si(OH)_4$ or $Ge(OH)_4$? As in part (b) above, z is the same but the $r + d$ parameter is different for these two compounds. The comparison here is also between species containing period 3 and period 4 central atoms in the same group, and the species containing the smaller central atom, $Si(OH)_4$, is more acidic.

(d) $HClO_3$ or $HClO_4$? These two acids are shown below. According to Pauling's rule 1 for mononuclear oxoacids, the species with more oxo groups has the lower pK_a and is the stronger acid. Thus, $HClO_4$ is a stronger acid than $HClO_3$. Note that the oxidation state of the central chlorine atom in the stronger acid (+7) is higher than in the weaker acid (+5).

$HClO_3$ $HClO_4$

(e) H_2CrO_4 or $HMnO_4$? As in part (d) above, the oxidation states of these two acids are different, VI for the chromium atom in H_2CrO_4 and VII for the manganese atom in $HMnO_4$. The species with the higher central-atom oxidation state, $HMnO_4$, is the stronger acid. Note that this acid has more oxo groups, three, than H_2CrO_4, which has two.

(f) H_3PO_4 or H_2SO_4? The oxidation state of sulfur in H_2SO_4 is VI, while the oxidation state of phosphorus in H_3PO_4 is only V. Furthermore, sulfuric acid has two oxo groups attached to the central sulfur atom, while phosphoric acid has only one oxo group attached to the central phosphorus atom. Therefore, on both counts (which by now you can see are really manifestations of the same thing) H_2SO_4 is a stronger acid than H_3PO_4.

4.15 **Arrange the following oxides in order of increasing basicity? Al_2O_3, B_2O_3, BaO, CO_2, Cl_2O_7, and SO_3?** First you pick out the intrinsically acidic oxides, since these will be the *least* basic. The compounds B_2O_3, CO_2, Cl_2O_7, and SO_3 are acidic, since the central element for each of them is found in the acidic region of the periodic table (see the *s* and *p* block diagram in the answer to Exercise **4.1**). The most acidic compound, Cl_2O_7, has the highest central-atom oxidation state, +7, while the least acidic, B_2O_3, has the lowest, +3. Of the remaining compounds, Al_2O_3 is amphoteric, which puts it on the borderline between acidic and basic oxides, and BaO is basic. Therefore, a list of these compounds in order of increasing basicity is $Cl_2O_7 < SO_3 < CO_2 < B_2O_3 < Al_2O_3 <$ BaO.

4.16 **Arrange the following in order of increasing acidity? HSO_4^-, H_3O^+, H_4SiO_4, CH_3GeH_3, NH_3, and HSO_3F?** The weakest acids, CH_3GeH_3 and NH_3, are easy to pick out of this group since they do not contain any –OH bonds. Ammonia is the weaker acid of the two, since it has a lower central-atom oxidation state, III, than that for the germanium atom in CH_3GeH_3, which is IV. Of the remaining species, note that HSO_3F is very similar to H_2SO_4 as far as structure and sulfur oxidation state (VI) are concerned, so it is reasonable to suppose that HSO_3F is a very strong acid, which it is. The anion HSO_4^- is a considerably weaker acid than HSO_3F, for the same reason that it is a considerably weaker acid than H_2SO_4, namely, Pauling's rule 2 for mononuclear oxoacids. Since HSO_4^- is not completely deprotonated in water, it is a weaker acid than H_3O^+, which is the strongest possible acidic species in water. Finally, it is difficult to place exactly $Si(OH)_4$ in this group. It is certainly more acidic than NH_3 and CH_3GeH_3, and it turns out to be *less* acidic than HSO_4^-, despite the negative charge of the latter species. Therefore, a list of these species in order of increasing acidity is $NH_3 < CH_3GeH_3 < H_4SiO_4 < HSO_4^- < H_3O^+ <$ HSO_3F.

The structures of H_2SO_4 and HSO_3F:

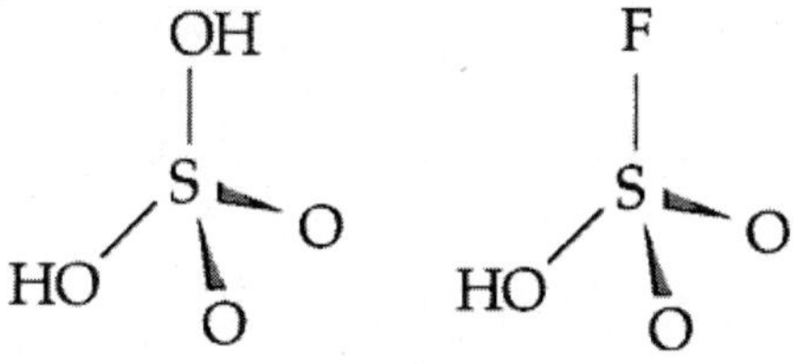

OH S O HO O   F S O HO O

4.17 **Which aqua ion is the stronger acid, Na^+ or Ag^+?** Even though these two ions have about the same ionic radius, Ag^+–OH_2 bonds are much more covalent than Na^+–OH_2 bonds, a common feature of the chemistry of *d*-block vs. *s*-block metal ions. The greater covalence of the Ag^+–OH_2 bonds has the effect of delocalizing the positive charge of the cation over the whole aqua complex. As a consequence, the departing proton is repelled more by the positive charge of $Ag^+(aq)$ than by the positive charge of $Na^+(aq)$, and the former ion is the stronger.

4.18 **Which of the following elements form oxide polyanions or polycations? Al, As, Cu, Mo, Si, B, Ti?** As discussed in Section 4.4, the aqua ions of metals that have amphoteric oxides generally undergo polymerization to polycations. The elements Al, Cu, and Ti fall into this category. On the other hand, polyoxoanions (oxide polyanions) are important for some of the early *d*-block metals, especially for V, Mo, and W in high oxidation states. Furthermore, many of the *p*-block elements form polyoxoanions, including As, B, and Si.

4.19 **The change in charge upon aqua ion polymerization?** One example of aqua ion polymerization is

$$2[Al(OH_2)_6]^{3+} + H_2O \rightarrow [(H_2O)_5Al\text{–}O\text{–}Al(OH_2)_5]^{4+} + 2H_3O^+$$

The charge per aluminum atom is +3 for the mononuclear species on the left-hand side of the equation but only +2 for the dinuclear species on the right-hand side. Thus, poly*cation* formation reduces the average positive charge per central M atom by +1 per M.

$$\left[(H_2O)_4Al{-}O{-}Al(OH_2)_5\right]^{4+}$$

H₂O OH₂ H₂O OH₂
H₂O—Al—O—Al—OH₂
H₂O OH₂ H₂O OH₂

4.20 Write balanced equations for the formation of $P_4O_{12}^{4-}$ from PO_4^{3-} and for the formation of $[(H_2O)_4Fe(OH)_2Fe(OH_2)_4]^{4+}$ from $[Fe(OH_2)_6]^{3+}$? The two balanced equations are shown below. Note that the condensation reactions involve a neutralization of charge, either by adding H^+ to a highly charged anion or by removing H^+ from a highly charged cation. The structure of $P_4O_{12}^{4-}$, which is called *cyclo*tetrametaphosphate, is also shown below.

$$4PO_4^{3-} + 8H_3O^+ \rightarrow P_4O_{12}^{4-} + 12H_2O$$

$$2[Fe(OH_2)_6]^{3+} \rightarrow [(H_2O)_4Fe(OH)_2Fe(OH_2)]^{4+} + 2H_3O^+$$

The structure of the $[P_4O_{12}]^{4-}$ ion in the salt $[NH_4]_4[P_4O_{12}]$ is shown below.

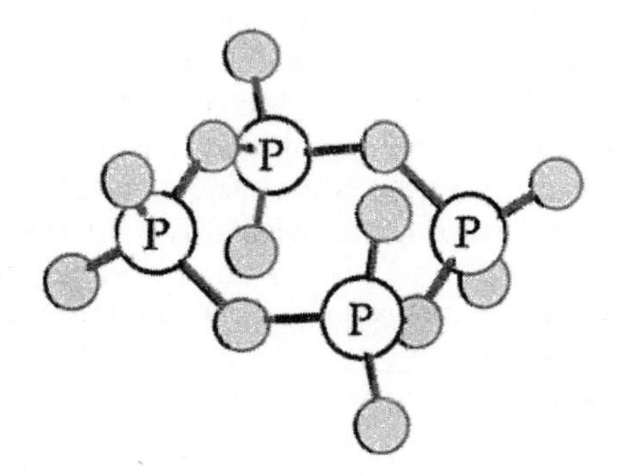

4.21 More balanced equations? (a) H_3PO_4 and Na_2HPO_4? You can use the successive K_a values for phosphoric acid to estimate the equilibrium constant for the equilibrium below:

$$H_3PO_4 + HPO_4^{2-} \rightleftharpoons 2H_2PO_4^- \qquad K = ?$$

The three K_a values can be found in Table 4.1 and are 7.5×10^{-3} (K_{a1}), 6.2×10^{-8} (K_{a2}), and 2.2×10^{-13} (K_{a3}). The equilibrium above is the sum of the two equilibria below, so K for the equilibrium above is the product $(K_{a1})(1/K_{a2})$ $= 1.2 \times 10^5$.

$$H_3PO_4 + H_2O \rightleftharpoons H_2PO_4^- + H_3O^+ \qquad K_{a1} = 7.5 \text{ X } 10^{-3}$$

$$HPO_4^{2-} + H_3O^+ \rightleftharpoons H_2PO_4^- + H_2O \qquad 1/K_{a2} = 1.6 \text{ X } 10^7$$

(b) CO_2 and $CaCO_3$? Successive K_a values can also be used to show that the equilibrium below lies to the right.

$$CO_2 + CaCO_3 + H_2O \rightleftharpoons Ca^{2+} + 2HCO_3^-$$

4.22 Give the equations for HF in H_2SO_4 and HF in liquid NH_3? HF behaves like an acid in anhydrous sulphuric acid, so the equation is:

$$H_2SO_4 + HF \Leftrightarrow H_3SO_4^+ + F^-$$

HF behaves like a base in liquid ammonia, so the equation is:

$$NH_3 + HF \rightleftharpoons NH_2^- + H_2F^+$$

4.23 **Why is H_2Se a stronger acid than H_2S?** As you go down a family in the periodic chart, the acidy of the homologous hydrogen compounds increases. This is do primarally to the fact that the bond dissociation energy is smaller as you go down a family, due to pure orbital overlap. Since the H-X bond is weaker for H_2Se, it will release protons more readily than H_2S in a given solvent.

4.24 **Identifying elements that form Lewis acids?** All of the *p*-block elements except nitrogen, oxygen, fluorine, and the lighter noble gases form Lewis acids in one of their oxidation states. Examples are as follows: BF_3, B_2O_3, and B_2H_6 are Lewis acids; CO_2, organic ketones, and carbonium ions are Lewis acids; Ga, In, and Tl all form +1 cations, which are Lewis acids, and $GaCl_3$, $InCl_3$, and $TlCl_3$ are Lewis acids; the dichlorides and tetrachlorides of Ge, Sn, and Pb are Lewis acids; the trifluorides and pentafluorides of P, As, Sb, and Bi are Lewis acids (PF_3 and AsF_3 are also Lewis bases toward *d*-block metals, so these compounds are amphoteric); the dioxides of S, Se, and Te are Lewis acids, as are the tetrafluorides and hexafluorides of Se and Te; the trifluorides and pentafluorides of Cl, Br, and I are Lewis acids, as is the heptafluoride IF_7; the tetrafluoride and hexafluoride of xenon, XeF_4 and XeF_6, are also Lewis acids.

4.25 **Identifying acids and bases: (a) $SO_3 + H_2O \rightarrow HSO_4^- + H^+$?** The acids in this reaction are the Lewis acids SO_3 and H^+ and the base is the Lewis base OH^-. The complex (or adduct) HSO_4^- is formed by the displacement of the proton from the hydroxide ion by the stronger acid SO_3. In this way, the water molecule is thought of as an adduct of H^+ and OH^-. Since the proton must be bound to a solvent molecule, even though this fact is not explicitly shown in the reaction, the water molecule exhibits Brønsted acidity. Note that it is easy to tell that this is a displacement reaction instead of just a complex formation reaction because, while there is only one base in the reaction, there are *two* acids. A complex formation reaction only occurs with a single acid and a single base. A double displacement, or metathesis, reaction only occurs with two acids and two bases.

(b) $Me[B_{12}]^- + Hg^{2+} \rightarrow [B_{12}] + MeHg^+$? (Note: $[B_{12}]$ designates the Co center of the macrocyclic complex called coenzyme B_{12}.) This is a displacement reaction. The Lewis acid Hg^{2+} displaces the Lewis acid $[B_{12}]$ from the Lewis base CH_3^-.

(c) $KCl + SnCl_2 \rightarrow K^+ + [SnCl_3]^-$? This is also a displacement reaction. The Lewis acid $SnCl_2$ displaces the Lewis acid K^+ from the Lewis base Cl^-.

(d) $AsF_3(g) + SbF_5(g) \rightarrow [AsF_2][SbF_6]$? Even though this reaction is the formation of an ionic substance, it is *not* simply a complex formation reaction. It is a displacement reaction. The very strong Lewis acid SbF_5 (one of the strongest known) displaces the Lewis acid $[AsF_2]^+$ from the Lewis base F^-.

(e) EtOH readily dissolves in pyridine? A Lewis acid–base complex formation reaction between EtOH (the acid) and py (the base) produces the adduct EtOH–py, which is held together by the kind of dative bond that you refer to as a hydrogen bond.

4.26 **Select the compound with the named characteristic? (a) Strongest Lewis acid: BF_3, BCl_3, or BBr_3?** The simple argument that more electronegative substituents lead to a stronger Lewis acid does not work in this case. Boron tribromide is observed to be the strongest Lewis acid of these three compounds. The shorter boron–halogen bond distances in BF_3 and BCl_3 than in BBr_3 are believed to lead to stronger halogen-to-boron *p–p* π bonding. According to this explanation, the acceptor orbital (empty *p* orbital) on boron is involved to a greater extent in π bonding in BF_3 and BCl_3 than in BBr_3; the acidities of BF_3 and BCl_3 are diminished relative to BBr_3.

$BeCl_2$ or BCl_3? Boron trichloride is expected to be the stronger Lewis acid of the two for two reasons. The first reason, which is more obvious, is that the oxidation number of boron in BCl_3 is +3, while for the beryllium atom in $BeCl_2$ it is only +2. The second reason has to do with structure. The boron atom in BCl_3 is only three-coordinate, leaving a vacant site to which a Lewis base can coordinate. Since $BeCl_2$ is polymeric, each beryllium atom is four-coordinate, and some Be–Cl bonds must be broken before adduct formation can take place.

A piece of the infinite linear chain structure of $BeCl_2$. Each Be atom is four-coordinate, and each Cl atom is two-coordinate. The polymeric chains are formed by extending this piece to the right and to the left:

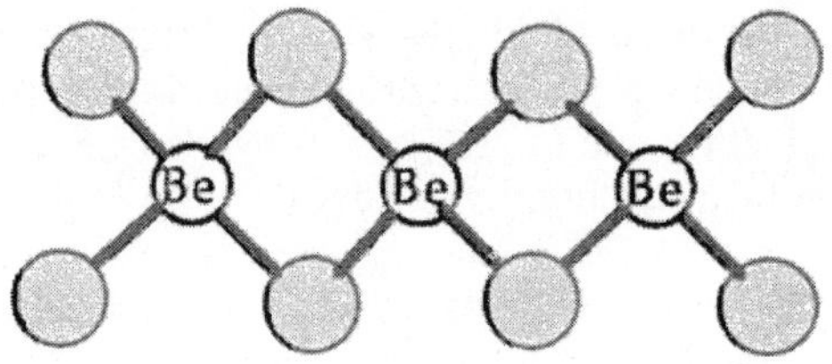

$B(n\text{-}Bu)_3$ or $B(t\text{-}Bu)_3$? The Lewis acid with the unbranched substituents, $B(n\text{-}Bu)_3$, is the stronger of the two because, once the complex is formed, steric repulsions between the substituents and the Lewis base will be less than with the bulky, branched substituents in $B(t\text{-}Bu)_3$.

(b) More basic toward BMe_3: NMe_3 or NEt_3? These two bases have nearly equal basicities toward the proton in aqueous solution or in the gas phase. Steric repulsions between the substituents on the bases and the proton are negligible, since the proton is very small. However, steric repulsions between the substituents on the bases and *molecular* Lewis acids like BMe_3 are an important factor in complex stability, and so the smaller Lewis base NMe_3 is the stronger in this case.

2-Me-py or 4-Me-py? As above, steric factors favor complex formation with the smaller of two bases that have nearly equal Brønsted basicities. Therefore, 4-Me-py is the stronger base toward BMe_3, since the methyl substituent in this base cannot affect the strength of the B–N bond by steric repulsions with the methyl substituents on the Lewis acid.

4.27 **Which of the following reactions have $K_{eq} > 1$? (a) $R_3P\text{–}BBr_3 + R_3N\text{–}BF_3 \rightarrow R_3P\text{–}BF_3 + R_3N\text{–}BBr_3$?** You know that phosphines are softer bases than amines. So, to determine the position of this equilibrium, you must decide which Lewis acid is softer, since the softer acid will preferentially form a complex with a soft base than with a hard base of equal strength. Boron tribromide is a softer Lewis acid than BF_3, a consequence of the relative hardness and softness of the respective halogen substituents. Therefore, the equilibrium position for this reaction will lie to the left, the side with the soft–soft and hard–hard complexes, so the equilibrium constant is less than 1. In general, it is found that soft substituents (or ligands) lead to a softer Lewis acid than for the same central element with harder substituents.

(b) $SO_2 + Ph_3P\text{–}HOCMe_3 \rightarrow Ph_3P\text{–}SO_2 + HOCMe_3$? In this reaction, the soft Lewis acid sulfur dioxide displaces the hard acid *t*-butyl alcohol from the soft base triphenylphosphine. The soft–soft complex is favored, so the equilibrium constant is greater than 1.

The adduct formed between triphenylphosphine and sulfur dioxide:

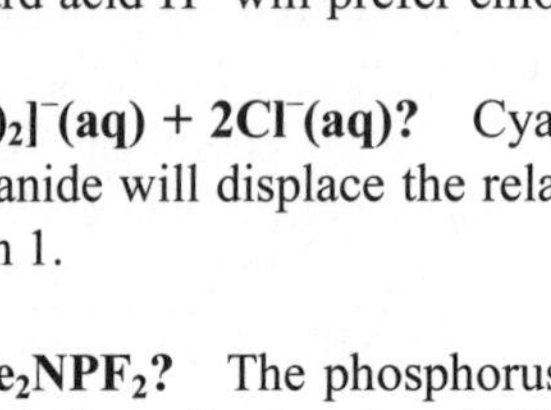

(c) $CH_3HgI + HCl \rightarrow CH_3HgCl + HI$? Iodide is a softer base than chloride, an example of the general trend that elements later in a group are softer than their progenors. The soft acid CH_3Hg^+ will form a stronger complex with iodide than with chloride, while the hard acid H^+ will prefer chloride, the harder base. Thus, the equilibrium constant is less than 1.

(d) $[AgCl_2]^-(aq) + 2\ CN^-(aq) \rightarrow [Ag\ (CN)_2]^-(aq) + 2Cl^-(aq)$? Cyanide is a softer and generally stronger base than chloride (see Table 4.5). Therefore, cyanide will displace the relatively harder base from the soft Lewis acid Ag^+. The equilibrium constant is greater than 1.

4.28 **Choose between the two basic sites in Me_2NPF_2?** The phosphorus atom in Me_2NPF_2 is the softer of the two basic sites, so it will bond more strongly with the softer Lewis acid BH_3 (see Table 4.5). The hard nitrogen atom will bond more strongly to the hard Lewis acid BF_3.

4.29 **Why is trimethylamine out of line?** As you add methyl groups to the nitrogen, the nitrogen becomes more basic due to the inductive effect (electron donating ability) of the methyl groups. So the trend

NH_3<$NH_2(CH_3)$<$NH(CH_3)_2$ makes since, but when you put three methyl groups around nitrogen, sterics come into play. Trimethyl amine is sterically large enough to fall out of line with the given enthalpies of reaction.

4.30 Discuss relative basicities? (a) Acetone and DMSO? Since both E_B and C_B are larger for DMSO than for acetone (see Table 4.6), DMSO is the stronger base regardless of how hard or how soft is the Lewis acid. The ambiguity for DMSO is that both the oxygen atom and sulfur atom are potential basic sites.
(b) Me_2S and DMSO? Dimethylsulfide has a C_B value that is two and a half times larger than that for DMSO, while its E_B value is only one quarter that for DMSO. Thus, depending on the E_A and C_A values for the Lewis acid, either base could be stronger. For example, DMSO is the stronger base toward BF_3, while SMe_2 is the stronger base toward I_2. This can be predicted by calculating the ΔH of complex formation for all four combinations:

$$\text{DMSO–BF}_3\text{: } \Delta H = -[(20.21)(2.76) + (3.31)(5.83)] = -75.1 \text{ kJ/mol}$$

$$\text{SMe}_2\text{–BF}_3\text{: } \Delta H = -[(20.21)(0.702) + (3.31)(15.26)] = -64.7 \text{ kJ/mol}$$

$$\text{DMSO–I}_2\text{: } \Delta H = -[(2.05)(2.76) + (2.05)(5.83)] = -17.6 \text{ kJ/mol}$$

$$\text{SMe}_2\text{–I}_2\text{: } \Delta H = -[(2.05)(0.702) + (2.05)(15.26)] = -32.7 \text{ kJ/mol}$$

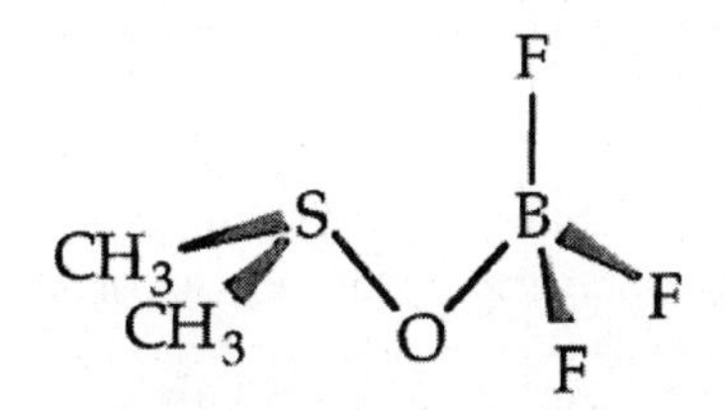

The stable complex of DMSO and boron trifluoride

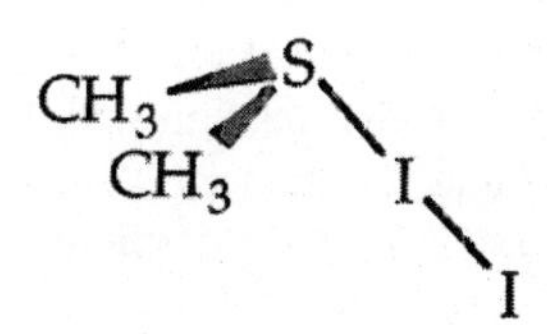

The stable complex of dimethyl sulfide and I_2

4.31 Write a balanced equation for the dissolution of SiO_2 by HF? The metathesis of two hard acids with two hard bases occurs as an equilibrium is established between solid, insoluble SiO_2 and soluble H_2SiF_6:

$$SiO_2 + 6HF \rightleftharpoons 2H_2O + H_2SiF_6$$

or

$$SiO_2 + 4HF \rightleftharpoons 2H_2O + SiF_4$$

This is both a Brønsted acid–base reaction and a Lewis acid–base reaction. The Brønsted reaction involves the transfer of protons from HF molecules to O^{2-} ions, while the Lewis reaction involves complex formation between Si^{4+} ions and F^- ions.

4.32 Write a balanced equation to explain the foul odor of damp Al_2S_3? The foul odor suggests that a volatile compound is formed when Al_2S_3 comes in contact with water. The only volatile species that could be present, other than odorless water, is H_2S, which has the characteristic odor of rotten eggs. Thus, an equilibrium is established between two hard acids, Al(III) and H^+, and the bases O^{2-} and S^{2-}:

$$Al_2S_3 + 3H_2O \rightleftharpoons Al_2O_3 + 3H_2S$$

4.33 Describe solvent properties? (a) Favour displacement of Cl^- by I^- from an acid centre? Since in this case you have no control over the hardness or softness of the acid centre, you must do something else that will favour the acid–iodide complex over the acid-chloride complex. If you choose a solvent that decreases the activity of chloride relative to iodide, you can shift the following equilibrium to the right:

$$\text{acid-Cl}^- + I^- \rightleftharpoons \text{acid-I}^- + Cl^-$$

Such a solvent should interact more strongly with chloride (i.e., form an adduct with chloride) than with iodide. Thus, the ideal solvent properties in this case would be *weak*, *hard*, and *acidic*. It is important that the solvent be a weak acid, since otherwise the activity of both halides would be rendered negligible. An example of a suitable solvent is anhydrous HF. Another suitable solvent is H_2O.

(b) Favour basicity of R_3As over R_3N? In this case you wish to enhance the basicity of the soft base trialkylarsine relative to the hard base trialkyl-amine. You can decrease the activity of the amine if the solvent is a hard acid, since the solvent-amine complex would then be less prone to dissociate than the solvent-arsine complex. Alcohols such as methanol or ethanol would be suitable.

The Lewis acid–Lewis base complex of a trialkylamine (a hard base) and an alcohol (a hard acid) is shown below.

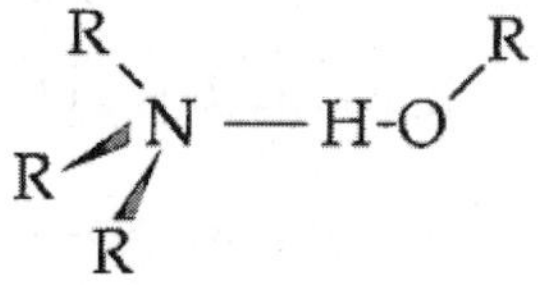

$R_3N—H\text{-}O\text{-}R$

(c) Favor acidity of Ag^+ over Al^{3+}? If you review the answers to parts (a) and (b) of this exercise, a pattern will emerge. In both cases, a hard acid solvent was required to favour the reactivity of a soft base. In this part of the exercise, you want to favor the acidity of a soft acid, so logically a solvent that is a hard base is suitable. Such a solvent will "tie up" (i.e., decrease the activity of) the hard acid Al^{3+} relative to the soft acid Ag^+. An example of a suitable solvent is diethyl ether. Another suitable solvent is H_2O.

(d) Promote the reaction $2FeCl_3 + ZnCl_2 \rightarrow Zn^{2+} + 2[FeCl_4]^-$? Since Zn^{2+} is a softer acid than Fe^{3+}, a solvent that promotes this reaction will be a softer base than Cl^-. The solvent will then displace Cl^- from the Lewis acid Zn^{2+}, forming $[Zn(solv)_x]^{2+}$. The solvent must also have an appreciable dielectric constant, since ionic species are formed in this reaction. A suitable solvent is acetonitrile, MeCN.

4.34 **Propose a mechanism for the acylation of benzene?** The mechanism described in Section 4.10(b), *Group 13 Lewis acids*, involves the abstraction of Cl^- from CH_3COCl by $AlCl_3$ to form $AlCl_4^-$ and the Lewis acid CH_3CO^+ (an acylium cation). This cation then attacks benzene to form the acylated aromatic product. An alumina surface, such as the partially dehydroxylated one shown below, would also provide Lewis acidic sites that could abstract Cl^-:

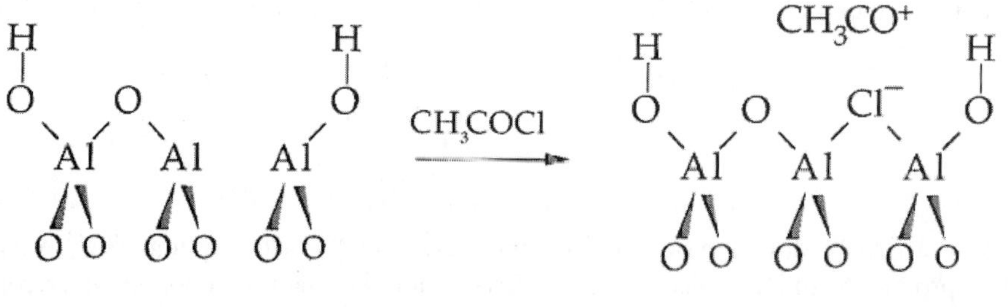

H H H CH$_3$CO$^+$ H
O O O $\xrightarrow{CH_3COCl}$ O O Cl$^-$ O
Al Al Al Al Al Al
O O O O O O O O O O O O

4.35 **Why does Hg(II) occur only as HgS?** Mercury(II) is a soft Lewis acid, and so is found in nature only combined with soft Lewis bases, the most common of which is S^{2-}. Sulfide can readily and permanently abstract Hg^{2+} from its complexes with harder bases in ore forming geological reaction mixtures. Zinc(II), which exhibits borderline behavior, is harder and forms stable compounds (i.e., complexes) with hard bases such as O^{2-}, CO_3^{2-}, and silicates, as well as with S^{2-}. The particular ore that is formed with Zn^{2+} depends on factors including the relative concentrations of the competing bases.

4.36 **Write Brønsted acid–base reactions in liquid HF? (a) CH_3CH_2OH?** Ethanol is a weaker acid than H_2O but a stronger base than H_2O, due to the electron donating property of the $–C_2H_5$ group. The balanced equation is:

$$CH_3CH_2OH + HF \rightleftharpoons CH_3CH_2OH_2^+ + F^-$$

(b) NH_3? The equation is

$$NH_3 + HF \rightleftharpoons NH_4^+ + F^-$$

(c) C_6H_5COOH? In this case, benzoic acid is a significantly stronger acid than water. Therefore, it will protonate hydrogen fluoride:

$$C_6H_5COOH + HF \rightleftharpoons C_6H_5COO^- + H_2F^+$$

4.37 **The dissolution of silicates by HF?** This is both a Brønsted acid–base reaction and a Lewis acid–base reaction. The reaction involves proton transfers from HF to the silicate oxygen atoms (a Brønsted reaction) and the formation of complexes such as SiF_6^{2-} from the Lewis acid Si^{4+} and the Lewis base F^- (a Lewis reaction).

4.38 **Are the *f*-block elements hard?** Since the trivalent lanthanides and actinides are found as complexes with hard oxygen bases (i.e., silicates) and not with soft bases such as sulfide, they must be hard. Since they are found *exclusively* as silicates, they must be considered very hard, unlike the borderline behaviour of Zn(II).

4.39 **Calculate the enthalpy change for I_2 with phenol?** From Table 4.6 we find $E = 2.05$ and $C = 2.05$ for I_2 and $E = 8.86$ and $C = 0.09$ for phenol. The Drago-Wayland equation gives:

$$\Delta_f H^{\ominus} = -[E_A E_B + C_A C_B]$$

$$\Delta_f H^{\ominus} = -20.0 \text{kJ/mol}$$

Indicating an exothermic reaction for the formation of an I_2 phenol adduct.

Chapter 5 Oxidation and Reduction

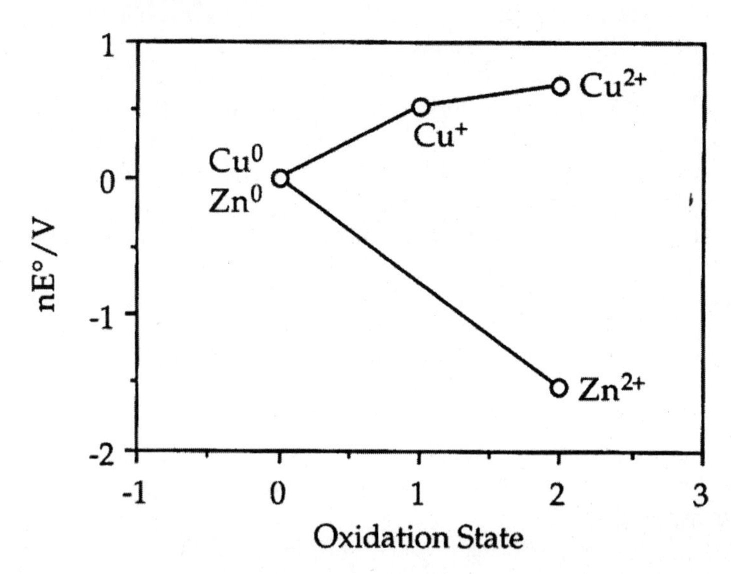

Frost diagrams, shown here for Cu and Zn in aqueous acid (pH 0), are useful representations of electrode potential data. These two graphs show that Cu^0 will not reduce $H^+(aq)$ to H_2, that Zn^0 will reduce $H^+(aq)$ to H_2 and will also reduce $Cu^{2+}(aq)$ to Cu^0, and that $Cu^+(aq)$ will disproportionate to Cu^0 and $Cu^{2+}(aq)$.

S5.1 Half-reactions and balanced reaction for oxidation of zinc metal by permanganate ions?

$$2[MnO_4^-(aq) + 8H^+(aq) + 5e^- \rightarrow Mn^{2+}(aq) + 4H_2O(l)] \quad \text{reduction}$$
$$5[Zn(s) \rightarrow Zn^{2+}(aq) + 2e^-] \quad \text{oxidation}$$

$$2MnO_4^-(aq) + 5Zn(s) + 16H^+(aq) \rightarrow 5Zn^{2+}(aq) + 2Mn^{2+}(aq) + 8H_2O(l)$$

S5.2 Does Cu metal dissolve in dilute HCl? No, because the $E°$ for the reaction $Cu^{2+}(aq) + 2e^- \rightarrow Cu(s)$, $E°(Cu^{2+},Cu)$ = +0.34 V is positive. The oxidation of Cu metal in HCl is not favored thermodynamically since the cell potential for the reaction is –0.34 V.

S5.3 Can $Cr_2O_7^{2-}$ be used to oxidize Fe^{2+}, and would Cl^- oxidation be a problem? The standard reduction potential for the $Cr_2O_7^{2-}/Cr^{3+}$ couple is 1.38 V, so it can oxidize any couple whose reduction potential is less than 1.38 V. In general terms, if the reduction potential for Ox + e^- → Red is *X*, then the overall potential *E* for the (unbalanced) reaction $Cr_2O_7^{2-}$ + Red → Cr^{3+} + Ox is (1.38 – *X*) V. Since $\Delta G = -NFE$, the reaction will have a negative free energy change as long as *E* is positive. Thus as long as the *reduction* potential *X* is less than 1.38 V, *E* will be positive. Returning to the specific question at hand, since the reduction potential for the Fe^{3+}/Fe^{2+} couple is 0.77 V, Fe^{2+} will be oxidized to Fe^{3+} by dichromate. Since the reduction potential for the Cl_2/Cl^- couple is 1.36 V, oxidation of chloride ion is only slightly favored and in practice it is not observed to occur at an appreciable rate.

S5.4 Fuel cell emf with oxygen and hydrogen gases at 5.0 bar? The standard emf of the cell from Example 5.5 was +1.23 V. In general terms, the Nernst equation is given by the formula:

$$E = E° - (RT/nF)(\ln Q) \text{ where } Q \text{ is the reaction quotient}$$

For the fuel-cell reaction, $O_2(g) + H_2(g) \rightarrow 2\, H_2O(1)$,

$$Q = 1/[p(O_2)p(H_2)] \text{ and } E = E° - [(0.059\text{ V})/4][\log(1/(5.0)^2)]$$

Therefore, the emf for the fuel cell is: E = 1.23 V + 0.02 V = 1.25 V.

S5.5 **The fate of SO_2 emitted into clouds?** You are told that the standard reduction potential for the SO_4^{2-}/SO_2 couple is 0.17 V. Since the standard reduction potential for the O_2/H_2O couple is 1.23 V, the potential for the coupled reaction

$$2SO_2\,(aq) + O_2\,(g) + 2H_2O(l) \rightarrow 2SO_4^{2-}\,(aq) + 4H^+\,(aq)$$

is $E^\circ = 1.23\text{ V} - 0.17\text{ V} = 1.06\text{ V}$. Since this potential is large and positive, this reaction will be driven nearly to completion. The aqueous solution of SO_4^{2-} and H^+ ions precipitates as acid rain, which can have a pH as low as 2 (the pH of rain water that is not contaminated with sulfuric or nitric acid is ~5.6).

S5.6 **Can Fe^{2+} disproportionate under standard conditions?** The disproportionation of Fe^{2+} involves the reduction of one equivalent of Fe^{2+} to Fe^0, a net gain of two equivalents of electrons, and the concomitant oxidation of two equivalents to Fe^{2+} to Fe^{3+}, a net loss of two equivalents of electrons:

$$3Fe^{2+}\,(aq) \rightarrow Fe^0\,(s) + 2Fe^{3+}\,(aq)$$

The value of E for this reaction can be calculated by subtracting E° for the Fe^{2+}/Fe couple (–0.44 V) from E° for the Fe^{3+}/Fe^{2+} couple (0.77 V),

$$E = -0.44 - 0.77\text{ V} = -1.21\text{ V}$$

This potential is large *and negative*, so the disproportionation will *not* occur.

S5.7 **bpy binding to Fe(III) or Fe(II)?** Compared to the aqua redox couple of +0.77 V, this value is higher by +0.25 V and we conclude that the lower oxidation state, Fe(II), binds preferentially.

S5.8 **Potential of AgCl/Ag,Cl⁻ couple?** We note that $E_{cell} = (RT/NF)\ \ln K_{sp}$ and $E_{ox} = E_{red} - E_{cell}$. We can use the first equation to calculate the overall cell potential and the second to find the half-cell potential for the anode (oxidation) reaction.

$$E_{cell} = (RT/NF)\ \ln(K_{sp})$$

$$E_{cell} = (0.025693\text{ V})\ \ln(1.77 \times 10^{-10})$$

$$E = -0.58\text{ V}$$

$$E_{ox} = E_{red} - E_{cell} = -0.58 - 0.80\text{V} = -1.38\text{ V}$$

Eox = 0.8 + 0.58 = 1.38V

S5.9 **Latimer diagram for Pu?** (a) Pu(IV) disproportionates to Pu(III) and Pu(V) in aqueous solution; (b) Pu(V) does not disproportionate into Pu(VI) and Pu(IV), though the potentials are so similar that solutions of Pu(IV) and Pu(V) will contain distributions of all Pu oxidation states between Pu(III) and Pu(VI).

S5.10 **Frost diagram for thallium in aqueous acid?** See the figure below. This plot was made using the potentials given, $NE^\circ = 0$ V for Tl^0 ($N = 0$), $NE^\circ = -0.34$ V for Tl^+ ($N = 1$), and $NE^\circ = 2.19$ V for Tl^{3+} ($n = 3$). Note that Tl^+ is stable with respect to disproportionation in aqueous acid. Note also that Tl^{3+} is a strong oxidant (i.e., it is very readily reduced), since the slope of the line connecting it with either lower oxidation state is large and positive.

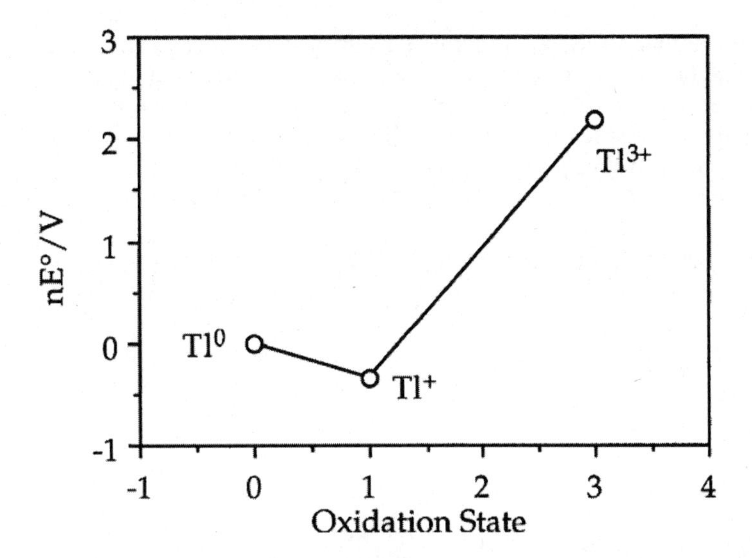

S5.11 **The oxidation number of manganese?** When permanganate, MnO_4^-, is used as an oxidant in aqueous solution, the manganese species that will remain is the most stable manganese species under the conditions of the reaction (i.e., acidic or basic). By "most stable" we mean most stable with respect to a reducing agent. Inspection of the Frost diagram for manganese, shown in Figure 5.10, shows that Mn^{2+}(aq) is the most stable species present, since it has the most negative $\Delta_f G$. Therefore, Mn^{2+}(aq) will be the product of the redox reaction when MnO_4^- is used as an oxidizing agent in aqueous acid.

S5.12 **Compare the strength of NO_3^- as an oxidizing agent in acidic and basic solution?** As suggested in Section 5.13, *Frost diagrams*, the reduction of nitrate ion usually proceeds to NO, which is evolved from the solution, instead of proceeding to N_2O or all the way to N_2. If you compare the portion of the Frost diagram for nitrogen, below, containing the NO_3^-/NO couple in acidic and in basic solution, you see that the slope for the couple in acidic solution is positive while the slope for basic solution is negative. Therefore, nitrate is a stronger oxidizing agent (i.e., it is more readily reduced) in acidic solution than in basic solution. Even if the reduction of NO_3^- proceeded all the way to N_2, the slope of that line is still less than the slope of the line for the NO_3^-/NO couple in acid solution.

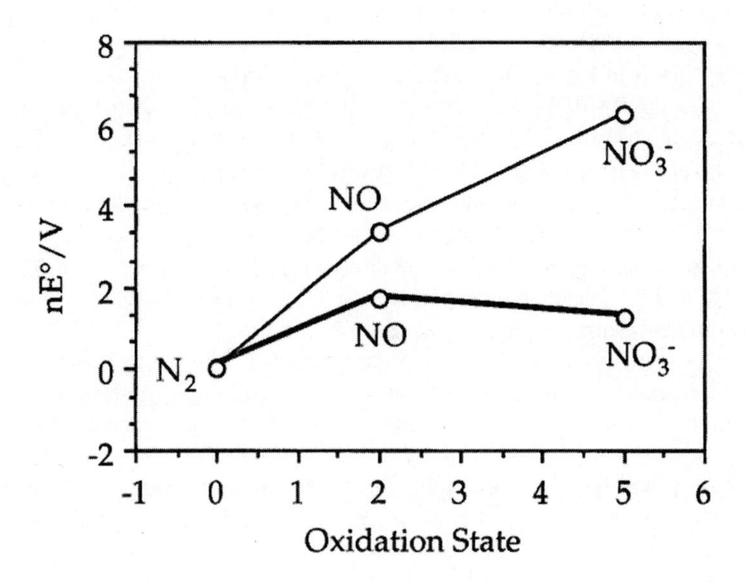

A portion of the Frost diagram for nitrogen is shown in the figure above. The points connected with the bold lines are for basic solution, and the other points are for acid solution.

S5.13 **The possibility of finding $Fe(OH)_3$ in a waterlogged soil?** According to Figure 5.12, a typical waterlogged soil (organic rich and oxygen depleted) has a pH of about 4 and a potential of about –0.1 V. If you find the point on the Pourbaix diagram for naturally occurring iron species, Figure 5.11, you see that $Fe(OH)_3$ is not stable and $Fe^{2+}(aq)$ will be the predominant species. In fact, as long as the potential remained at –0.1 V, Fe^{2+} remains the predominant species below pH = 8. Above pH = 8, Fe^{2+} is oxidized to Fe_2O_3 at this potential. Note that for a potential of –0.1 V and a pH below 2, Fe^{2+} is still the predominant iron species in solution, but water is reduced to H_2.

S5.14 **The minimum temperature for reduction of MgO by carbon?** According to the Ellingham diagram shown in Figure 5.16, at about 1800°C the line for the reducing agent (C) dips below the MgO line, which means that the reactions $2Mg(l) + O_2(g) \rightarrow 2MgO(s)$ and $2C(s) + O_2(g) \rightarrow 2CO(g)$ have the same free energy change at that temperature. Thus, coupling the two reactions (i.e., subtracting the first from the second) yields the overall reaction $MgO(s) + C(s) \rightarrow Mg(l) + CO(g)$ with $\Delta G = 0$. At 1800°C or above, MgO can be conveniently reduced to Mg by carbon.

5.1 **Oxidation numbers?**

$2\,NO(g)$	+	$O_2(g)$	→	$2\,NO_2(g)$				
+1 -2		0		+4 -2				
$2Mn^{3+}(aq)$	+	$2H_2O$	→	MnO_2	+	Mn^{2+}	+	$4H^+(aq)$
+3		+1 -2		+4 -2		+2		+1
$LiCoO_2(s)$	+	$C(s)$	→	$LiC(s)$	+	$CoO_2(s)$		
+1 +3 -2	0			+1-1		+4 -2		
$Ca(s)$	+	$H_2(g)$	→	$CaH_2(s)$				
0		0		+2 -1				

5.2 **Suggest chemical reagents for redox transformations? (a) Oxidation of HCl to Cl_2?** Assume that all of these transformations are occurring in acid solution (pH 0). Referring to Resource Section 3, you will find that the oxidation potential (not the reduction potential) for the Cl^-/Cl_2 couple is –1.358 V. To oxidize Cl^- to Cl_2, you need a couple with a reduction potential more positive than 1.358 V, because then the net potential will be positive and DG will be negative. Examples include the $S_2O_8^{2-}/SO_4^{2-}$ couple (E° = 1.96 V), the H_2O_2/H_2O couple (E° = 1.763 V), and the α–PbO_2/Pb^{2+} couple (E° = 1.468 V). Therefore, since the *oxidized* form of the couple will get *reduced* by *oxidizing* Cl^-, you would want to use either $S_2O_8^{2-}$, H_2O_2, or α–PbO_2 to oxidize Cl^- to Cl_2.
(b) Reducing $Cr^{3+}(aq)$ to $Cr^{2+}(aq)$? In this case, the reduction potential is –0.424 V. Therefore, to have a net E° > 0, you need a couple with an oxidation potential more positive than 0.424 V (or a couple with a reduction potential more negative than –0.424 V). Examples include Mn^{2+}/Mn (E° = –1.18 V), Zn^{2+}/Zn (E° = –0.7626 V), and NH_3OH^+/N_2 (E° = –1.87 V). Remember that it is the reduced form of the couple that you want to use to reduce Cr^{3+} to Cr^{2+}, so in this case you would choose metallic manganese, metallic zinc, or NH_3OH^+.
(c) Reducing $Ag^+(aq)$ to Ag(s)? The reduction potential is 0.799 V. As above, the reduced form of any couple with a reduction potential less than (i.e., less positive than) 0.799 V will reduce Ag^+ to silver metal.
(d) Reducing I_2 to I^-? The reduction potential is 0.535 V. As above, the reduced form of any couple with a reduction potential less than 0.535 V will reduce iodine to iodide.

5.3 **Write balanced equations, if a reaction occurs, for the following species in aerated aqueous acid? (a) Cr^{2+}?** For all of these species, you must determine whether they can be oxidized by O_2. The standard potential for the reduction $O_2 + 4\,H^+ + 4\,e^- \rightarrow 2\,H_2O$ is 1.23 V. Therefore, only redox couples with a reduction potential less positive than 1.23 V will be driven to completion to the oxidized member of the couple by the reduction of O_2 to H_2O. Since the Cr^{3+}/Cr^{2+} couple has E° = –0.424 V, Cr^{2+} *will* be oxidized to Cr^{3+} by O_2. The balanced equation is:

$$4Cr^{2+}(aq) + O_2(g) + 4H^+(aq) \rightarrow 4Cr^{3+}(aq) + 2H_2O(l) \qquad E^\circ = 1.65\text{ V}$$

(b) Fe^{2+}? Since the Fe^{3+}/Fe^{2+} couple has E° = 0.771 V, Fe^{2+} *will* be oxidized to Fe^{3+} by O_2. The balanced equation is:

$$4Fe^{2+}(aq) + O_2(g) + 4H^+(aq) \rightarrow 4Fe^{3+}(aq) + 2H_2O(l) \qquad E^\circ = 0.46\ V$$

(c) Cl^-? Both of the following couples have E° values, shown in parentheses, more positive than 1.23 V, so there will be no reaction when O_2 is mixed with aqueous chloride ion in acid solution: ClO_4^-/Cl^- (1.287 V); Cl_2/Cl^- (1.358 V). The appropriate equation is $Cl^-(aq) + O_2(g) \rightarrow$ NR (NR = no reaction).
(d) HOCl? Since the $HClO_2/HClO$ couple has E° = 1.701 V, HClO *will not* be oxidized to $HClO_2$ by O_2. The standard potential for the oxidation of HOCl by O_2 is –0.47 V.
(e) Zn(s)? Since the Zn^{2+}/Zn couple has E° = –0.763 V, metallic zinc *will* be oxidized to Zn^{2+} by O_2. The balanced equation is:

$$2Zn(s) + O_2(g) + 4H^+(aq) \rightarrow 2Zn^{2+}(aq) + 2H_2(l) \qquad E^\circ = 1.99\ V$$

A competing reaction is $Zn(s) + 2H^+(aq) \rightarrow Zn^{2+}(aq) + H_2(g)$ (E° = 0.763 V).

5.4 **Balanced equations for redox reactions? (a) Fe^{2+}?** When a chemical species is dissolved in aerated acidic aqueous solution, you need to think about four possible redox reactions. These are: (i) the species might oxidize water to O_2, (ii) the species might reduce water (hydronium ions) to H_2, (iii) the species might be oxidized by O_2, and (iv) the species might undergo disproportionation. In the case of Fe^{2+}, the two couples of interest are Fe^{3+}/Fe^{2+} (E° = 0.77 V) and Fe^{2+}/Fe (E° = –0.44 V). Consider the four possible reactions (refer to Resource Section 2 for the potentials you need): (i) The O_2/H_2O couple has E° = 1.229 V, so the oxidation of water would only occur if the Fe^{2+}/Fe potential were *greater* than positive 1.229 V. In other words, the net reaction below is not spontaneous because the net E° is less than zero:

$$2Fe^{2+} + 4e^- \rightarrow 2Fe \qquad E^\circ = -0.44\ V$$

$$2H_2O \rightarrow O_2 + 4H^+ + 4e^- \qquad E^\circ = -1.67\ V$$

$$2Fe^{2+} + 2H_2O \rightarrow 2Fe + O_2 + 4H^+ \qquad E^\circ = -1.67\ V$$

Therefore, Fe^{2+} will not oxidize water. (ii) The H_2O/H_2 couple has E° = 0 V (by definition), so the reduction of water would only occur if the potential for the oxidation of Fe^{2+} to Fe^{3+} was positive, and it is –0.77 V (note that it is the *reduction* potential for the Fe^{3+}/Fe^{2+} couple that is *positive* 0.77 V). Therefore, Fe^{2+} will not reduce water. (iii) Since the O_2/H_2O has E° = 1.229 V, the reduction of O_2 would occur as long as the potential for the oxidation of Fe^{2+} to Fe^{3+} was less negative than –1.229 V. Since it is only –0.77 V (see above), Fe^{2+} will reduce O_2 and in doing so will be oxidized to Fe^{3+}. (iv) The disproportionation of a chemical species will occur if it can act as its own oxidizing agent and reducing agent, which will occur when the potential for reduction minus the potential for oxidation is *positive*. For the Latimer diagram for iron in acidic solution (see Resource Section 3), the difference (–0.44 V) – (0.771 V) = –1.21 V, so disproportionation will not occur.
(b) Ru^{2+}? Consider the Ru^{3+}/Ru^{2+} (E° = 0.249 V) and Ru^{2+}/Ru (E° = 0.81 V) couples. Since the Ru^{2+}/Ru couple has a potential that is not more positive than 1.229 V, Ru^{2+} will not oxidize water. Also, since the Ru^{3+}/Ru^{2+} couple has a positive potential, Ru^{2+} will not reduce water. However, since the potential for the oxidation of Ru^{2+} to Ru^{3+}, –0.249 V, is less negative than –1.229 V, Ru^{2+} will reduce O_2 and in doing so will be oxidized to Ru^{3+}. Finally, since the difference (0.81 V) – (0.249 V) is positive, Ru^{2+} will disproportionate in aqueous acid to Ru^{3+} and metallic ruthenium. It is not possible to tell from the potentials whether the reduction of O_2 by Ru^{2+} or the disproportionation of Ru^{2+} will be the faster process.
(c) $HClO_2$? Consider the $ClO_3^-/HClO_2$ (E° = 1.181 V) and $HClO_2/HClO$ (E° = 1.674 V) couples. Since the $HClO_2/HClO$ couple has a potential that is more positive than 1.229 V, $HClO_2$ *will* oxidize water. However, since the $ClO_3^-/HClO_2$ couple has a positive potential, $HClO_2$ will not reduce water. Since the potential for the oxidation of $HClO_2$ to ClO_3^-, –1.181 V, is less negative than –1.229 V, $HClO_2$ will reduce O_2 and in doing so will be oxidized to ClO_3^-. Finally, since the difference (1.674 V) – (1.181 V) is positive, $HClO_2$ will disproportionate in aqueous acid to ClO_3^- and HClO. As above, it is not possible to tell from the potentials whether the reduction of O_2 by $HClO_2$ or the disproportionation of $HClO_2$ will be the faster process.

(d) Br_2? Consider the Br_2/Br^- (E^o = 1.065 V) and $HBrO/Br_2$ (E^o = 1.604 V) couples. Since the Br_2/Br^- couple has a potential that is less positive than 1.229 V, Br_2 will not oxidize water. Also, since the $HBrO/Br_2$ couple has a positive potential, Br_2 will not reduce water. Since the potential for the oxidation of Br_2 to HBrO, –1.604 V, is more negative than –1.229 V, Br_2 will not reduce O_2. Finally, since the difference (1.065 V) – (1.604 V) is negative, Br_2 will not disproportionate in aqueous acid to Br^- and HBrO. In fact, the equilibrium constant for the following reaction is 7.2×10^{-9}.

$$Br_2(aq) + H_2O(l) \Leftrightarrow Br^-(aq) + HBrO(aq) + H^+(aq)$$

5.5 **Standard potentials vary with temperature in opposite directions?** The two complexes have different stabilities with respect to the preferred oxidation state for binding that varies with temperature. As noted in Section 5.10, the cyano complex favors Ru(III) at room temperature. These complexation reactions involve dynamic equilibria that shift with varying temperature. The amino and cyano complexes must have different equilibrium shifts with respect to changes in temperature that results in the opposite directions of change for the cell potential.

5.6 **Balance redox reaction in acid solution: $MnO_4^- + H_2SO_3 \rightarrow Mn^{2+} + HSO_4^-$? pH dependence?**

$$2\,[MnO_4^-(aq) + 8H^+(aq) + 5e^- \rightarrow Mn^{2+}(aq) + 4H_2O(l)] \qquad \text{reduction}$$

$$5\,[H_2O(l) + H_2SO_3(aq) \rightarrow HSO_3^-(aq) + 2e^- + 3H^+(aq)] \qquad \text{oxidation}$$

$$2MnO_4^-(aq) + 5H_2SO_3(aq) + H^+(aq) \rightarrow 2Mn^{2+}(aq) + 5HSO_3^-(aq) + 3H_2O(l)$$

The potential decreases as the pH increases and the solution becomes more basic (See Section 5.6 for further clarification).

5.7 **Write the Nernst equation for (a) The reduction of O_2?** In general terms, the Nernst equation is given by the formula: $E = E^o - (RT/nF)(\ln Q)$ where Q is the reaction quotient.

For the reduction of oxygen, $O_2(g) + 4H^+(aq) + 4e^- \rightarrow 2H_2O(1)$,

$$Q = 1/(p(O_2)[H^+]^4) \text{ and } E = E^o - [(0.059\text{ V})/4][\log(1/(p(O_2)[H^+]^4)]$$

Therefore, the potential for O_2 reduction at pH = 7 and $p(O_2)$ = 0.20 bar is:

$$E = 1.229\text{ V} - 0.42\text{ V} = 0.81\text{ V}$$

(b) The reduction of $Fe_2O_3(s)$? For the reduction of solid iron(III) oxide, $Fe_2O_3(s) + 6H^+(aq) + 6e^- \rightarrow 2\,Fe(s) + 3H_2O(1)$, we have:

$$Q = 1/[H^+]^6 \text{ and } E = E^o - (RT/nF)(13.8\text{ pH})$$

$$\text{since pH} = -\log[H^+] = -2.3\ln[H^+] \text{ and } \log[H^+]^6 = 6\log[H^+].$$

5.8 **Using Frost diagrams? (a) What happens when Cl_2 is dissolved in aqueous basic solution?** The Frost diagram for chlorine in basic solution is shown in Figure 5.18. If the points for Cl^- and ClO_4^- are connected by a straight line, Cl_2 lies above it. Therefore, Cl_2 is thermodynamically susceptible to disproportionation to Cl^- and ClO_4^- when it is dissolved in aqueous base. In practice, the oxidation of ClO^- is slow, so a solution of Cl^- and ClO^- is formed when Cl_2 is dissolved in aqueous base.

(b) What happens when Cl_2 is dissolved in aqueous acid solution? The Frost diagram for chlorine in acidic solution is shown in Figure 6.16. If the points for Cl^- and any positive oxidation state of chlorine are connected by a straight line, the point for Cl_2 lies below it (if only slightly). Therefore, Cl_2 will not disproportionate. However, E^o for the Cl_2/Cl^- couple, 1.36 V, is more positive than E^o for the O_2/H_2O couple, 1.23 V. Therefore, Cl_2 is thermodynamically capable of oxidizing water as follows, although the reaction is very slow:

$$Cl_2(aq) + H_2O(l) \rightarrow 2Cl^-(aq) + 2H^+(aq) + \tfrac{1}{2}O_2(g) \qquad E = 0.13\ V$$

(c) Should $HClO_3$ disproportionate in aqueous acid solution? The point for ClO_3^- in acidic solution on the Frost diagram lies above the single straight line connecting the points for Cl_2 and ClO_4^-. Therefore, since ClO_3^- is thermodynamically unstable with respect to disproportionation in acidic solution (i.e., it *should* disproportionate), the failure of it to exhibit any observable disproportionation must be due to a kinetic barrier.

5.9 **Write equations for the following reactions: (a) N_2O is bubbled into aqueous NaOH solution?** The Frost diagram for nitrogen in basic solution, showing some pertinent species, is shown below:

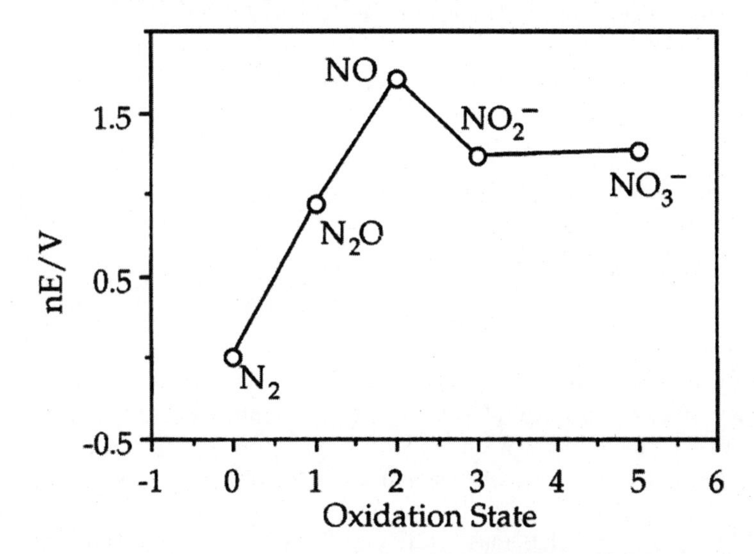

Inspection of the diagram shows that N_2O lies above the line connecting N_2 and NO_3^-. Therefore, N_2O is thermodynamically susceptible to disproportionation to N_2 and NO_3^- in basic solution:

$$5N_2O(aq) + 2OH^-(aq) \rightarrow 2NO_3^-(aq) + 4N_2(g) + H_2O(l)$$

However, the redox reactions of nitrogen oxides and oxyanions are generally slow. In practice, N_2O has been found to be inert.

(b) Zinc metal is added to aqueous acidic sodium triiodide? The overall reaction is:

$$Zn(s) + I_3^-(aq) \rightarrow Zn^{2+}(aq) + 3I^-(aq)$$

Since E° values for the Zn^{2+}/Zn and I_3^-/I^- couples are –0.76 V and 0.54 V, respectively (these potentials are given in Resource Section 3), E° for the net reaction above is 0.54 V + 0.76 V = 1.30 V. Since the net potential is positive, this is a favorable reaction, and it should be kinetically facile if the zinc metal is finely divided and thus well exposed to the solution.

(c) I_2 is added to excess aqueous acidic $HClO_3$? Since E° values for the I_2/I^- and ClO_3^-/ClO_4^- couples are 0.54 V and –1.20 V, respectively (see Resource Section 2), the net reaction involving the reduction of I_2 to I^- and the oxidation of ClO_3^- to ClO_4^- will have a net negative potential, E° = 0.54 V + (–1.20 V) = –0.66 V. Therefore, this net reaction will not occur. However, since E° values for the IO_3^-/I_2 and ClO_3^-/Cl^- couples are 1.19 V and 1.47 V, respectively, the following net reaction will occur, with a net E° = 0.28 V:

$$5\,[ClO_3^-(aq) + 6H^+(aq) + 6e^- \rightarrow Cl^-(aq) + 3H_2O(l)] \qquad \text{reduction}$$

$$3\,[I_2(s) + 6H_2O(l) \rightarrow 2IO_3^-(aq) + 12H^+(aq) + 10e^-] \qquad \text{oxidation}$$

$$3I_2(s) + 5ClO_3^-(aq) + 3H_2O(l) \rightarrow 6IO_3^-(aq) + 5Cl^-(aq) + 6H^+(aq)$$

5.10 **Electrode potential for Ni^{2+}/Ni couple at pH = 14?** The standard potential for the reduction of Ni^{2+} is 0.25 V. The cell potential, *E*, is given by the Nernst equation as follows:

$$E = E^{o} - (RT/\nu F)\ \ln Q \text{ where } Q = 1/[Ni^{2+}(aq)]$$

$$\text{At pH} = 14 \text{ the } [OH^{-}] = 1\ M \text{ and } [Ni^{2+}] = K_{sp},$$

$$E = 0.25\ V - ((8.3145\ J\ K^{-1}mol\text{-}1)(298\ K) / (2)(9.6485 \times 10^{4}\ C\ mol^{-1}))\ln(1/1.5 \times 10^{-16})$$

$$E = 0.25\ V - (0.01284)\ln(1/1.5 \times 10^{-16})\ V$$

$$E = 0.25\ V - (0.01284)(36.4)\ V$$

$$E = -0.21\ V$$

5.11 **Will acid or base most favour the following half-reactions? (a) $Mn^{2+} \rightarrow MnO_4^-$?** You can answer questions such as this by applying Le Chatelier's principle to the complete, balanced half-reaction, which in this case is:

$$Mn^{2+}(aq) + 4H_2O(l) \rightarrow MnO_4^-(aq) + 8H^+(aq) + 5e^-$$

Since hydrogen ions are produced by this oxidation half-reaction, raising the pH will favor the reaction. (Alternatively, you could come to the same conclusion by using the Nernst equation.) Thus this reaction is favored in basic solution. At a sufficiently high pH, Mn^{2+} will precipitate as $Mn(OH)_2$. The rate of oxidation for this solid will be slower than for dissolved species.
(b) $ClO_4^- \rightarrow ClO_3^-$? The balanced half-reaction is:

$$ClO_4^-(aq) + 2H^+(aq) + 2e^- \rightarrow ClO_3^-(aq) + H_2O(l)$$

Since hydrogen ions are consumed by this reduction half-reaction, lowering the pH will favor the reaction. Thus this reaction is favored in acidic solution. This is the reason that perchloric acid is a dangerous oxidizing agent, even though salts of the perchlorate ion are frequently stable in neutral or basic solution.
(c) $H_2O_2 \rightarrow O_2$? The balanced half-reaction is:

$$H_2O_2(aq) \rightarrow O_2(g) + 2H^+(aq) + 2e^-$$

Since hydrogen ions are produced by this oxidation half-reaction, raising the pH will favor the reaction. Thus, the reaction is favored in basic solution. This means that hydrogen peroxide is a better reducing agent in base than in acid.
(d) $I_2 \rightarrow 2I^-$? Since protons are neither consumed nor produced in this reduction half-reaction, and since I^- is not protonated in aqueous solution (because HI is a very strong acid), this reaction has the same potential in acidic or basic solution, 0.535 V (this can be confirmed by consulting Resource Section 3).

5.12 **Determine the standard potential for the reduction of ClO_4^- to Cl_2?** The Latimer diagram for chlorine in acidic solution is given in Resource Section 3.To determine the potential for any couple, you must calculate the *weighted average* of the potentials of intervening couples. In general terms it is:

$$(n_1E^{o}_1 + n_2E^{o}_2 + \ldots + n_nE^{o}_n) / (n_1 + n_2 + \ldots + n_n)$$

and in this specific case it is:

$$[(2)(1.201\ V) + (2)(1.181\ V) + (2)(1.674) + (1)(1.630\ V)]/(2 + 2 + 2 + 1) = 1.392\ V$$

Thus, the standard potential for the ClO_4^-/Cl_2 couple is 1.392 V. The half-reaction for this reduction is:

$$2ClO_4^-(aq) + 16H^+(aq) + 14e^- \rightarrow Cl_2(g) + 8H_2O(l)$$

5.13 Calculate the equilibrium constant for $Au^+(aq) + 2CN^-(aq) \rightarrow [Au(CN)_2]^-(aq)$? To calculate an equilibrium constant using thermodynamic data, you can make use of the expressions $\Delta G = -RT\ln K$ and $\Delta G = -nFE$. You can use the given potential data to calculate ΔG for each of the two half-reactions, then you can use Hess's Law to calculate ΔG for the overall process, and finally calculate K from the net ΔG.

The net reaction $Au^+(aq) + 2\,CN^-(aq) \rightarrow [Au(CN)_2]^-(aq)$ is the following sum:

$$Au^+(aq) + e^- \rightarrow Au(s)$$

$$Au(s) + 2CN^-(aq) \rightarrow [Au(CN)_2]^-(aq)$$

$$Au^+(aq) + 2\,CN^-(aq) \rightarrow [Au(CN)_2]^-(aq)$$

ΔG for the first reaction is $-nFE = -(1)(96.5\text{ kJ mol}^{-1}\text{ V}^{-1})(1.69\text{ V}) = -163\text{ kJ mol}^{-1}$. ΔG for the second reaction is $-(1)(96.5\text{ kJ mol}^{-1}\text{ V}^{-1})(0.6\text{ V}) = -58\text{ kJ mol}^{-1}$. The net ΔG is the sum of these two values, -221 kJ mol^{-1}. Therefore, assuming that $T = 298$ K:

$$K = \exp[(221\text{ kJ mol}^{-1}) / (8.31\text{ J K}^{-1}\text{mol}^{-1})(298)] = \exp(89.2) = 5.7 \times 10^{38}$$

5.14 Find the approximate potential of an aerated lake at pH = 6, and predict the predominant species? (a) Fe? According to Figure 5.12, the potential range for surface water at pH 6 is 0.5 – 0.6 V, so a value of 0.55 V can be used as the approximate potential of an aerated lake at this pH. Inspection of the Pourbaix diagram for iron (Figure 5.11) shows that at pH 6 and $E = 0.55$ V, the stable species of iron is $Fe(OH)_3$. Therefore, this compound of iron would predominate.

(b) Mn? Inspection of the Pourbaix diagram for manganese (Figure 5.13) shows that at pH 6 and $E = 0.55$ V, the stable species of manganese is Mn_2O_3. Therefore, this compound of manganese would predominate.

(c) S? At pH 0, the potential for the HSO_4^-/S couple is 0.387 V (this value was calculated using the weighted average of the potentials given in the Latimer diagram for sulfur in Resource Section 3), so the lake will oxidize S_8 all the way to HSO_4^-. At pH 14, the potentials for intervening couples are all negative, so SO_4^{2-} would again predominate. Therefore, HSO_4^- is the predominant sulfur species at pH 6.

5.15 Frost diagram and standard potential for the $HSO_4^-/S_8(s)$ couple? Consult Figure 16.2 for the Frost diagram for sulphur. To determine the potential for any couple, you must calculate the *weighted average* of the potentials of intervening couples. In general terms it is:

$$(n_1E°_1 + n_2E°_2 + \ldots + n_nE°_n) / (n_1 + n_2 + \ldots + n_n)$$

and in this specific case it is:

$$[(2)(0.16\text{ V}) + (2)(0.40\text{ V}) + (2)(0.60\text{ V})] / (2 + 2 + 2) = 0.387\text{ V}$$

5.16 Equilibrium constant for the reaction $Pd^{2+}(aq) + 4\,Cl^-(aq) \equiv [PdCl_4]^{2-}(aq)$ in 1 M HCl(aq)? To calculate an equilibrium constant using thermodynamic data, you can make use of the expressions $\Delta G = -RT\ln K$ and $\Delta G = -nFE$. You can use the given potential data to calculate ΔG for each of the two half-reactions, then you can use Hess's Law to calculate ΔG for the overall process, and finally calculate K from the net ΔG.
The net reaction $Pd^{2+}(aq) + 4\,Cl^-(aq) \rightarrow [Pd(Cl)_4]^{2-}(aq)$ is the following sum:

$$Pd^{2+}(aq) + 2e^- \rightarrow Pd(s)$$

$$Pd(s) + 4\,Cl^-(aq) \rightarrow [PdCl_4]^{2-}(aq)$$

$$Pd^{2+}(aq) + 4\,Cl^{-}(aq) \rightarrow [PdCl_4]^{2-}(aq)$$

ΔG for the first reaction is $-nFE = -(2)(96.5\text{ kJ mol}^{-1}\text{ V}^{-1})(0.915\text{ V}) = -176.6\text{ kJ mol}^{-1}$. ΔG for the second reaction is $-(2)(96.5\text{ kJ mol}^{-1}\text{ V}^{-1})(-0.6\text{ V}) = +115.8\text{ kJ mol}^{-1}$. The net ΔG is the sum of these two values, -60.8 kJ mol^{-1}. Therefore, assuming that $T = 298$ K:

$$K = \exp[(60.8\text{ kJ mol}^{-1}) / (8.31\text{ J K}^{-1}\text{mol}^{-1})(298)] = \exp(24.5) = 4.37 \times 10^{10}$$

5.17 **Reduction potential for MnO_4^- to $MnO_2(s)$ at pH = 9.00?** The standard potential for the reduction $MnO_4^-(aq) + 4H^+(aq) + 3e^- \rightarrow MnO_2(s) + 2H_2O(l)$ is 1.69 V. The pH, or $[H^+]$, dependence of the potential E is given by the Nernst equation ($[MnO_4^-] = 1$) as follows:

$$E = E^\circ - ((0.059\text{ V})/3)(\log(1/[H^+]^4)$$

Note that $n = 3$ for the reduction of MnO_4^-. Note also that the factor 0.059 V/n can only be used at 25°C. Since at pH = 9, $[H^+] = 10^{-9}$ M,

$$E = 1.69\text{V} - (0.01967\text{ V})(\log(1/10^{-36}) = 1.69\text{ V} - (0.01967\text{ V})(36)$$

$$E = 1.69\text{ V} - 0.71\text{ V} = 0.98\text{ V}$$

5.18 **Tendency of mercury species to act as an oxidizing agent, a reducing agent, or to undergo disproportionation?** Consult Section 5.11 and Self-test 5.11 for the construction of the Frost diagram. Hg^{2+} and Hg_2^{2+} are both oxidizing agents as they have large positive standard reduction potentials are therefore likely to undergo reduction encouraging oxidation of the other species in the reaction. None of these species are likely to be good reducing agents as their standard reduction potentials are positive and oxidation is not favored thermodynamically. Hg_2^{2+} is not likely to undergo disproportionation as the cell potential for this reaction would be negative ($E = 0.796\text{ V} - (+0.911\text{V}) = -0.115\text{V}$) and not spontaneous.

5.19 **Thermodynamic tendency of HO_2 to undergo disproportionation?** The standard cell potential for the disproportionation of HO_2 can be calculated from the standard reduction potentials in the Latimer diagram provided. $E = E_{red} - E_{ox} = +1.150 - (-0.125)\text{ V} = +1.275$ V. Since the E value is positive, HO_2 will undergo disproportionation.

5.20 **Dissolved carbon dioxide corrosive toward iron?** The air oxidation of iron is favored in acidic solution. Increased levels of carbon dioxide and water generate carbonic acid which encourages the corrosion process by lowering the pH of the solution.

5.21 **What is the maximum E for an anaerobic environment rich in Fe^{2+} and H_2S?** Any species capable of oxidizing either Fe^{2+} or H_2S at pH 6 cannot survive in this environment. According to the Pourbaix diagram for iron (Figure 5.11), the potential for the $Fe(OH)_3/Fe^{2+}$ couple at pH 6 is approximately 0.3 V. Using the Latimer diagrams for sulfur in acid and base (see Resource Section 3), the H_2S, S potential at pH 6 can be calculated as follows:

$$0.14\text{ V} - (6/14)[0.14 - (-0.45)] = -0.11\text{ V}$$

Any potential higher than this will oxidize hydrogen sulfide to elemental sulfur. Therefore, as long as H_2S is present, the maximum potential possible is approximately -0.1 V.

5.22 **How will edta^{4-} complexation affect $M^{2+} \rightarrow M^0$ reductions?** Since edta^{4-} forms very stable complexes with $M^{2+}(aq)$ ions of period 4 *d*–block elements but *not* with the zerovalent metal atoms, the reduction of a $M(edta)^{2-}$ complex will be more difficult than the reduction of the analogous M^{2+} aqua ion. Since the reductions are more difficult, the reduction potentials become less positive (or more negative, as the case may be). The reduction of the

$M(edta)^{2-}$complex includes a decomplexation step, with a positive free energy change. The reduction of $M^{2+}(aq)$ does not require this additional expenditure of free energy.

5.23 **Which of the boundaries depend on the choice of $[Fe^{2+}]$?** Any boundary between a soluble species and an insoluble species will change as the concentration of the soluble species changes. For example, consider the line separating $Fe^{2+}(aq)$ from $Fe(OH)_3(s)$ in Figure 5.11. As shown in the text, $E = E^o - (0.059\ V/2)\log([Fe^{2+}]^2/[H_3O^+]^6)$ (see Section 5.14, *Pourbaix diagrams*). Clearly, the potential at a given pH is dependent on the concentration of soluble $Fe^{2+}(aq)$. The boundaries between the two soluble species, $Fe^{2+}(aq)$ and $Fe^{3+}(aq)$, and between the two insoluble species, $Fe(OH)_2(s)$ and $Fe(OH)_3(s)$, will not depend on the choice of $[Fe^{2+}]$.

5.24 **Under what conditions will Al reduce MgO?** The lines for Al_2O_3 and MgO on the Ellingham diagram (Figure 5.16) represent the change in ΔG^o with temperature for the following reactions:

$$\tfrac{4}{3}Al(l) + O_2(g) \rightarrow \tfrac{2}{3}Al_2O_3(s) \text{ and } 2Mg(l) + O_2(g) \rightarrow 2MgO(s)$$

At temperatures below about 1400°C, the free energy change for the MgO reaction is more negative than for the Al_2O_3 reaction. This means that under these conditions MgO is more stable with respect to its constituent elements than is Al_2O_3, and that Mg will react with Al_2O_3 to form MgO and Al. However, above about 1400°C the situation reverses, and Al will react with MgO to reduce it to Mg with the concomitant formation of Al_2O_3. This is a rather high temperature, achievable in an electric arc furnace (compare the extraction of silicon from its oxide, discussed in Section 5.16).

Chapter 6 Molecular Symmetry

The H atom 1*s* orbitals in H_2O form the two linear combinations shown. The unshaded lobes are + throughout and the shaded lobe is – throughout. These symmetry-adapted orbitals are labeled a_1 (top) and b_2 (bottom). The molecular orbitals of H_2O are formed by making suitable linear combinations of these symmetry-adapted orbitals with O atomic orbitals of a_1 and b_2 symmetry.

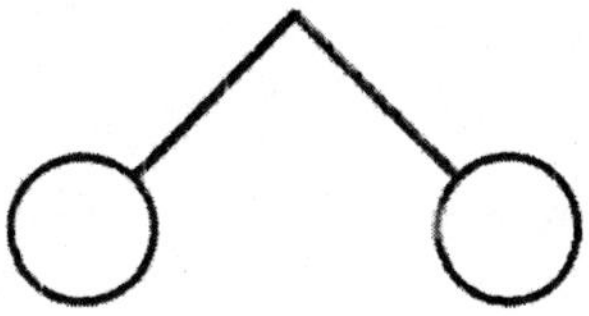

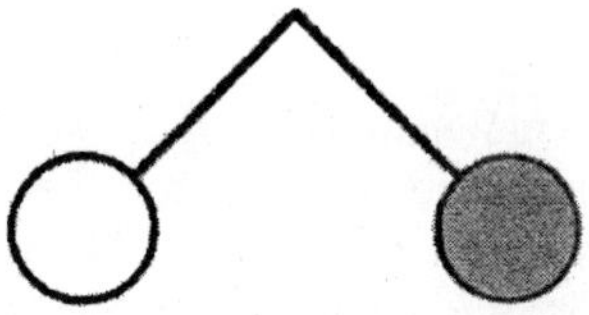

S6.1 **Sketch the S_4 axis of an NH_4^+ ion. How many of these axes are there in the ion?**
The S_4 axis is best portrayed by separating the components. There are three S_4 axes.

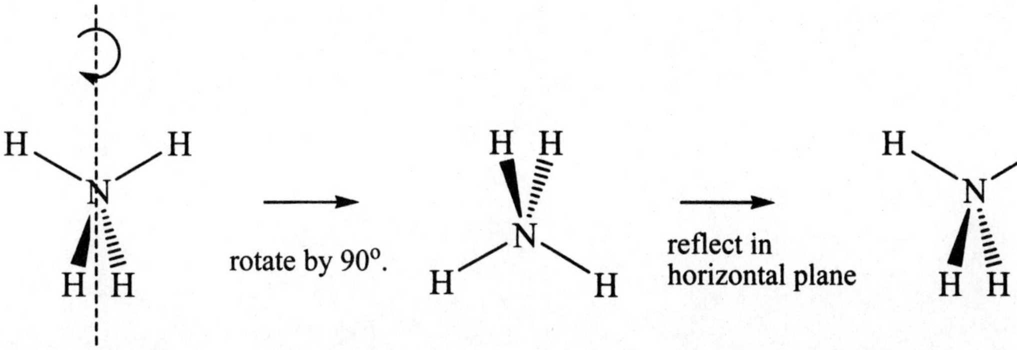

rotate by 90°.

reflect in horizontal plane

S6.2 **(a) BF_3 point group?** All molecules possess the identity, *E*. Since BF_3 is trigonal planar, it is obvious that it possesses a 3-fold rotation axis (C_3) and a mirror plane of symmetry that coincides with the molecular plane (σ_h, since it is perpendicular to the C_3 axis, which is the "major" axis). There are also three 2-fold axes (C_2) that coincide with the three B–F bond vectors, and three more mirror planes, each of which contains a B–F bond and is perpendicular to σ_h (these are called σ_v since they are parallel to the major axis). The set of elements (E, C_3, $3C_2$, σ_h, $3\sigma_v$) corresponds to the group D_{3h}. Refer to Table 6.2 and note that the D_{3h} point group also contains an S_3 improper rotation axis. However, in many cases it is not necessary to find the complete set of symmetry elements to determine uniquely the point group of a molecule or ion. In fact, if you use the decision tree shown in Figure 6.9 to determine the point group of BF_3, you will find that the smaller set of elements (C_3, $3C_2$, σ_h) uniquely corresponds to D_{3h}.

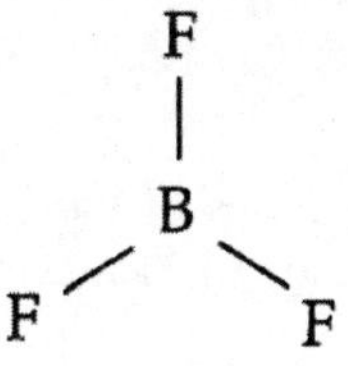

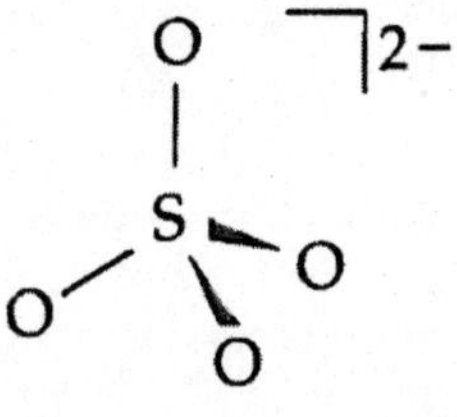

(b) SO_4^{2-} point group? Using the decision tree in Figure 6.9 is generally the easiest way to determine the point group of a molecule or ion. The sulfate ion (i) is nonlinear, (ii) possesses four 3-fold axes (C_3), like NH_4^+ (see the answer to S6.1), and (iii) does not have a center of symmetry. The sequence of "no, yes, no" on the decision tree leads to the conclusion that SO_4^{2-} belongs to the T_d point group.

S6.3 **Symmetry species of all five d orbitals of the central Xe atom in XeF_4 (D_{4h}, Fig. 6.3)?** In order to determine the orbital symmetry, we need to see how the orbitals behave under specific symmetry operations for D_{4h} point group. We can use the D_{4h} character table to determine symmetry species. We note that: d_{x2-y2} has a symmetry species of B_{1g}; d_{xy} is B_{2g}; d_{xz} and d_{yz} are E_g; and d_{z2} is A_{1g}.

S6.4 **What is the maximum possible degeneracy for an O_h molecule?** This question, although asked about SF_6, could have been asked about any molecule with rigorous octahedral symmetry, i.e., a molecule that belongs to the O_h point group. If you refer to the character table for this group, which is given in Resource Section 4, you find that there are characters of 1, 2, *and* 3 in the column headed by the identity element, E. Therefore, the *maximum* possible degree of degeneracy of the orbitals in SF_6 is 3 (although nondegenerate and twofold degenerate orbitals are allowed). As an example, the sulfur atom valence p orbitals, and any molecular orbitals formed using them, are triply degenerate in SF_6.

S6.5 **A conformation of the ferrocene molecule that lies 4 kJ mol^{-1} above the lowest energy configuration is a pentagonal antiprism. Is it polar?** Like ruthenocene, the staggered conformation of ferrocene has a C_5 axis passing through the centroids of the C_5H_5 rings and the Fe atom. It also has five C_2 axes that pass through the Fe atom but are perpendicular to the major C_5 axis, so it belongs to one of the D point groups, in this case D_{5d} (it lacks the σ_h plane of symmetry that a D_{5h} molecule like ruthenocene possesses). Since the D_{5d} conformation of ferrocene has a C_5 axis *and* perpendicular C_2 axes, it is not polar (see Section 6.3). You may find it difficult to find the n C_2 axes for a D_{nd} structure. However, if you draw the mirror planes, the C_2 axes lie between them. In this case, one C_2 axis interchanges the front vertex of the top ring with one of the two front vertices of the bottom ring, while a second C_2 axis, rotated exactly 36° from the first one, interchanges the same vertex on top with the other front bottom one.

S6.6 **Is the skew form of H_2O_2 chiral?** Except for the identity, E, the only element of symmetry that this conformation of hydrogen peroxide possesses is a C_2 axis that passes through the midpoint of the O–O bond and bisects the two O–O–H planes (these are *not* mirror planes of symmetry). Hence this form of H_2O_2 belongs to the C_2 point group, and it is chiral since this group does not contain any S_n axes. In general, any structure that belongs to a C_n or D_n point group is chiral, as are molecules that are asymmetric (C_1 symmetry). Optically active H_2O_2 might be observable, even transiently, if bound in a chiral host such as the active site of an enzyme (see Chapter 26).

S6.7 **Can the bending mode of N_2O be Raman active?** The Lewis structure of nitrous oxide is shown below. Based on this structure, you can predict that its geometry should be linear. However, unlike CO_2, with which it is isoelectronic, N_2O does not have a centre of symmetry. Therefore, the exclusion rule does not apply, and a band that is IR active *can* be Raman active as well.

$$\ddot{\underset{..}{N}}=N=\ddot{\underset{..}{O}}$$

S6.8 **Confirm that the symmetric mode is A_g?** All of the operations of this group leave the displacement vectors unchanged. Therefore, all of the operations have a character of +1. This corresponds to the first row in the D_{2h} character table, which is the A_g symmetry type.

S6.9 **Show that the four CO displacements in the square-planar (D_{4h}) $[Pt(CO)_4]^{2+}$ cation transform as $A_{1g} + B_{1g} + E_u$. How many bands would you expect in the IR and Raman spectra for the $[Pt(CO)_4]^{2+}$ cation?**
The reducible representation is obtained as follows:

D_{4h}	E	$2C_4$	C_2	$2C_2'$	$2C_2''$	i	$2S_4$	σ_h	$2\sigma_v$	$2\sigma_d$
Γ_{3N}	4	0	0	2	0	0	0	4	2	0

This reduces to $A_{1g} + B_{1g} + E_u$
$A_{1g} + B_{1g}$ are Raman active modes as they have the same symmetry as a component of the molecular polarizability. E_u is IR active as it has the same symmetry as a component of the electric dipole vector.

S6.10 **Orbital symmetry for a tetrahedral array of H atoms in methane?** The molecule CH_4 has T_d symmetry, and the combination of H atom 1*s* orbitals given also has T_d symmetry. The group theory jargon that is used in this case is to say that the combination has the *total* symmetry of the molecule. This is true because each time the H atom array of orbitals is subjected to an operation in the T_d point group, the array changes into itself. In each case, the character is 1. Each point group has one symmetry label (one row) for which all the characters are one, and for the T_d point group it is called A_1 (see Resource Section 3). Thus the symmetry label of the given combination of H atom 1*s* orbitals is A_1.

S6.11 **Orbital symmetry for a square-planar array of H atoms?** You must adopt some conventions to answer this one question. First, you assume that the combination of H atom 1*s* orbitals given looks like the figure shown below. This array of H atoms has D_{4h} symmetry. Inspection of the character table for this group, which is given in Resource Section 4, reveals that there are three different types of C_2 rotation axes, i.e., there are three columns labeled C_2, C_2'', and C_2''. The first of these is the C_2 axis that is coincident with the C_4 axis; the second type, C_2', represents two axes in the H_4 plane that do not pass through any H atoms; the third type, C_2'', represents two axes in the H_4 plane that pass through pairs of opposite H atoms. Now, instead of applying operations from all ten columns to this array, to see if it changes into itself (i.e., the +/– signs of the lobes stay the same) or if it changes sign, you can make use of a shortcut. Notice that the array changes into itself under the inversion operation through the centre of symmetry. Thus the character for this operation, *i*, is 1. This means that the symmetry label for this array is one of the first four in the character table, A_{1g}, A_{2g}, B_{1g}, or B_{2g}. Notice also that for these four, the symmetry type is uniquely determined by the characters for the first five columns of operations, which are:

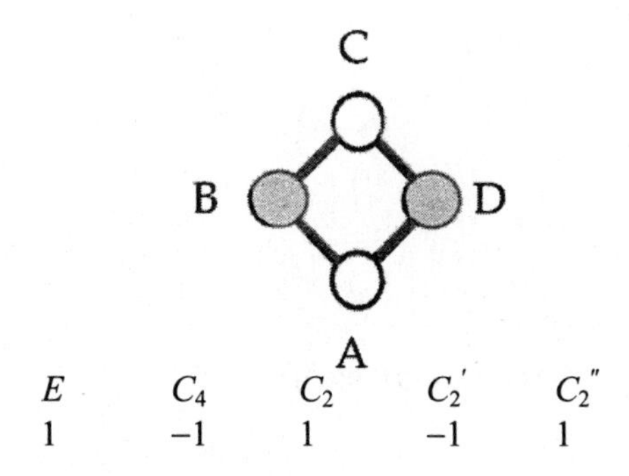

E	C_4	C_2	C_2'	C_2''
1	–1	1	–1	1

These match the characters of the B_{2g} symmetry label.

S6.12 **Which Pt atomic orbitals can combine with which of these SALCs? The atomic orbitals must have matching symmetries to generate SALCs.** By consulting Resource Section 5 and the D_{4h} character table, we note that 5s and $4d_{z2}$ have A1g symmetry; the $d_{x2\text{-}y2}$ has B_{1g} symmetry; and $5p_x$ and $5p_y$ have E_u symmetry. Therefore, these Pt orbitals with matching symmetries can be used to generate SALCs.

S6.13 **Predict how the IR and Raman spectra of SF_5Cl differ from that of SF_6?** Recall that symmetry types with the same symmetry as the function $x^2 + y^2 + z^2$ are Raman active, not IR active. On the other hand, symmetry types with the same symmetry as the functions *x*, *y*, or *z* are IR active, not Raman active.

SF_6 has O_h symmetry. Analysis of the stretching vibrations leads to:

$$\Gamma_{str} = A_{1g} \text{ (Raman, polarized)} + E_g \text{ (Raman)} + T_{1u} \text{ (IR)}.$$

SF_5Cl has C_{4v} symmetry. Analysis of the stretching vibrations leads to:

$$\Gamma_{str} = 3A_1 \text{ (IR and Raman, polarized)} + 2B_1 \text{ (Raman)} + E \text{ (IR, Raman)}.$$

Of the A_1 vibrations, one (S-Cl) will be at much lower frequency than the other two.

S6.14 Symmetries of all the vibration modes of $[PdCl_4]^{2-}$? For the molecule with D4h symmetry,

D_{4h}	E	$2C_4$	C_2	$2C_2'$	$2C_2''$	i	$2S_4$	σ_h	$2\sigma_v$	$2\sigma_d$
Γ_{3N}	15	1	−1	−3	−1	−3	−1	5	3	1

This reduces to $A_{1g} + A_{2g} + B_{1g} + B_{2g} + E_g + 2A_{2u} + B_{2u} + 3E_u$

The translations span A_{2u} and E_u and the rotations span A_{2g} and E_g. Subtracting these terms gives:

$$A_{1g} + B_{1g} + B_{2g} + A_{2u} + B_{2u} + 2E_u$$

as the symmetries for the vibration modes.

S6.15 SALCs for sigma bonding in O? $A_{1g} + E_g + T_{1u}$, see S6.13 and Resource Section 5.

6.1 Symmetry elements? (a) a C_3 axis and a σ_v plane in the NH_3 molecule?

C_3 σ_v

(b) a C_4 axis and a σ_h plane in the square-planar $[PtCl_4]^{2-}$ ion?

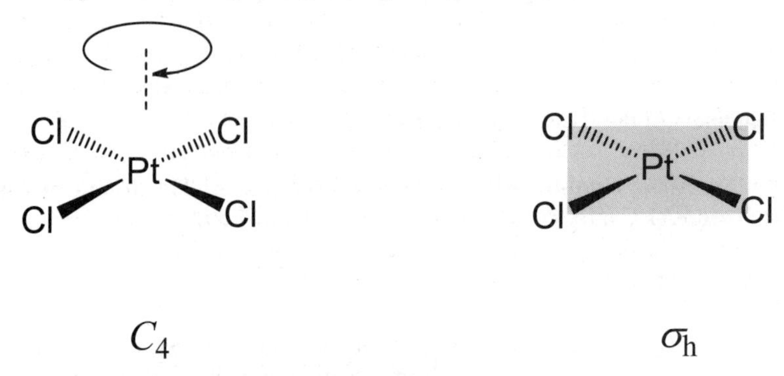

C_4 σ_h

6.2 **S_4 or i? (a) CO_2?** This linear molecule has a center of inversion. There are only two point groups for linear molecules, $D_{\infty h}$ (has i) or $C_{\infty v}$ (does not have i). Therefore, the point group is $D_{\infty h}$.
(b) C_2H_2? This linear molecule also has i. The point group is $D_{\infty h}$.
(c) BF_3? This molecule possesses neither i nor S_4. It belongs to the D_{3h} point group.
(d) SO_4^{2-}? This ion has three different S_4 axes, which are coincident with three C_2 axes, but there is no i. The point group is T_d.

6.3 **Assigning point groups: (a) NH_2Cl?** The only element of symmetry that this molecule possesses other than E is a mirror plane that contains the N and Cl atoms and bisects the H–N–H bond angle. The set of symmetry elements (E, σ) corresponds to the point group C_s (the subscript comes from the German word for mirror, *Spiegel*).

N
Cl H
H

(b) CO_3^{2-}? The carbonate anion is planar, so it possesses at least one plane of symmetry. Since this plane is perpendicular to the major proper rotation axis, C_3, it is called σ_h. In addition to the C_3 axis, there are three C_2 axes coinciding with the three C–O bond vectors. There are also other mirror planes and improper rotation axes, but the elements listed so far (E, C_3, σ_h, $3C_2$) uniquely correspond to the D_{3h} point group (note that CO_3^{2-} has the same symmetry as BF_3). A complete list of symmetry elements is E, C_3, $3C_2$, S_3, σ_h, and $3\sigma_v$.

O 2–
C
O O

(c) SiF_4? This molecule has four C_3 axes, one coinciding with each of the four Si–F bonds. In addition, there are six mirror planes of symmetry (any pair of F atoms and the central Si atom define a mirror plane, and there are always six ways to choose a pair of objects out of a set of four). Furthermore, there is no center of symmetry. Thus, the set (E, $4C_3$, 6σ, no i) describes this molecule and corresponds to the T_d point group. A complete list of symmetry elements is E, $4C_3$, $3C_2$, $3S_4$, and $6\sigma_d$.

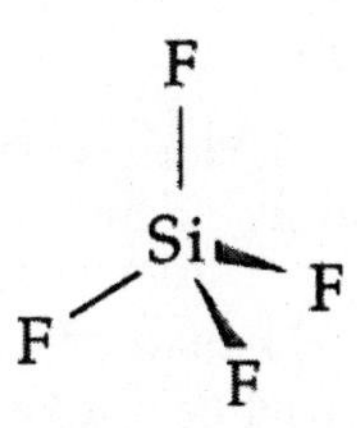

(d) HCN? Hydrogen cyanide is linear, so it belongs to either the $D_{\infty h}$ or the $C_{\infty v}$ point group. Since it does not possess a center of symmetry, which is a requirement for the $D_{\infty h}$ point group, it belongs to the $C_{\infty v}$ point group.
(e) SiFClBrI? This molecule does not possess any element of symmetry other than the identity element, E. Thus it is asymmetric and belongs to the C_1 point group, the simplest possible point group.
(f) BrF_4^-? This anion is square planar. It has a C_4 axis and four perpendicular C_2 axes. It also has a σ_h mirror plane. These symmetry elements uniquely correspond to the D_{4h} point group. A complete list of symmetry elements is E, C_4, a parallel C_2, four perpendicular C_2, S_4, i, σ_h, $2\sigma_v$, and $2\sigma_d$.

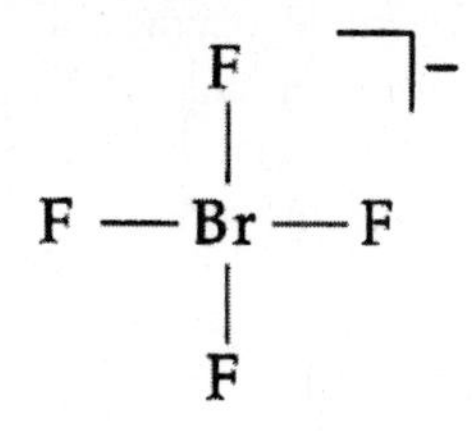

6.4 **How many planes of symmetry does a benzene molecule possess? What chloro-substituted benzene has exactly four planes of symmetry?** The point group symmetry of benzene is D_{6h}. It has a C_6 axis perpendicular to six C_2 axes and a horizontal mirror plane that contains all the atoms in the ring. Overall, it has seven mirror planes ($3\sigma_v$, $3\sigma_d$, and $1\sigma_h$). The chloro-substituted, $C_6H_3Cl_3$, with chlorines on alternating carbons around the ring, has D_{3h} symmetry and four mirror planes ($3\sigma_v$ and $1\sigma_h$).

6.5 **The symmetry elements of orbitals? (a) An *s* orbital?** An *s* orbital, which has the shape of a sphere, possesses an infinite number of C_n axes where n can be any number from 1 to ∞, plus an infinite number of mirror planes of symmetry. It also has a centre of inversion, *i*. A sphere has the highest possible symmetry.
(b) A *p* orbital? The + and – lobes of a *p* orbital are not equivalent and therefore cannot be interchanged by potential elements of symmetry. Thus a *p* orbital does not possess a mirror plane of symmetry perpendicular to the long axis of the orbital. It does, however, possess an infinite number of mirror planes that pass through both lobes and include the long axis of the orbital. In addition, the long axis is a C_n axis, where n can be any number from 1 to ∞ (in group theory this is referred to as a C_∞ axis).
(c) A d_{xy} orbital? The two pairs of + and – lobes of a d_{xy} orbital are interchanged by the center of symmetry that this orbital possesses. It also possesses three mutually perpendicular C_2 axes, each one coincident with one of the three Cartesian coordinate axes. Furthermore, it possesses three mutually perpendicular mirror planes of symmetry, which are coincident with the *xy* plane and the two planes that are rotated by 45° about the *z* axis from the *xz* plane and the *yz* plane.
(d) A d_z2 orbital? Unlike a p_z orbital, a d_{z2} orbital has two large + along its long axis, and a – torus (or doughnut) around the middle. In addition to the symmetry elements possessed by a *p* orbital (see above), the infinite number of mirror planes that pass through both lobes and include the long axis of the orbital as well as the C_∞ axis, a d_{z2} orbital also possesses (i) a center of symmetry, (ii) a mirror plane that is perpendicular to the C_∞ axis, (iii) an infinite number of C_2 axes that pass through the center of the orbital and are perpendicular to the C_∞ axis, and (iv) an S_∞ axis.

6.6 **SO_3^{2-} ion? (a) Point group?** Using the decision tree shown in Figure 6.9, you will find the point group of this anion to be C_{3v} (it is nonlinear, it only has one proper rotation axis, a C_3 axis, and it has three σ_v mirror planes of symmetry).

(b) Degenerate MOs? Inspection of the C_{3v} character table (Resource Section 4) shows that the characters under the column headed by the identity element, *E*, are 1 and 2. Therefore, the maximum degeneracy possible for molecular orbitals of this anion is 2.
(c) Which *s* and *p* orbitals have the maximum degeneracy? According to the character table, the S atom 3*s* and $3p_z$ orbitals are each singly degenerate (and belong to the A_1 symmetry type), but the $3p_x$ and $3p_y$ orbitals are doubly degenerate (and belong to the *E* symmetry type). Thus the $3p_x$ and $3p_y$ atomic orbitals on sulfur can contribute to molecular orbitals that are two-fold degenerate.

6.7 **PF_5? (a) Point group?** As above, you use the decision tree to assign the point group, in this case concluding that PF_5 has D_{3h} symmetry (it has a trigonal bipyramidal structure, by analogy with PCl_5; it is nonlinear; it has only one high-order proper rotation axis, a C_3 axis; it has three C_2 axes that are perpendicular to the C_3 axis; and it has a σ_h mirror plane of symmetry).
(b) Degenerate MOs? Inspection of the D_{3h} character table (Resource Section 3) reveals that the characters under the *E* column are 1 and 2, so the maximum degeneracy possible for a molecule with this symmetry is 2.
(c) Which *p* orbitals have the maximum degeneracy? The P atom $3p_x$ and $3p_y$ atomic orbitals, which are doubly degenerate and are of the E' symmetry type (i.e., they have E' symmetry), can contribute to molecular orbitals that are twofold degenerate. In fact, if they contribute to molecular orbitals at all, they *must* contribute to twofold degenerate ones.

6.8 **$AsCl_5$ Raman spectrum consistent with a trigonal bipyramidal geometry?** From VSEPR, $AsCl_5$ should be trigonal pyramidal with symmetry D_{3h}. We first obtain the representation Γ_{3N}:

D_{3h}	E	$2C_3$	$3C_2$	σ_h	$2S_3$	$3\sigma_v$
Γ_{3N}	18	0	−2	4	−2	4

Γ_{3N} reduces to: $2A_1' + A_2' + 4E' + 3A_2'' + 2E''$; subtracting Γ_{trans} $(E' + A_2'')$ and Γ_{rot} $(A_2' + E'')$, we obtain Γ_{vib}: $2A_1' + 3E' + 2A_2'' + E''$. Thus we expect 6 Raman bands: $2A_1'$ (polarized) + $3E' + E''$ (note that A_2'' is inactive in Raman since symmetry type does not contain the same symmetry as the function $x^2 + y^2 + z^2$).

6.9 **Vibrational modes of SO_3? (a) In the plane of the nuclei?** SO_3 has a trigonal-planar structure. If you consider the C_3 axis to be the z axis, then each of the four atoms has two independent displacements in the xy plane, namely along the x and along the y axis. The product (4 atoms)(2 displacement modes/atom) gives 8 displacement modes, not all of which are vibrations. There are two translation modes in the plane of the nuclei, one each along the x and y axes. There is also one rotational mode around the z axis. Therefore, if you subtract these 3 nonvibrational displacement modes from the total of 8 displacement modes, you arrive at a total of 5 vibrational modes in the plane of the nuclei.

D_{3h}	E	$2C_3$	$3C_2$	σ_h	$2S_3$	$3\sigma_v$
Γ_{3N}	12	0	−2	4	−2	2

Γ_{3N} reduces to: $A_1' + A_2' + 3E' + 2A_2'' + E''$.
Subtracting Γ_{trans} $(E' + A_2'')$ and Γ_{rot} $(A_2' + E'')$, we obtain Γ_{vib}: $A_1' + 2E' + A_2''$.

In the plane: A_1' is a symmetric stretch; $2E'$ are modes consisting of (i) mainly asymmetric stretching and (ii) deformation:

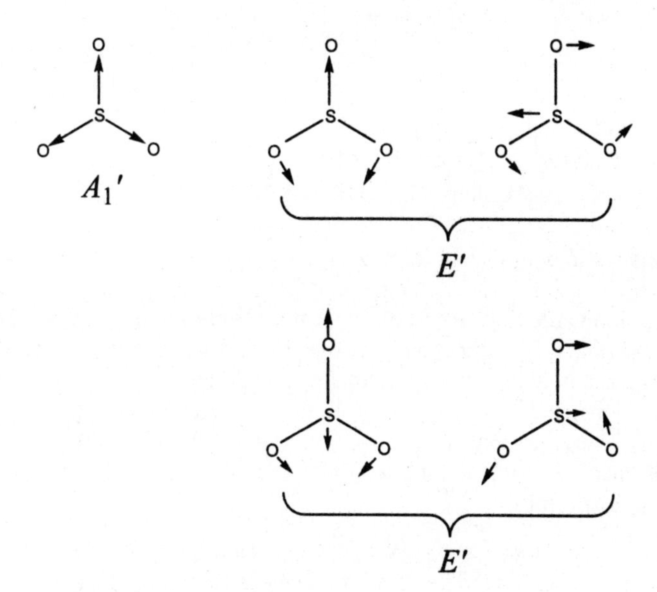

(b) Perpendicular to the molecular plane? You can use your answer to part (a), above, to answer this question. Since there are four atoms in the molecule, there are 3(4) – 6 = 6 vibrational modes. You discovered that there are 5 vibrational modes in the plane of the nuclei for SO_3, so there must be only 1 vibrational mode perpendicular to the molecular plane. A_2'' is the deformation in and out of the plane.

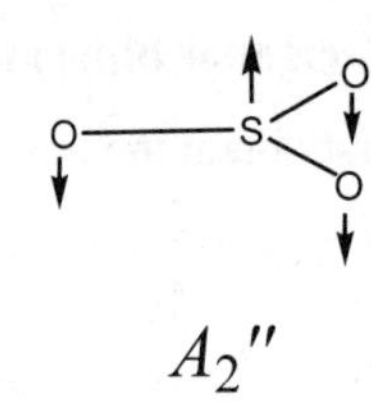

A_2''

6.10 **Vibrations that are IR and Raman active? (a) SF_6?** Since sulfur hexafluoride has a center of symmetry, the mutual exclusion rule applies. Therefore, none of the vibrations of this molecule can be *both* IR and Raman active. A quick glance at the O_h character table in Resource Section 4 confirms that the functions x, y, and z (required for IR activity) have the T_{1u} symmetry type and that all of the binary product functions such as x^2, xy, etc. (required for Raman activity) have different symmetry types.

O_h	E	$8C_3$	$6C_2$	$6C_4$	$6C_2$	i	$6S_4$	$8S_6$	$3\sigma_h$	$6\sigma_d$
Γ_{3N}	21	0	−1	3	−3	−3	−1	0	5	3

Γ_{3N} reduces to: $A_{1g} + E_g + T_{1g} + T_{2g} + 3T_{1u} + T_{2u}$.
Then subtracting Γ_{trans} (T_{1u}) and Γ_{rot} (T_{1g}),
we obtain Γ_{vib}: $A_{1g} + E_g + T_{2g} + 2T_{1u} + T_{2u}$.

None of these modes are both IR and Raman active as there is a centre of inversion.

(b) BF_3? Boron trifluoride does not have a center of symmetry. Therefore, it is possible that some vibrations are both IR and Raman active. You should consult the D_{3h} character table in Resource Section 4. Notice that the pairs of functions (x, y) and ($x^2 - y^2$, xy) have the E' symmetry type. Therefore, any E' symmetry vibration will be observed as a band in both IR and Raman spectra.

D_{3h}	E	$2C_3$	$3C_2$	σ_h	$2S_3$	$3\sigma_v$
Γ_{3N}	12	0	−2	4	−2	2

Γ_{3N} reduces to: $A_1' + A_2' + 3E' + 2A_2'' + E''$.
Subtracting Γ_{trans} ($E' + A_2''$) and Γ_{rot} ($A_2' + E''$),
we obtain Γ_{vib}: A_1' (Raman) + $2E'$ (IR and Raman) + A_2'' (IR).

The E' modes are active in both IR and Raman.

6.11 **Vibrations of a C_{6v} molecule that are neither IR nor Raman active?** You will need to consult the C_{6v} character table in Resource Section 4. If a vibration is neither IR nor Raman active, then the symmetry type of the vibration must be different than the symmetry types for the functions x, y, and z (required for IR activity) and all of the binary product functions such as x^2, xy, etc. (required for Raman activity). The symmetry types A_2, B_1, and B_2 satisfy this criterion. Therefore, any A_2, B_1, or B_2 vibrations of a C_{6v} molecule will not be observed in either the IR spectrum or the Raman spectrum. Examples of species having C_{6v} symmetry are arene complexes of Group 3 elements, such as $[Tl(benzene)]^+$.

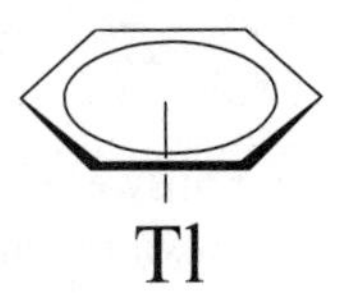

For this species the vibrational modes are determined as follows:

C_{6v}	E	$2C_6$	$2C_3$	C_2	$3\sigma_v$	$3\sigma_d$
Γ_{3N}	39	2	0	−1	5	1

Γ_{3N} reduces to: $5A_1 + 2A_2 + 4B_1 + 2B_2 + 7E_1 + 6E_2$, and subtracting Γ_{trans} $(A_1 + E_1)$ and Γ_{rot} $(A_2 + E_1)$, we obtain:

Γ_{vib}: $4A_1$ (IR, Raman) + A_2 (inactive) + $4B_1$ (inactive) + $2B_2$ (inactive) + $5E_1$ (IR, Raman) + $6E_2$ (Raman).

6.12 $[AuCl_4]^-$ ion? Γ of all 3*N* displacements and irreducible representations?

D_{4h}	E	$2C_4$	C_2	$2C_2'$	$2C_2''$	i	$2S_4$	σ_h	$2\sigma_v$	$2\sigma_d$
Γ_{3N}	15	1	−1	−3	−1	−3	−1	5	3	1

This reduces to $A_{1g} + A_{2g} + B_{1g} + B_{2g} + E_g + 2A_{2u} + B_{2u} + 3E_u$

6.13 IR and Raman to distinguish between: (a) planar and pyramidal forms of PF_3, (b) planar and 90°-twisted forms of B_2F_4 (D_{2h} and D_{2d} respectively)?

(a) Planar PF_3 has D_{3h} symmetry:
The vibrations are A_1' (Raman, polarized) + $2E'$ (IR and Raman) + A_2'' (IR).

Pyramidal PF_3 has C_{3v} symmetry:
The vibrations are $2A_1$ (IR and Raman, polarized) + $2E'$ (IR and Raman)

(b) For the planar form of B_2F_4 (D_{2h}), we obtain:

D_{3h}	E	$C_2(z)$	$C_2(y)$	$C_2(x)$	i	$\sigma(xy)$	$\sigma(xz)$	$\sigma(yz)$
Γ_{3N}	18	−2	0	0	0	0	6	2

which reduces to: $3A_g + B_{1g} + 3B_{2g} + 2B_{3g} + A_u + 3B_{1u} + 2B_{2u} + 3B_{3u}$.

Subtracting Γ_{trans} $(B_{1u} + B_{2u} + B_{3u})$ and Γ_{rot} $(B_{1g} + B_{2g} + B_{3g})$, we obtain Γ_{vib}.

The vibrations are:

$3A_g$ (Raman, polarized) + $2B_{2g}$ (Raman) + B_{3g} (Raman) + A_u(inactive) + $2B_{1u}$ (IR) + B_{2u} (IR) + $2B_{3u}$ (IR).

For the 90°-twisted form of B_2F_4 (D_{2d})

D_{2d}	E	$2S_4$	C_2	$2C_2'$	$2\sigma_d$
Γ_{3N}	18	0	−2	0	4

which reduces to: $3A_1 + A_2 + B_1 + 3B_2 + 5E$.

Subtracting Γ_{trans} $(B_2 + E)$ and Γ_{rot} $(A_2 + E)$, we obtain Γ_{vib}.

The vibrations are: $3A_1$ (Raman, polarized) + B_1 (Raman) + $2B_2$ (IR and Raman) + $3E$ (IR and Raman).

6.14 (a) Take the 4 hydrogen 1s orbitals of CH_4 and determine how they transform under T_d. (b) Confirm that it is possible to reduce this representation to $A_1 + T_2$. (c) Which atomic orbitals on C can form MOs with H1s SALCs?

Using symmetry T_d

T_d	E	$8C_3$	$3C_2$	$6S_4$	$6\sigma_d$
Γ_{3N}	4	1	0	0	2

Γ_{3N} reduces to: $A_1 + T_2$.

The 2s atomic orbitals have symmetry species A_1 and the 2p atomic orbitals have symmetry species T_2. Therefore, the MOs would be constructed from SALCs with H1s and 2s and 2p atomic orbitals on C.

6.15 Use the projection operator method to construct the SALCs of $A_1 + T_2$ symmetry that derive from the four H1s orbitals in methane. The SALCs for methane have the same forms as the expressions for the AOs as SALCs of sp^3 hybrid orbitals.

$$s = (1/2)(\varphi_1 + \varphi_2 + \varphi_3 + \varphi_3) \quad (= A_1)$$

$$p_x = (1/2)(\varphi_1 - \varphi_2 + \varphi_3 - \varphi_3) \quad (= T_2)$$

$$p_y = (1/2)(\varphi_1 - \varphi_2 - \varphi_3 + \varphi_3) \quad (= T_2)$$

$$p_z = (1/2)(\varphi_1 + \varphi_2 - \varphi_3 - \varphi_3) \quad (= T_2)$$

6.16 SALCs for σ-bonds

(a) BF_3?

$$(1/\sqrt{3})(\varphi_1 + \varphi_2 + \varphi_3) \quad (= A_1')$$

$$(1/\sqrt{6})(2\varphi_1 - \varphi_2 - \varphi_3) \text{ and } (1/\sqrt{2})(\varphi_2 - \varphi_3) \quad (= E')$$

note that $(1/\sqrt{2})(\varphi_2 - \varphi_3)$ is obtained by combining the other two E'-type SALCs,

i.e., $(2\varphi_2 - \varphi_3 - \varphi_1) - (2\varphi_3 - \varphi_1 - \varphi_2)$

(b) PF_5?

(axial F atoms are $\varphi_4 + \varphi_5$)

$$(1/\sqrt{2})(\varphi_4 + \varphi_5) \quad (= A_1')$$

$$(1/\sqrt{2})(\varphi_4 - \varphi_5) \quad (= A_2'')$$

$$(1/\sqrt{3})(\varphi_1 + \varphi_2 + \varphi_3) \quad (= A_1')$$

$$(1/\sqrt{6})(2\varphi_1 - \varphi_2 - \varphi_3) \text{ and } (1/\sqrt{2})(\varphi_2 - \varphi_3) \quad (= E')$$

Chapter 7 An Introduction to Coordination Compounds

Many of *d*-metal complexes can be understood on the basis of Lewis acid/base theory. The Lewis acid being the transition metal, and the Lewis base being the ligand bound to it. Most of the transition metal ions have empty *s*, *p*, and some *d* orbital available for bonding, making them an exceptional Lewis acid. The bond between the ligand and the metal is formally referred to as a coordinate covalent bond, in which both electrons in the chemical bond come from the ligand (Lewis base). A couple of examples are shown below, incorporating both anion and neutral ligands.

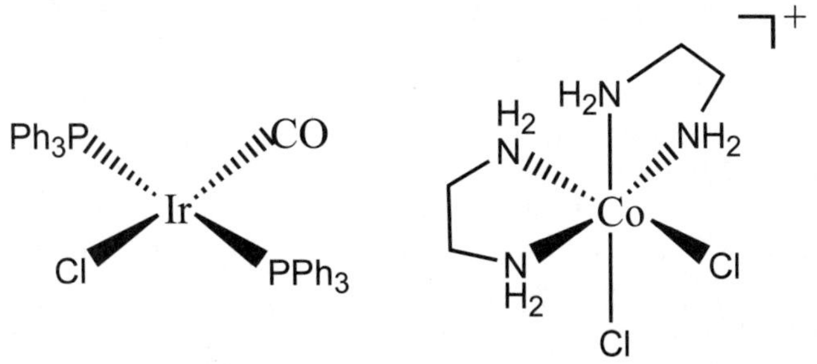

S7.1 **Give formulas corresponding to the following names? (a) *Cis*-diaquadichloroplatinum(II)?** As with most Pt(II) complexes, this complex is square-planar. The *cis* prefix indicates adjacent positions for the two aqua and two chloro ligands. The formula is *cis*-$[PtCl_2(OH_2)_2]$. Note the order used in naming complexes of the *d*-block metals (the metal ion is listed; then ligands are listed in alphabetical order). For the other isomer, ***trans*-diaquadichloroplatinum(II)**, the formula is *trans*-$[PtCl_2(OH_2)_2]$. Read section 7.7 for more explanation of the isomers of square planar complexes.

(b) Diammine*tetra*(isothiocyanato) chromate(III)? This is an octahedral complex of Cr(III). The *ate* suffix indicates an overall negative charge. Two NH_3 molecules and four NCS^- anions are bonded to the Cr(III) ion. The NCS^- ligands use their N atoms to bond to Cr(III) (-isothiocyanato-, not -thiocyanato-). The formula is $[Cr(NCS)_4(NH_3)_2]^-$. Note that a complex with this composition can exist as two structural isomers, *cis*-$[Cr(NCS)_4(NH_3)_2]^-$ or *trans*-$[Cr(NCS)_4(NH_3)_2]^-$.

(c) *Tris*(ethylenediamine)rhodium (III)? This is also an octahedral complex. The metal ion is Rh(III), there are three symmetric bidentate ligands, and the complex is not anionic (-rhodium(III), not -rhodate(III)). Since the accepted abbreviation for ethylenediamine ($H_2NCH_2CH_2NH_2$) is en, the formula may be written as $[Rh(en)_3]^{3+}$.

(d) Bromo*penta*carbonylmanganese (I)? This is also an octahedral complex. The manganese has five CO bonded to it along with one bromide. There are no isomers for this complex, therefore the formula is $[MnBr(CO)_5]$.

(e) Chloro*tris*(triphenylphosphine)rhodium (I)? This is a d^8, square-planar complex. Given that the complex has three phosphine ligands that are the same, there are no isomers for it, therefore the formula is $[RhCl(PPh_3)_3]$.

S7.2 **What type of isomers are possible for $[Cr(NO_2)_2{\bullet}6H_2O]$?** The hydrate isomers $[Cr(NO_2)(H_2O)_5]NO_2{\bullet}H_2O$ and $[Cr(H_2O)_6](NO_2)_2$ are possible, as are linkage isomers of the NO_2 group. One further isomer is $[Cr(ONO)(H_2O)_5]NO_2{\bullet}H_2O$.

S7.3 **Identifying isomers?** The two square-planar isomers of $[PtBrCl(PR_3)_2]$ are shown below. The NMR data indicate that isomer A is the *trans* isomer since the two trialkylphosphine ligands occupy opposite corners of the square plane. Isomer B is the *cis* isomer. Note that the two phosphine ligands in the *trans* isomer are related by symmetry elements that this C_{2v} molecule possesses, namely the C_2 axis (the Cl-Pt-Br axis) and the σ_v mirror plane that is perpendicular to the molecular plane. Therefore, they exhibit the same chemical shift in the ^{31}P NMR spectrum of this compound. The two phosphine ligands in the *cis* isomer are not related by the σ mirror plane that this C_s molecule possesses. Since they are chemically nonequivalent, they give rise to separate groups of ^{31}P resonances.

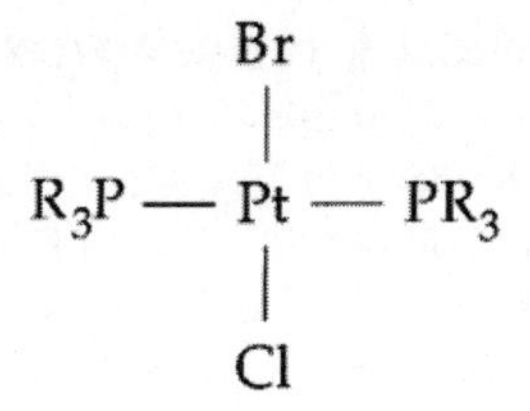

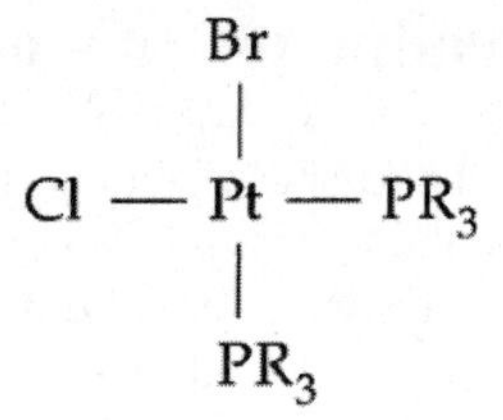

trans– $[PtBrCl(PR_3)_2]$ *cis*– $[PtBrCl(PR_3)_2]$

A B

S7.4 **Sketches of the *mer* and *fac* isomers of $[Co(gly)_3]$?** The anion of glycine is an unsymmetrical bidentate ligand (it has a neutral amine nitrogen donor atom and a negatively charged carboxylate oxygen donor atom). In the *fac* isomer, the three possible N–Co–N bond angles are ~90°, while in the *mer* isomer, two N–Co–N bond angles are ~90° and one is 180°. If you imagine that the complex is a sphere, the three N atoms in the *mer* isomer lie on a *meridian* of the sphere (the largest circle that can be drawn on the surface of the sphere). In contrast, the three N atoms in the *fac* isomer form one of the eight triangular *faces* of the $[Co(gly)_3]$ octahedron.

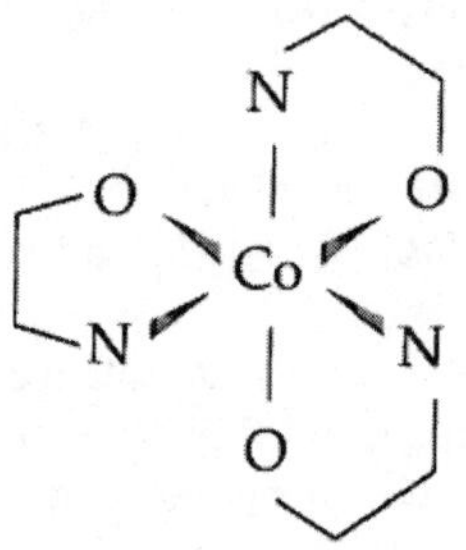

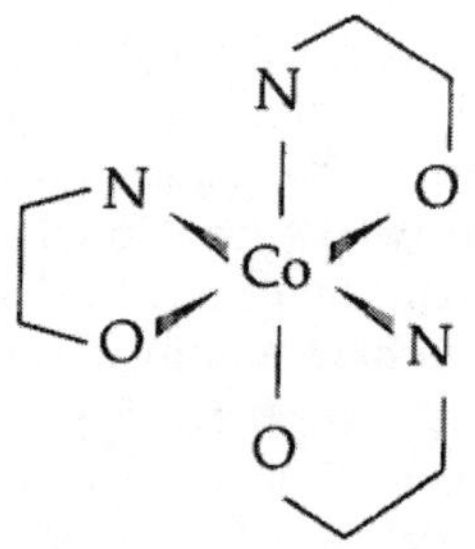

fac– $[Co(gly)_3]$ *mer*– $[Co(gly)_3]$

S7.5 **Which of the following are chiral? (a) *cis*-$[CrCl_2(ox)_2]^{3-}$?** Drawings of two mirror images of this complex are shown below. They are not superimposable, and therefore represent two enantiomers. Therefore, this complex is chiral. Note that it does not possess an S_n axis, only a single C_2 axis. The point group of both enantiomers is C_2, so they are dissymmetric, not asymmetric.

(b) *trans*-$[CrCl_2(ox)_2]^{3-}$? Drawings of two mirror images of this complex are also shown below. They *are* superimposable, and therefore do not represent two enantiomers but only a single isomer. Since this complex is achiral, it must possess at least one S_n axis. In fact, it possesses three different σ planes of symmetry, each of which is an S_1 axis.

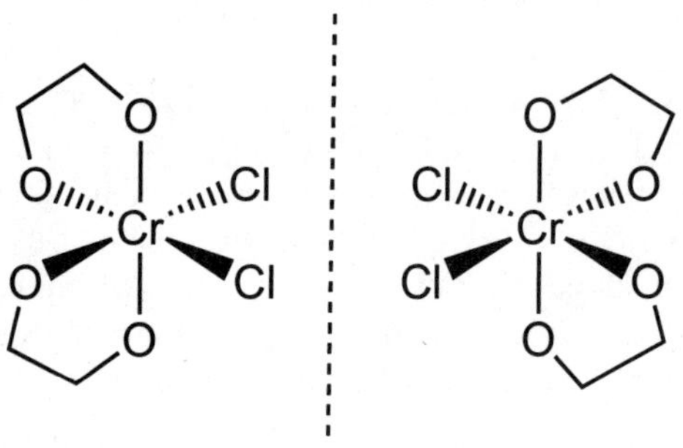

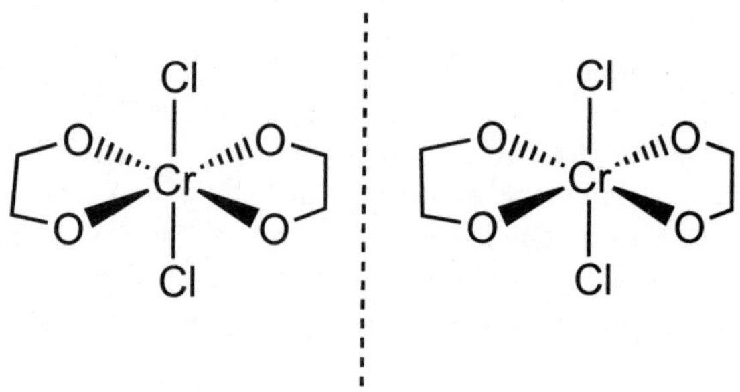

cis *trans*

(c) *cis*-[RhH(CO)(PR_3)$_2$]? This is a complex of Rh(I), which is a d^8 metal ion. Four-coordinate d^8 complexes of period 5 and period 6 metal ions are almost always square-planar, and [RhH(CO)(PR_3)$_2$] is no exception. The bulky PR_3 ligands are *cis* to one another. This compound has C_s symmetry, with the mirror plane coincident with the rhodium atom and the four ligand atoms bound to it. A planar complex cannot be chiral, whether it is square-planar, trigonal planar, etc.

S7.6 **Calculate all of the stepwise formation constants?** By study Table 7.4, one can see that each successive formation constant is essentially 30% less than the former. Given this, we can predict what the successive formation constants should be given that $K_{f1} = 1 \times 10^5$. K_{f2} will be 30% less or 30000, $K_{f3} = 9000$, $K_{f4} = 2700$, $K_{f5} = 810$, and finally $K_{f6} = 243$. The overall formation constant (β_n), is the product of the stepwise formation constants. Therefore $\beta_n = 100000 \times 30000 \times 9000 \times 2700 \times 810 \times 243 = 1.43 \times 10^{22}$.

7.1 **Name and draw the structures of the complexes? (a) [Ni(CO)$_4$]?** Nickel tetracarbonly or tetracarbonyl nickel(0) has a tetrahedral geometry, shown below.

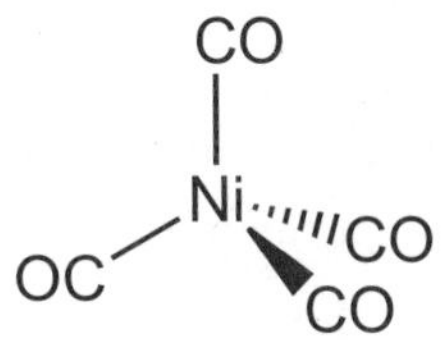

(b) [Ni(CN)$_4$]$^{2-}$? Tetracyanonickelate (II), like most d^8 metal complexes with four ligands, has square-planar geometry, shown below.

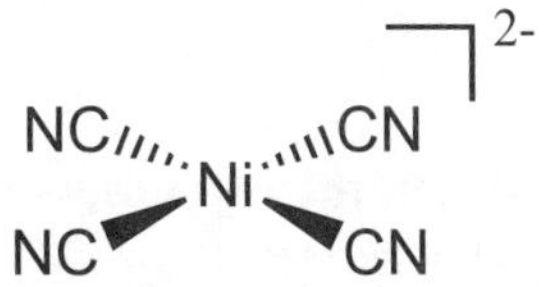

(c) [CoCl$_4$]$^{2-}$? Tetrachlorocobaltate (II) is tetrahedral, like most of the first-row transition metal chlorides.

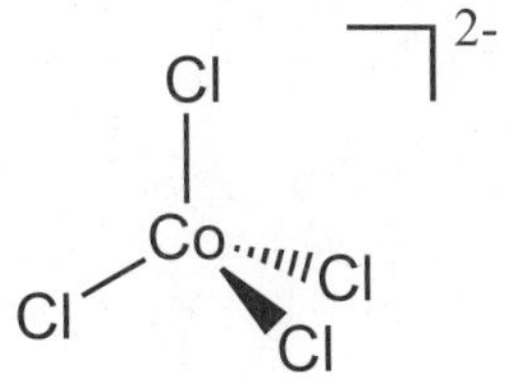

(d) [Mn(NH$_3$)$_6$]$^{2+}$? Hexaamminemanganesium (II) has octahedral geometry, shown below.

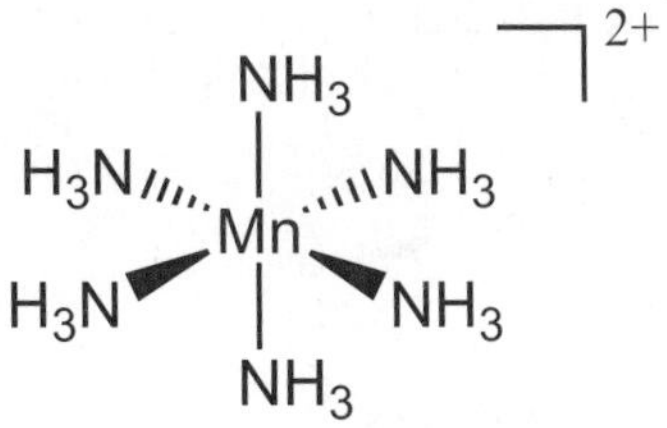

7.2 **Write the formulas for the following complexes?** The key to writing formulas from a name is keeping the balance of charge between the metal and its ligands. To do this, you need to know if the ligands that directly bond to the metal are neutral or ionic; this will help you determine what the oxidation state is on the metal. The chemical name gives the oxidation state of the metal as a roman numeral; simply balance charge with the ligands that directly bond to the metal and the counter-ions, if necessary.

(a) $[CoCl(NH_3)_5]Cl_2$, since it is cobalt (III), and the ammonia ligands are neutral and the chloride bond to the metal is anionic, you know that you need two chlorides as your counter-ion.

(b) $[Fe(OH_2)_6](NO_3)_3$, again, since the water's ligands are neutral, you need three nitrates as your counter-ion to balance charge.

(c) *cis*-$[FeCl_2(en)_2]$ has a divalent metal centre, the chlorides are anionic, and the ethylene diamines are neutral; therefore you do not need a counter-ion to achieve a 2+ charge for the iron atom, rendering the complex neutral.

(d) $[Cr(NH_3)_5\mu\text{–}OH\text{–}Cr(NH_3)_5]Cl_5$; the Greek symbol μ means you have a bridging ligand between two metal complexes, so you need to balance charge for the entire dimeric complex. The ammonia's ligands are neutral, and the bridging hydroxide is an anionic ligand, giving one of the chromium atoms a 1+ charge; thus you need five chloride counter-ions to bring each chromium atom up to a 3+ charge. This gives us an overall 6+ charge for the dimeric complex.

7.3 Name the following complexes?

(a) *cis*-$[CrCl_2(NH_3)_4]^+$? *cis*-tetra(ammine)di(chloro)chromium(III)

(b) *trans*-$[Cr(NCS)_4(NH_3)_2]^-$? *trans*-di(ammine)tetrakis(isothiocyanato)chromate (III)

(c) $[Co(C_2O_4)(en)_2]^+$? bis(ethylenediamine)oxalatocobalt(III), which is neither *cis* nor *trans* but does have one optical isomer, shown below.

7.4 Four-coordinate complexes? (a) Sketch the two observed structures? With four-coordinate complexes, the two possible geometries are tetrahedral or square planar. Most of the the first-row divalent transition metal halides are tetrahedral, while metals that have a d^8 electronic configuration tend to be square-planar.

(b) Isomers expected for MA_2B_2? For a tetrahedral complex there are no isomers for MA_2B_2; however, for a square-planar complex, there are two isomers, *cis* and *trans*, shown below.

7.5 **For five-coordinate complexes, sketch the two observed structures?** With five-coordinate complexes, the two possible geometries are trigonal bipyramidal and square-based pyramidal, as shown below.

Trigonal Bipyramidal

A = axial ligands

E = equatorial ligands

Square based pyramid

A = axial

B = basal

7.6 **Six-coordinate complexes? (a) Sketch the two observed structures?** Most six-coordinate complexes are either octahedral or trigonal-prismatic. Drawings of these are shown below.

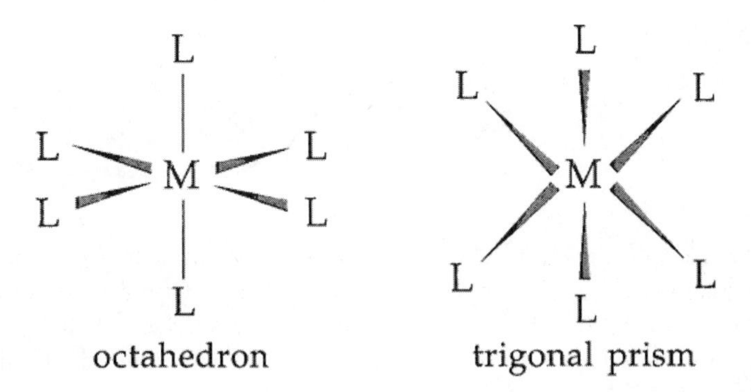

(b) Which one of these is rare? The trigonal prism is rare. Nearly all six-coordinate complexes are octahedral.

7.7 **Explain the difference between monodentate, bidentate, and quadridentate?** These terms describe how many Lewis bases you have on your ligand and whether they can physically bind to the metal. A monodentate ligand can bond to a metal atom only at a single atom, a bidentate ligand can bond through two atoms, a quadridentate ligand can bond through four atoms. Examples of monodentate and bidentate ligands are shown below. Tetraazacclotetradecane, from Table 7.1, is an example of a quadridentate ligand, all four nitrogen atoms can donate a lone pair to the metal.

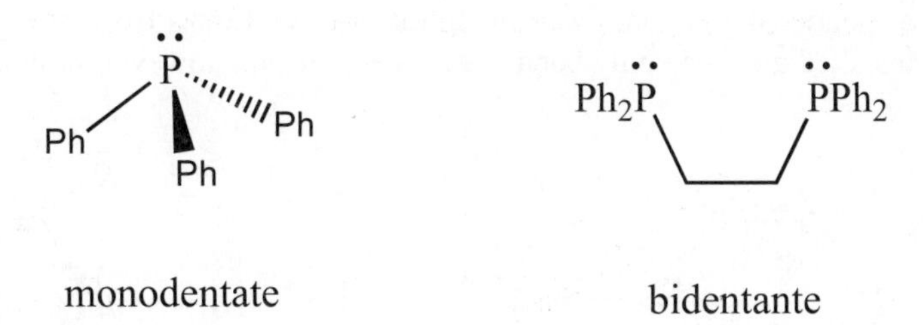

monodentate

bidentante

7.8 **What type of isomers do you get with ambidentate ligands?** Ambidentate ligands are ligands that have two different atoms within the molecule that can serve as a Lewis base. An example is the thiocyanide anion NCS^-. It can bond to a metal either through the nitrogen's lone pair or through the sulphur's lone pair, depending on the electronics of the metal. If the metal is soft, then the softer-base sulphur is preferred; if the metal is hard, then the hard-base nitrogen is preferred. These two isomers are known as linkage isomers.

NCS–M (thiocyanato) SCN–M (isothiocyanato)

7.9 Which ligand could act like a chelating ligand? A chelating ligand has to have at least two Lewis basic sites on the molecule that can physically bond to the metal. **(a)** Triphenylphosphite is a monodentate ligand, which bonds to a metal through the lone pair on the phosphorous atom.

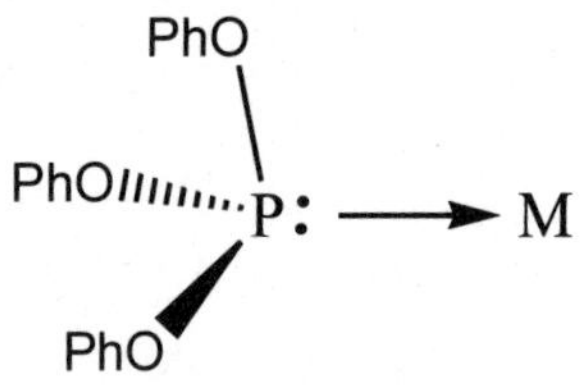

PhO
PhO P: → M
PhO

(b) Bis(dimethyl)phosphino ethane (dmpe) is a chelating ligand; it is able to bond to a metal through both phosphorous atoms.

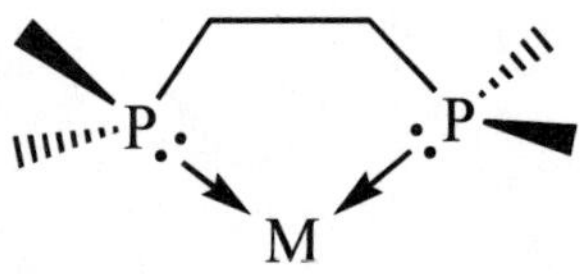

P P
M

(c) Bipyridine (biby) is a chelating ligand that, like bis(dimethyl)phosphino ethane, is able to bond to a metal through both of its nitrogen atoms.

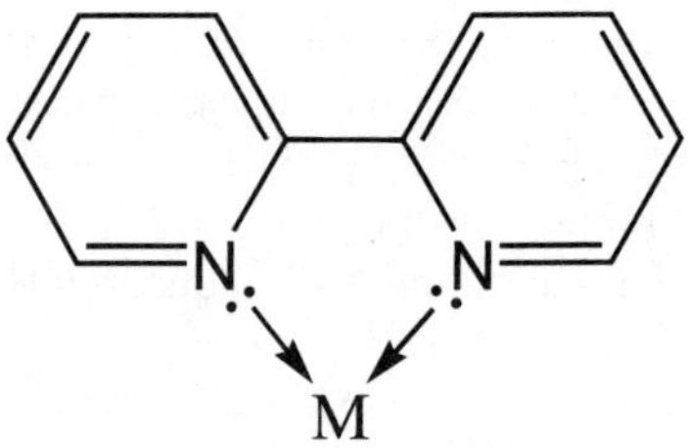

N N
M

(d) Pyrazine is a monodentate ligand even though it has two Lewis basic sites. Because of the location of the nitrogen atoms, the ligand can only bond to one metal; it can, however, bridge two metals, as shown below.

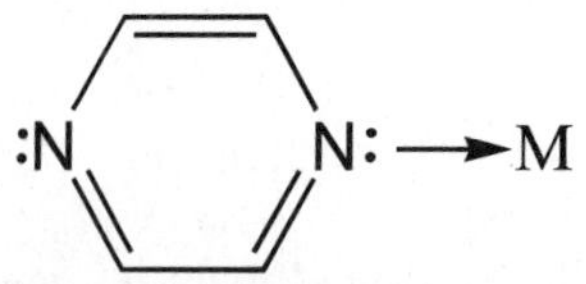

:N N: → M   M ← :N N: → M

monodentate ligand    bridging ligand

7.10 Draw structures of complexes that contain the ligands (a) en, (b) ox, (c) phen, and (d) edta? Generic drawings of octahedral complexes of metal M for (a), (b), and (c) are shown below. In (a), the bidentate ligand ethylenediamine (en = $H_2NCH_2CH_2NH_2$) takes up two coordination sites. For clarity, the carbon atoms of the ethylene bridge are not explicitly shown. The five-membered ring that is formed is not planar—one carbon atom is above and one is below the plane formed by M and the two N atoms. In (b), the bidentate ligand oxalate dianion (ox = $C_2O_4^{2-}$) also takes up two coordination sites. Once again, the carbon atoms of the ligand are shown simply as vertices. Due to the delocalized *p* system of the ligand, the five-membered ring in this case *is* planar. In (c), the bidentate ligand phenanthroline takes up two coordination sites. An example of a complex of ethylenediaminetetraacetate (edta^{4-} = $(O_2CCH_2)_2NCH_2CH_2N(CH_2CO_2)_2^{4-}$) is shown below. Note that the complex $Mg(edta)(OH_2)^{2-}$ is seven-coordinate.

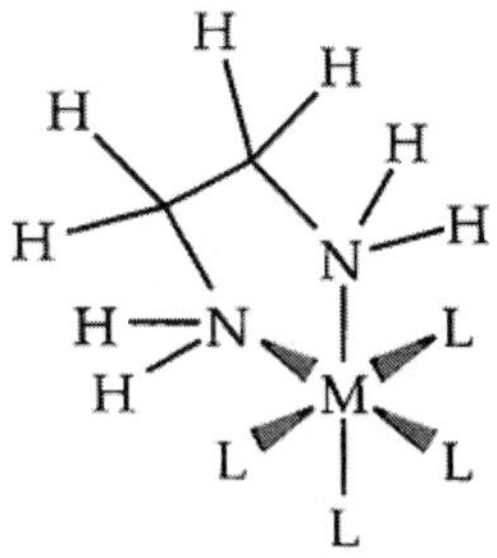

ML_4(en) 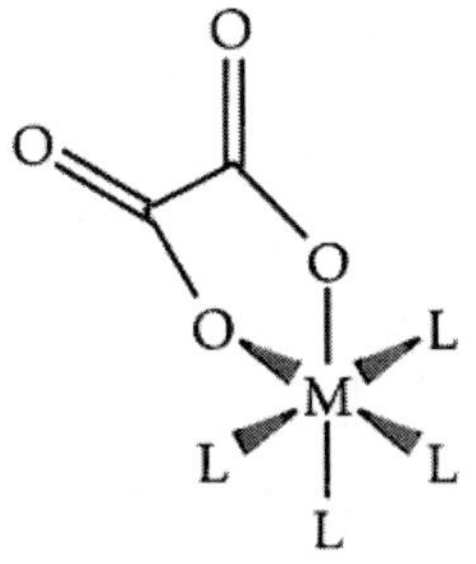 ML_4(ox) 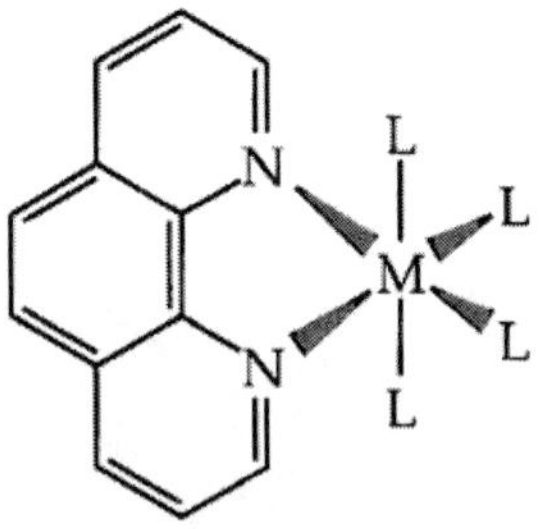 ML_4(phen)

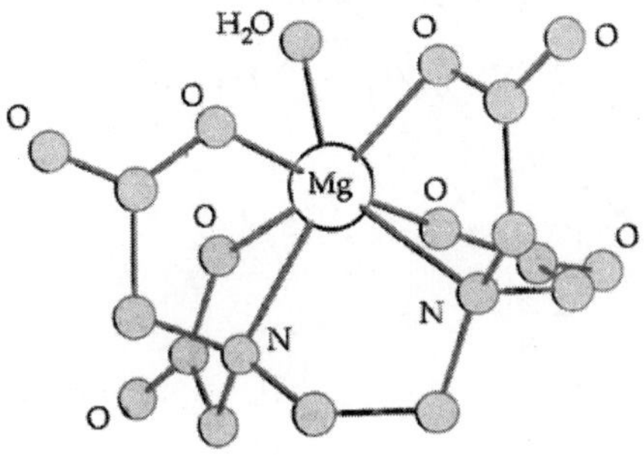

The structure of the $[Mg(edta)(OH_2)]^{2-}$ dianion. Hydrogen atoms have been omitted for clarity, and the carbon atoms of the edta^{4-} ligand are left unlabeled. The hexadentate edta^{4-} ligand nearly surrounds the metal ion.

7.11 What types of isomers are $[RuBr(NH_3)_5]Cl$ and $[RuCl(NH_3)_5]Br$? These complexes differ as to which halogen is bonded, and as to which one is the counter ion. These types of isomers are known as ionization isomers. When dissolved into solution, $[RuBr(NH_3)_5]Cl$ will release chloride ions, while $[RuCl(NH_3)_5]Br$ will release bromide ions.

7.12 Which complexes have isomers? For tetrahedral complexes, the only isomers found are when you have four different ligands bound to the metal. Therefore, $[CoBr_2Cl_2]^-$ and $[CoBrCl_2(OH_2)]$ have no isomers. However, $[CoBrClI(OH_2)]$ has an optical isomer shown below.

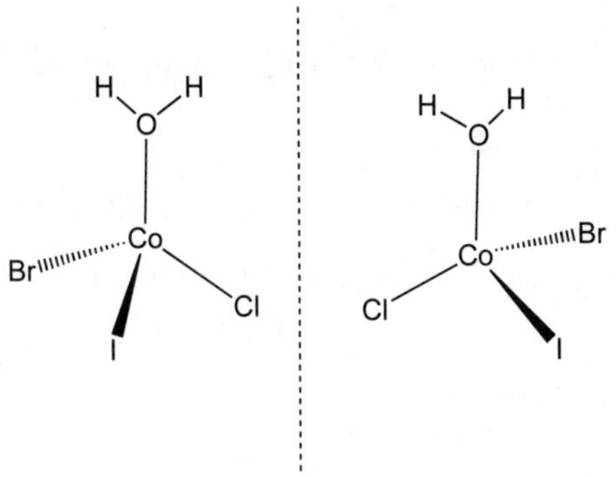

7.13 Which complexes have isomers? For square-planar complexes, depending on the ligands, several isomers are possible. **(a) [Pt(ox)(NH_3)$_2$]** has no isomers because of the chelating oxalato ligand. The oxalate forces the ammonia molecules to be *cis* only.

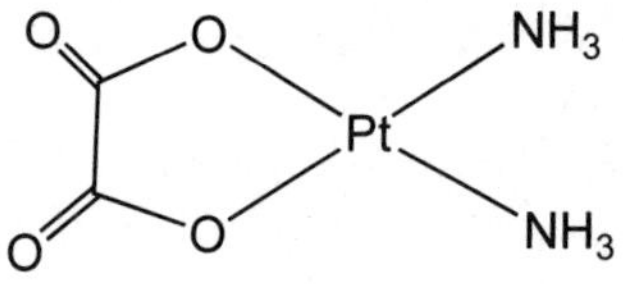

(b) [PdBrCl(PEt_3)$_2$] has two isomers, *cis* and *trans*, as shown below.

cis trans

(c) [IrHCO(PR_3)$_2$] has two isomers, *cis* and *trans*, as shown below.

cis trans

(d) [Pd(gly)$_2$] has two isomers, *cis* and *trans*, as shown below.

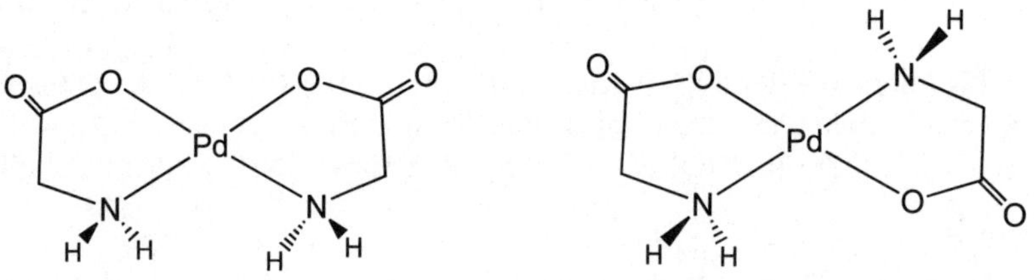

cis trans

7.14 How many isomers are possible for the following complexes? (a) [FeCl(OH_2)$_5$]$^{2+}$? This complex has no isomers.

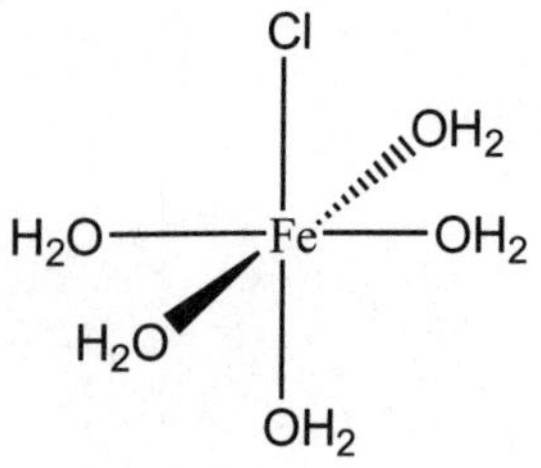

(b) [$IrCl_3(PEt_3)_2$]? There are two isomers for this complex, shown below.

facial (*fac*) meridianal (*mer*)

(c) [$Ru(biby)_3$]$^{2+}$? This complex contains optical isomers, shown below.

(d)

Λ isomer Δ isomer

[$CoCl_2(en)(NH_3)_2$]$^{+}$? This complex contains *cis* and *trans*, as well as optical isomers, shown below.

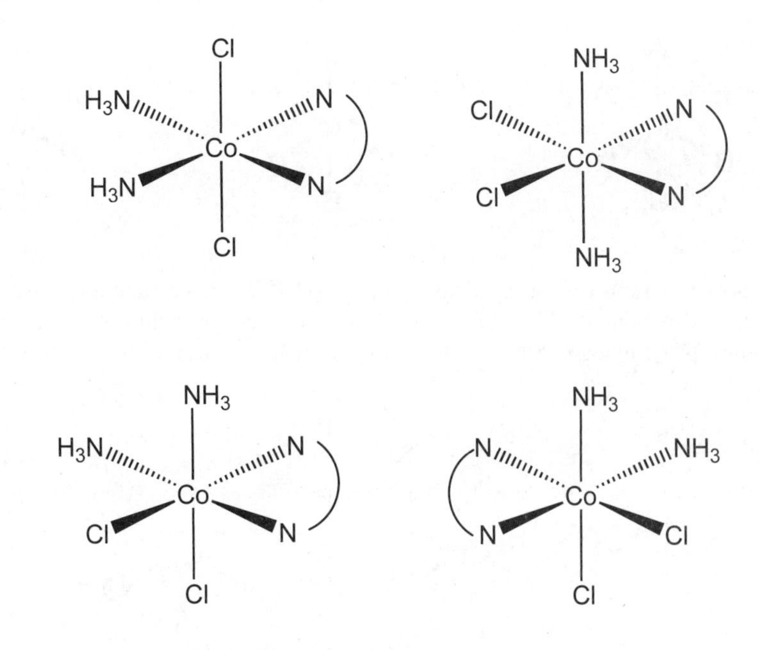

(e) [$W(CO)_4(py)_2$] has two isomers, *cis* and *trans*, shown below.

7.15 Draw all possible isomers for [MA_2BCDE]? Ignoring optical isomers, nine isomers are possible. Including optical isomers, 15 isomers are possible! Read section 7.10(a) *Geometrical isomerism*, for a better understanding of all the possible isomers.

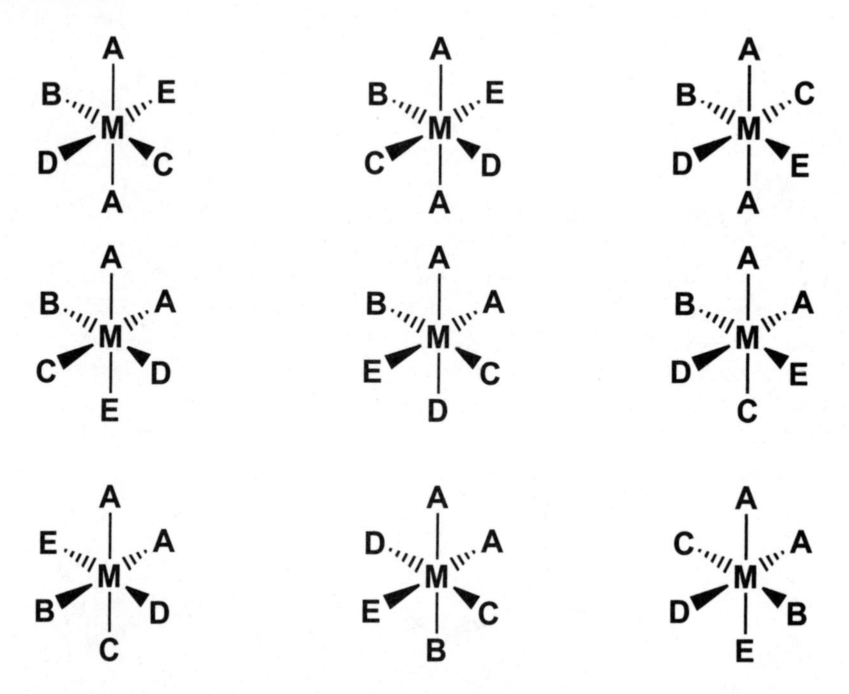

7.16 Which of the following complexes are chiral? (a) [$Cr(ox)_3]^{3-}$? All octahedral *tris*(bidentate ligand) complexes are chiral, since they can exist as either a right-hand or a left-hand propeller, as shown in the drawings of the two nonsuperimposable mirror images (the oxalate ligands are shown in abbreviated form and the 3– charge on the complexes has been omitted).

(b) *cis*-[$PtCl_2$(en)]? This is a four-coordinate complex of a period 6 d^8 metal ion, so it is undoubtedly square-planar. You will recall from Chapter 3 that any planar complex contains at least one plane of symmetry and must be achiral. In this case the five-membered chelate ring formed by the ethylenediamine ligand is not planar, so, strictly speaking, the complex is not planar. It *can* exist as two enantiomers, depending on the conformation of the chelate ring, as shown below. However, the conformational interconversion of the ethylene linkage is extremely rapid, so the two enantiomers cannot be separated.

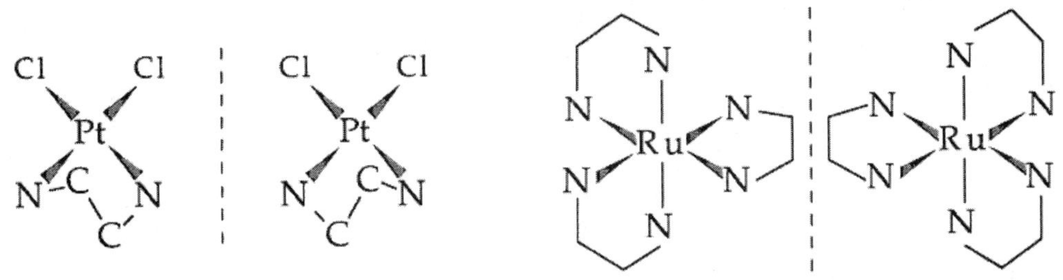

d and *l* [$PtCl_2$(en)]

Δ and Λ [Ru(bipy)$_3$]$^{3+}$

(c) *cis*-[$RhCl_2(NH_3)_4$]$^+$? This complex has C_{2v} symmetry, so it is not chiral. The C_2 axis is coincident with the bisector of the Cl–Rh–Cl bond angle, one σ_v plane is coincident with the plane formed by the Rh atom and the two Cl atoms, and the other σ_v plane is perpendicular to the first.

(d) [Ru(bipy)$_3$]$^{2+}$? As stated in the answer to part (a) above, all octahedral *tris* (bidentate ligand) complexes are chiral, and this one is no exception. The two nonsuperimposable mirror images are shown above (the bipyridine ligands are shown in abbreviated form, and the 3+ charge on the complex has been omitted).

(e) *fac*-[$Co(NO_2)_3$(dien)]? There is a mirror plane through the metal, bisecting the dien ligand shown below.

(f) *mer*-[$Co(NO_2)_3$(dien)]? There are two mirror planes: one through all three coordinating N atoms of the dien ligand and the metal atom; the other through all three nitro groups and the metal atom.

If you take into account the various conformations of the ethylene linkages, the stereoisomer possibilities are much more complicated. As explained in the text for en, the ethylene linkages undergo *rapid* conformational interconversion.

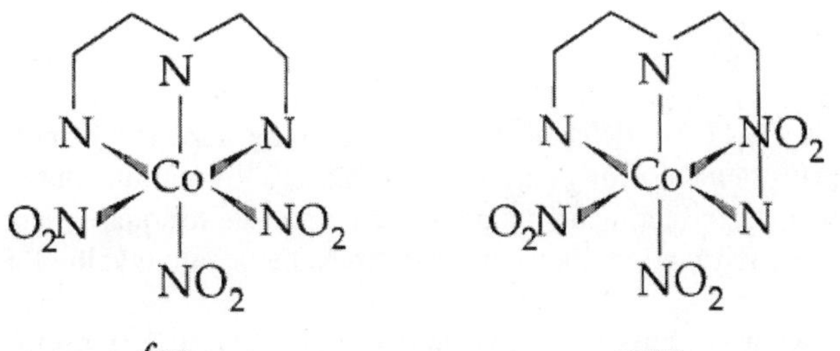

fac

mer

7.17 Which isomer, Λ or Δ, is the complex Mn(acac)$_3$, shown in the exercise? Below is a picture of both isomers. The best way to do this problem is to draw both isomers as mirror images of each other, and think of the chelating ligand as a propellor blade. Which way would it rotate if photons hit it? If it rotates clockwise, it is the Δ isomer, if it rotates counter clockwise, it is the Λ isomer. When you do this, it is obvious which isomer you have, in the case of this problem, the complex drawn in the exercise is the Λ isomer. For more help with this concept read section 7.10(b) *Chirality and Optical Isomers.*

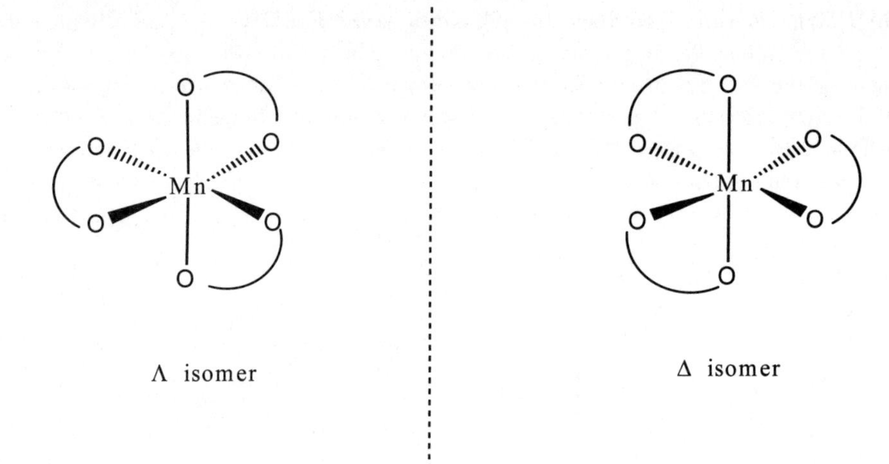

7.18 Draw both isomers, Λ or Δ, of the complex $[Ru(en)_3]^{+2}$? Below is a picture of both isomers. As in problem 7.17, the best way to do this problem is to draw both isomers as mirror images of each other, and think of the chelating ligand as a propellor blade. Which way would it rotate if photons hit it? Clockwise would be the Δ isomer, and counter clockwise would be the Λ isomer.

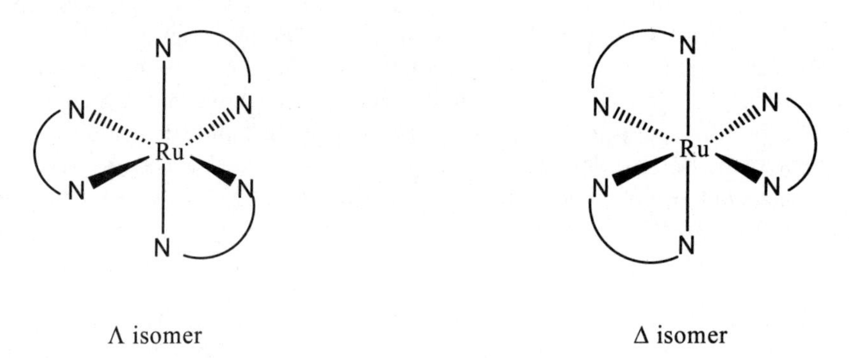

Λ isomer

Δ isomer

7.19 Suggest a reason why K_{f5} is so different? The value of the stepwise formation constants drop from 1 to 4, as expected on statistical grounds. However, the fifth stepwise formation constant is substantially lower, suggesting a change in coordination. In fact, what happens is the very stable square planar complex $[Cu(NH_3)_4]^{2+}$ complex ion forms, and further coordination does not occur, thus the negative value for K_{f5}.

7.20 Compare these values with those of ammonia given in exercise 7.19 and suggest why they are different? The salient comparison is of log β_2 (7.65) for the ammonia reaction with log K_{f2} (10.72) for the en reaction. The value for the en reaction is substantially higher, indicating more favourable complex formation, and this can be attributed to the chelate effect. A comparison can also be drawn between the 3rd and 4th stepwise formation constants with ammonia (5.02) and the 2nd one for en (9.31). Read section 7.14 *The chelate and marcrocylic effects*, for better understanding of this phenomenon.

Chapter 8 Physical Techniques in Inorganic Chemistry

The Bragg relation is derived from considering the layers of atoms in solid-state structures as arrays of reflecting planes (shown in the schematic below). X-rays constructively interfere with these layers when the additional path length 2dsinθ is equal to the wavelength of the X-ray, λ. Powder X-ray diffraction uses the Bragg relation to determine the lattice parameters and lattice types for the crystal structures of solids described in Chapter 3. It is one of a suite of physical techniques described in this chapter to investigate structure-property relationships within inorganic compounds and complexes.

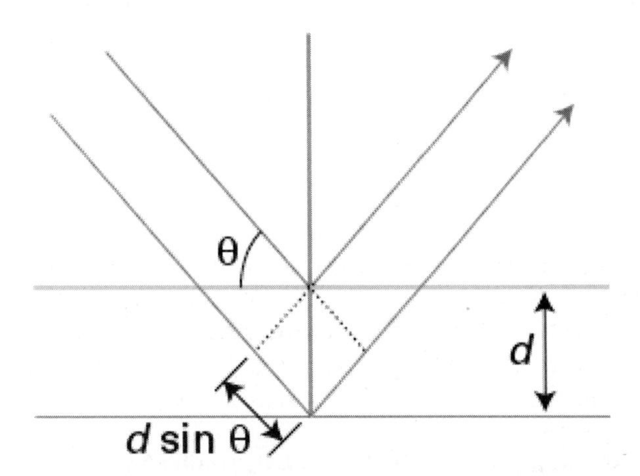

S8.1 **Main features of the CrO_2 powder XRD pattern?** The ionic radius of Cr(IV), 69 pm, is smaller than that of Ti(IV), 75 pm. The unit cell and *d* spacings will shrink as a result of the smaller radius of chromium. The XRD pattern for CrO_2 will show identical reflections to those of rutile TiO_2 (see Figure 8.4) but shifted to slightly higher diffraction angles.

S8.2 **TiO_2 in sunscreens?** As we can clearly see from Figure 8.12, titania articles absorb this ultraviolet radiation and prevent the waves from reaching the skin offering protection from the suns rays that can cause skin cancer.

S8.3 **Molecular shape and vibrational modes for XeF_2?** This molecule has a steric number of five and will assume a trigonal bipyramidal electron pair arrangement with the fluorines in the axial positions and the lone pairs in the equatorial positions to minimize overall repulsion. The molecule is linear with D?h point group symmetry. As a linear molecule, we expect it to have 3N-5=3(3)-5=4 total vibrational modes. These modes cannot be both IR and Raman active as the molecule has a center of symmetry.

S8.4 **(a) ^{77}Se-NMR spectrum consists of a triplet of triplets?** The triplet of triplets arises from the coupling of the two non-equivalent F nuclei with the Se nucleus. For n equivalent X (spin ½) nuclei, the A resonance is split into n +1 equally spaced lines with relative intensities given by Pascal's triangle. **(b) Proton resonance of the hydrido ligand consist of eight equal intensity lines?** The hydride resonance couples to three nuclei (two different ^{31}P nuclei and the ^{103}Rh nucleus) that are 100% abundant and all of which have I = ½ (see Table 8.4). Since the three coupling constants are different, the effect is to split the signal into a doublet of doublet of doublets, thus generating eight lines in the NMR of equal intensity.

S8.5 **EPR signal of new material arises from W sites?** Consulting Table 8.4, we see that 14% of the naturally occurring tungsten is ^{183}W, which has I = ½. Owing to this spin, the signal is split into 2 lines. This will be superimposed on a nonsplit signal that arises from the 86% of tungsten that does not have a spin.

S8.6 **Isomer shift for iron in Sr_2FeO_4?** The oxidation state for iron in this compound is +4. The outermost electron configuration is $3d^4$. We would expect the isomer shift to be smaller and less positive (below 0.2 mm s^{-1}).

S8.7 **Why does the mass spectrum of $ClBr_3$ have five peaks separated by 2 u?** Both chlorine (^{35}Cl 76% and ^{37}Cl 24%) and bromine (^{79}Br 51% and ^{81}Br 49%) exist as two isotopes. Consider the differences in mass numbers for the isotopes: any compound containing either Cl or Br will have molecular ions 2 u apart. The lightest isotopomer of $ClBr_3$ is $^{35}Cl^{79}Br_3$ at 272 u and the heaviest is $^{37}Cl^{81}Br_3$ at 280 u. Three other molecular masses are possible, giving rise to a total of 5 peaks in the mass spectrum shown in Figure 8.38. The differences in the relative intensities of these peaks are a consequence of the differences in the percent abundance for each isotope.

S8.8 **Why are percentages of hydrogen in 3d compounds less accurate than 5d compounds?** CHN analysis is used to determine percent composition (mass percentages of C, H, and N). Because the atomic masses of 5d metals are significantly higher than the atomic masses of 3d metals, the hydrogen percentages will be less accurate as they correspond to smaller fractions of the overall molecular masses of the compounds.

S8.9 **Stoichiometry of the tin oxide?** 7.673 mg of tin corresponds to 0.0646 mmol of Sn. The 10.000 mg sample will contain 2.327 mg of oxygen or 0.145 mmol of O. These molar ratios correspond to the formula, Sn_2O_5.

8.1 **How would you determine crystalline components in mineral sample?** As the mineral contains crystalline components, you could use powder X-ray diffraction to collect data. You could consult the JCPDS files for databases containing the main Bragg peaks in XRD patterns for these inorganic materials.

8.2 **Why are there no diffraction maxima in borosilicate glass?** Glass is a network covalent solid with no long-range periodicity or order. This periodicity is required for Bragg diffraction in order to have constructive interference of the X-rays that leads to observable peaks in the XRD pattern (see Figure 8.1).

8.3 **What is the minimum size of a cubic crystal that can be studied?** If the laboratory single-crystal diffractometer can study crystals that are 50 μm by 50 μm by 50 μm , with a new synchrotron source that has a millionfold increase in source intensity, one can reduce the size of the crystal area studied by $1/10^6$ or 0.5 μm by 0.5 μm by 0.5 μm.

8.4 **Wavelength of neutron at 2.20 km/s?** By using the de Broglie relation given in Section 8.2, we have:

$$\lambda = h/mv = (6.626 \times 10^{-34}\ \text{J s})/(1.675 \times 10^{-27}\ \text{kg})(2200\ \text{m/s}) = 1.80 \times 10^{-12}\ \text{m}$$

or 180 pm. This wavelength is very useful for diffraction, since it is comparable to interatomic spacings.

8.5 **Order of stretching frequencies?** The bond orders for CN^- and CO are the same, but N is lighter than O, hence the reduced mass for CN^- is smaller than for CO. From equation 8.4a, the smaller effective mass of the oscillator for CN^- causes the molecule to have the higher stretching frequency, since they are inversely proportional. The bond order for NO is 2.5, and N is heavier than C, hence CO has a higher stretching frequency than NO.

8.6 **Wavenumber for O–O in O_2^+?** The bond in O_2^+ (bond order = 2.5) is stronger than in O_2 (bond order = 2). Therefore, O_2^+ has the higher force constant. The stretching frequency is expected to be in the region of 1800 cm^{1}. See Section 2.8 for more on bond orders in homonuclear diatomics.

8.7 **UV photoelectron spectrum of NH_3?** The band at 11 eV is due to the lone pair. This might be expected to be sharp as it is nonbonding. However, the occupancy of the lone-pair orbital is linked strongly to the pyramidal angle. The ionised molecule has greater planarity, giving rise to the long progression.

8.8 **Raman bands assignments?** $N(SiH_3)_3$ is planar and thus N is at the center of the molecule and does not move in the symmetric stretch. $N(CH_3)_3$ is pyramidal and the N–C symmetric stretch involves displacement of N. We will more about these assignments in Chapter 7 as we consider the influence of the symmetry of the molecule on the observed spectroscopy.

8.9 **Single ^{13}C peak in NMR?** Even though there are chemically distinct carbonyls in $Co_2(CO)_9$, at room temperature they are exchanging position sufficiently quickly that an average signal is seen and only one peak is observed.

8.10 **Form of ^{19}F-NMR and ^{77}Se-NMR spectra of $^{77}SeF_4$?** The ^{19}F NMR spectrum of SeF_4 reveals two 1:3:3:1 quartets as the two distinct fluorine resonances will each be coupled to three spin ½ nuclei. (2N+1 = 2(3/2)+1 = 4). The ^{77}Se-NMR spectrum is a triplet of triplets (see S8.4 for details).

8.11 **NMR spectral features for XeF_5^-?** XeF_5^- has a pentagonal planar geometry, so all 5 of the F atoms are chemically equivalent and thus the molecule shows a single ^{19}F resonance. Approximately 25% of the Xe is present as ^{129}Xe, which has I = 1/2, and in this case the ^{19}F resonance is split into a doublet. The final result is a composite: two lines each of 12.5% intensity from the ^{19}F coupled to the ^{129}Xe, and one line of 75% intensity for the remainder.

8.12 **g-values?** Rearranging equation 8.5 we have $g = \Delta E / \mu_B B_0$. In order to calculate the g-values, we obtain the B_0 values from the EPR spectrum shown in Figure 6.33 (the three lines drawn to the x-axis correspond to 348, 387, and 432 mT) and set the $\Delta E = h\nu$ where ν is equal to the microwave frequency. The three g values are thus 1.94, 1.74, and 1.56, respectively. See below for an example of the calculation for the first line,

$$g = [(6.626 \times 10^{-34} \text{ J s})(9.43 \times 10^{9} \text{ s}^{-1})] / (9.27401 \times 10^{-24} \text{ J/T})(0.348 \text{ T}) = 1.94$$

8.13 **Slower process, NMR or EPR?** NMR, which uses typical frequencies of 0.5 GHz, will respond to slower chemical processes than EPR, with typical higher frequencies of 10 GHz. Each technique has key advantages and disadvantages, but both are powerful to probe localized spin environments and inorganic structures.

8.14 **Differences in EPR spectrum for *d*-metal with one electron in solution versus frozen?** In aqueous solution at room temperature, molecular tumbling is fast compared to the timescale of the EPR transition (approximately gm_bB_0/h), and this removes the effect of the g-value anisotropy: we expect a derivative-type spectrum, possibly exhibiting hyperfine structure that is centered at the average g value. In frozen solution, g-value anisotropy can be observed because the spectrum records the values of g projected along each of the three axes, and the averaging from molecular tumbling does not apply.

8.15 **Isomer shift for iron in $BaFe^{(VI)}O_4$?** The oxidation for iron results in a $3d^2$ configuration. As the 3d electrons are removed the isomer shift becomes less positive as the 3d electrons partly screen the nucleus from the inner s electrons. Compared to Fe(III) with isomer shift between +0.2 and +0.5 mm s^{-1}, we would expect a smaller positive shift for Fe(VI) well below +0.2 mm s^{-1}.

8.16 **Charge on Fe atoms in $Fe_4[Fe(CN)_6]_3$?** The total charge on Fe atoms must add up to +18 owing to the −18 negative charges on the 18 cyanide anions. Thus we can have four Fe(III) and three Fe(II). EPR can detect Fe(III) but not Fe(II) because all of the electrons are paired in the Fe(II) octahedral complex. Mössbauer spectroscopy can distinguish between Fe(II) and Fe(III) and between "complexed" Fe(II) coordinated by $6CN^-$ and "complexed" Fe(III) coordinated by $6CN^-$ because uncomplexed Fe(II) will have a larger quadrupole splitting than uncomplexed Fe(III), whereas complexed Fe(II) will have a smaller quadrupole splitting than complexed Fe(III).

8.17 **No quadrupole splitting in Mössbauer spectrum of SbF_5?** From VSEPR theory, SbF_5 is expected to have trigonal bipyramidal geometry, which may be subject to an electric field gradient. However, the absence of quadrupole splitting suggests that the geometry must be closer to cubic in the solid state; an octahedral geometry can be achieved if each Sb accepts a lone pair from an F atom on an adjacent molecule. This highly symmetrical cubic arrangement would not show the quadrupole splitting.

8.18 **No peak in the mass spectrum of Ag at 108 u?** Silver has two isotopes, ^{107}Ag (51.82%) and ^{109}Ag (48.18%). Thus the average mass is near 108, but no peak exists at this location since there is no isotope with this mass number. Compounds that contain silver will have two mass peaks flanking this average mass.

8.19 **Peaks in mass spectrum of $Mo(C_6H_6)(CO)_3$?** Considering systematic loss of individual ligands on the organometallic complex, the major peaks would be at 258 ($Mo(C_6H_6)(CO)_3^+$), 230 ($Mo(C_6H_6)(CO)_2^+$), 200 ($Mo(C_6H_6)(CO)^+$), 186 ($Mo(C_6H_6)^+$), and 174 ($Mo(CO)_3^+$).

8.20 **Cyclic voltammogram of Fe(III) complex?** The complex undergoes a reversible one-electron reduction with a reduction potential of 0.21 V. This can be calculating by taking the mean of E_{pa}(0.240 V) and E_{pc}(0.18 V), see Figure 8.46. Above 0.720 V the complex is oxidized to a species that undergoes a further chemical reaction, and thus is not rereduced. This step is not reversible.

8.21 **Zeolite of composition $CaAl_2Si_6O_{16}.nH_2O$, determine *n*?** The molar mass of the dehydrated zeolite is 518.5 g/mol. The molar mass of water is 18.0 g/mol. We can solve for n using the mass percentage of water.

Mass % H_2O = (mass water/mass sample)(100%) = 18n/(518.5 +18n)(100%) = 20%.

$18n = 0.2\ (518.5 + 18n)$

$14.4\ n = 103.7$

$n = 7.2$

Isolating n as an integer we can estimate that n = 7.

8.22 **Ratio of cobalt to acetylacetonate in the product?** The ratio is 3:1. By considering the mass percents given and converting to moles of Co and C, for every one mole of Co in the product we have 15 moles of C. This ratio holds for Co(III) having 3 bidentate, anionic acac ligands attached, and is consistent with the formula $Co(acac)_3$ (Consult Section 7.1 for more detail on the acetylacetonate ligand and cobalt coordination complexes).

Chapter 9 Periodic Trends

Periodicity is the regular manner in which the physical and chemical properties of the elements vary with atomic number. In this chapter we summarize some of the trends in the physical and chemical properties of the elements (including atomic radii in picometres shown below) and interpret them in terms of the underlying principles presented in Chapter 1.

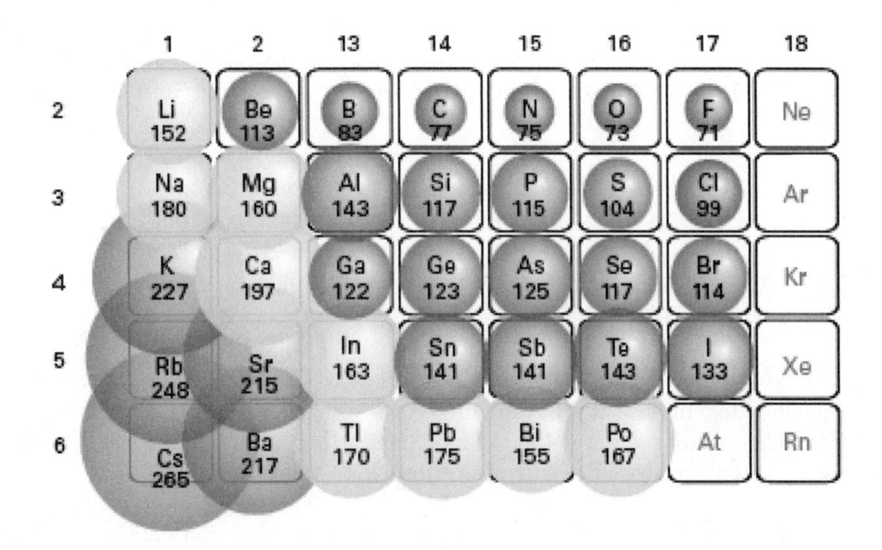

S9.1 **Found in aluminosilicate minerals or sulfides?** Cd and Pb are chalcophiles and will be found as sulfides. Rb and Sr are lithophiles are can be found in aluminosilicate minerals. Cr and Pd are siderophiles and can be found in both oxides and sulfides but are more likely to be found in their elemental states.

S9.2 **Sulfur forms catenated polysulfides whereas polyoxygen anions are unknown?** Oxygen forms a double bond that is three times more stable than its single bond. Owing to this strong tendency to form strong double bonds, it is more likely that these polyoxygen anions will form pi bonds that limit extended bonding owing to restrictions on pi orbital overlap through multiple bridging centres. Sulfur is much less likelty to form pi bonds and therefore more likely to generate catenated polysulfides.

S9.3 **Shape of XeO_4 and identify the Z + 8 compound with the same structure?** Xe is the central atom. With 8 valence electrons from Xe and 24 electrons (6 from each O) we have 32 total electrons and 16 electron pairs. We would predict a tetrahedral geometry to minimize electron pair repulsions. Note that in order to minimize formal charge, the xenon will form double bonds with each oxygen. Because the atomic number of Xe is 54, the Z + 8 is Sm. Therefore, the compound with the same structure is SmO_4.

S9.4 **Comment on $\Delta_f H^o$ values?** It is evident from the values that as we move down the group from S to Se to Te, the steric crowding of the fluorines is minimized owing to the increasing radius of the central atom. As a result, the enthalpy values become larger and more negative (more exothermic) and the higher steric number compounds (such as TeF_6) are more likely to form.

S9.5 **Further data useful when drawing comparisons with the value for V_2O_5?** We would have to know the products formed upon decomposition. That is, we need a balanced chemical reaction for the decomposition to determine which bond energies to use for the products of decomposition. We need comparable thermodynamic data for phosphorus as provided for other elements in Table 9.8.

9.1 **Maximum stable oxidation state?** (a) Ba; +2, (b) As; +5, (c) P; +5, (d) Cl; +7.

9.2 **Form saline hydrides, oxides and peroxides, and all the carbides react with water to liberate a hydrocarbon?** The group is the alkaline earth metals or Group 2 elements. Consult Sections 12.8 and 12.9 for detailed reactions.

9.3 **Elements vary from metals through metalloids to non-metals; form halides in oxidation states +5 and +3 and toxic gaseous hydrides?** These elements are Group 15. This group is very diverse. N and P are nonmetals; As and Sb are metalloids; and Bi is metallic. The +5 and +3 oxidation states are common for the group electron configuration of ns^2p^3. Phoshine and arsine are well known toxic gases. Consult Chapter 15 for more details on this diverse group of elements.

9.4 **Born–Haber cycle for the formation of the hypothetical compound $NaCl_2$? Which thermochemical step is responsible for the fact that $NaCl_2$ does not exist?** The key steps in the Born–Haber cycle are:

Na(s)	$\rightarrow$	Na(g)	sublimation	$\Delta_{sub}H^{\circ}$
Cl_2(g)	$\rightarrow$	2Cl(g)	dissociation	$\Delta_{dis}H^{\circ}$
Na(g)	$\rightarrow$	$Na^+(g) + 1e^-$	first ionization	$\Delta_{ion1}H^{\circ}$
Na^+(g)	$\rightarrow$	$Na^{2+}(g) + 1e^-$	second ionization	$\Delta_{ion2}H^{\circ}$
$2Cl(g) + 2e^-$	$\rightarrow$	$2Cl^-$ (g)	electron gain	$\Delta_{eg}H^{\circ}$
$Na^{2+}(g) + 2Cl^-(g)$	$\rightarrow$	$NaCl_2$(s)	lattice enthalpy	$-\Delta_L H^{\circ}$

We can use Figure 3.44 to construct a comparable cycle. You can calculate the lattice enthalpy by moving around the cycle and noting that the enthalpy of formation $\Delta_f H^{\circ} = \Delta_{sub}H^{\circ} + \Delta_{dis}H^{\circ} + \Delta_{ion1}H^{\circ} + \Delta_{ion2}H^{\circ} - \Delta_{eg}H^{\circ} - \Delta_L H^{\circ}$. The second ionization energy of sodium is 4562 kJ mol^{-1} and is responsible for the fact that the compound does not exist as it encourages a large, positive enthalpy of formation value.

9.5 **Inert pair effect beyond Group 15?** The relative stability of an oxidation state in which the oxidation number is 2 less than the group number is an example of the inert pair effect. This is a recurring theme in the heavier p-block elements. Beyond Group 15 where we see inert pair effect influencing the favored 3+ oxidation state for elements such as Bi and Sb, we also find stable +5 oxidation states for the halogens in Group 17. Examples include BrO_3^- and IO_3^-. As a result of this intermediate oxidation state, these compounds can function as both oxiding and reducing agents. Group 16 elements (except oxygen) also form several stable compounds with +4 oxidation states including SF_4 and SeF_4. There are only a few examples of the inert pair effect in Group 18 including the compound .XeO_3.

9.6 **Ionic radii, ionization energy, and metallic character?** Metallic character and ionic radii decrease across a period and down a group. Ionization energy increases across a period and decreases down a group. Large atoms typically have low ionization energy and are more metallic in character.

9.7 **Names of ores?** (a) **Mg**; $MgCO_3$ magnesite, (b) **Al**; Al_2O_3 bauxite, and (c) **Pb**; PbS galena.

9.8 **Identify the *Z* + 8 element for P. similarities?** The atomic number of P is 15. The Z + 8 element has an atomic number of 23 and is V (vanadium). Both form compounds with varying oxidation states up to a maximum value of +5. Both form stable oxides including ones in +5 oxidation state (V_2O_5 and P_2O_5). Like phosphorus, vanadium forms oxoanions including ortho-, pyro- and meta-anions. Consult Section 15.5 for analogous phosphorus oxoanions.

9.9 **Calculate $\Delta_f H^{\circ}$ for SeF_6?** If we assume that the B(Se–Se) for diatomic gas is 332 kJ mol^{-1} we can calculate $\Delta_f H^{\circ}$ for SeF_6 from the difference between the enthalpies of bonds broken and the bonds formed by the reaction of:

$$\tfrac{1}{2}Se_2(g) + 3F_2(g) \rightarrow SeF_6(g).$$

The enthalpy change accompanying bond breaking is ½(332) + 3(159) = 643 kJ mol^{-1}. The enthalpy change accompanying bond making is –6(340) = –2040 kJ mol^{-1}. Therefore, $\Delta_f H^{\circ}$ = –1397 kJ mol^{-1}.

Chapter 10 Hydrogen

Diborane, B_2H_6, is the simplest member of a large class of compounds, the electron-deficient boron hydrides. Like all boron hydrides, it has a positive standard free energy of formation, and so cannot be prepared directly from boron and hydrogen. The bridge B–H bonds are longer and weaker than the terminal B–H bonds (1.32 vs. 1.19 Å).

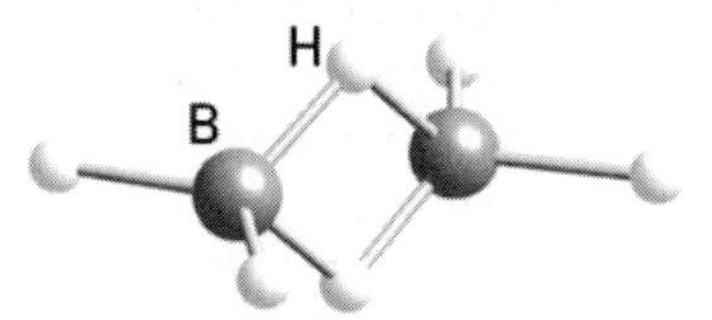

17 Diborane, B_2H_6

S10.1 **Which of the following CH_4, SiH_4, or GeH_4 would best H^+ or H^- donor?** To decide if the cleavage of a E-H bond is going to result in a H^- or H^+, you must first look at the electronegativity differences between H and E. If E is more electronegative than H, heterolytic cleavage accurs, and you realease a proton, H^+. If, E is less elecgtronegative than H, heterolytic cleavage occurs resulting in release of a hydride, H^-. Given the following compounds, CH_4, SiH_4, and GeH_4; carbon is the only E that is more electronegative than H, thus it would be the strongest Bronsted acid. In other words, CH_4 would be more likely to release protons than SiH_4 and GeH_4. Ge is the least electronegative E, thus GeH_4 would be the best hydride donor of the three.

S10.2 **Reactions of hydrogen compounds? (a)** $Ca(s) + H_2(g) \rightarrow CaH_2(s)$. This is the reaction of an active *s*-metal with hydrogen, which is the way that saline metal hydrides are prepared.
(b) $NH_3(g) + BF_3(g) \rightarrow H_3N{-}BF_3(g)$. This is the reaction of a Lewis base and a Lewis acid. The product is a Lewis acid–base complex.
(c) $LiOH(s) + H_2(g) \rightarrow$ NR. Although dihydrogen can behave as an oxidant (e.g., with Li to form LiH) or as a reductant (e.g., with O_2 to form H_2O), it does not behave as a Brønsted or Lewis acid or base. It does not react with strong bases, like LiOH, or with strong acids.

S10.3 **A procedure for making Et_3MeSn?** A possible procedure is as follows:

$$2Et_3SnH + 2Na \rightarrow 2Na^+Et_3Sn^- + H_2$$

$$Na^+Et_3Sn^- + CH_3Br \rightarrow Et_3MeSn + NaBr$$

10.1 **Where does Hydrogen fit in the periodic chart? (a) Hydrogen in group 1?** Hydrogen has one valence electron like the group 1 metals and is stable as H^+, especially in aqueous media. The other group 1 metals have one valence electron and are quite stable as M^+ cations in solution and in the solid state as simple ionic salts. In most periodic charts, hydrogen is generally put with this group, given the above information. However, physically, hydrogen does not fit in this group, it is a diatomic gas, while the group 1 metals are just that, metals. Your text does mention that hydrogen may exit as a solid under extreme pressures.
(b) Hydrogen in group 17? Hydrogen can fill its 1s orbital and make a hydride H^-. Hydrides are isoelectronic to He, a noble gas configuration, thus are relatively stable. Group 1 and group 2 metals, as well as transition metals, stabilize hydrides. The halogens form stable X^- anions, obtaining a noble gas configuration, both in solution and in the solid state as simple ionic salts. Some periodic charts put hydrogen both in group 1 and in group 17 for the reasons stated above. The halogens are diatomic gases just like hydrogen, so physically hydrogen fits well in group 17, but chemically it fits well in both group 1 and group 17.
(c) Hydrogen in group 14? There is no reason, chemical or physical, for hydrogen to be placed in this group.

10.2 **Low reactivity of hydrogen?** Hydrogen exists as a diatomic molecule (H_2). It has a high bond enthalpy (see Table 2.8); thus it takes a lot of energy to break the bond. It also only has two electrons shared between two protons, which is energetically very stable, leading to its chemical stability.

10.3 **Assign oxidation numbers to elements? (a) H_2S?** When hydrogen is less electronegative than the other element in a binary compound (H is 2.20, S is 2.58; see Table 1.7), its oxidation number is chosen to be +1. Therefore, the oxidation number of sulfur in hydrogen sulfide (sulfane) is –2.

(b) KH? In this case, hydrogen (2.20) is more electronegative than potassium (0.82), so its oxidation number is chosen to be –1. Therefore, the oxidation number of potassium in potassium hydride is +1.

(c) $[ReH_9]^{2-}$? The electronegativity of rhenium is not given in Table 1.7. However, since it is a metal, it is reasonable to conclude that it is less electronegative than hydrogen. Therefore, if hydrogen is counted as –1, then the rhenium atom in the $[ReH_9]^{2-}$ ion has an oxidation number of +7.

(d) H_2SO_4? The structure of sulfuric acid is shown below. Since the hydrogen atoms are bound to very electronegative oxygen atoms, their oxidation number is +1. Furthermore, since oxygen is always assigned an oxidation number of –2 (except for O_2 and peroxides), sulfur has an oxidation number of +6.

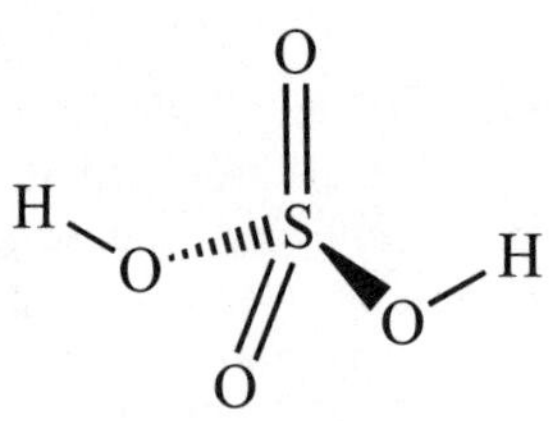

H_2SO_4 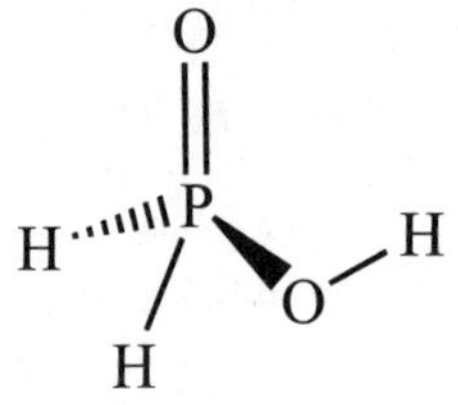 $\qquad$ $H_2PO(OH)$

(e) $H_2PO(OH)$? The structure of hypophosphorous acid (also called phosphinic acid) is shown above. There are two types of hydrogen atoms. The one that is bonded to an oxygen atom has an oxidation number of +1. The two that are bonded to the phosphorus atom present a problem, since phosphorus (2.19) and hydrogen (2.20) have nearly equal electronegativities. If these two hydrogen atoms are assigned an oxidation number of +1, and oxygen, as above, is assigned an oxidation number of –2, then the phosphorus atom in $H_2PO(OH)$ has an oxidation number of +1.

10.4 **Preparation of hydrogen gas?** As discussed in Section 10.4, *Production of dihydrogen*, the three industrial methods of preparing H_2 are (i) steam reforming, (ii) the water-gas reaction, and (iii) the shift reaction (also called the water-gas shift reaction). The balanced equations are:

$$\text{(i) } CH_4(g) + H_2O \rightarrow CO(g) + 3H_2(g) \ (1000°C)$$

$$\text{(ii) } C(s) + H_2O \rightarrow CO(g) + H_2(g) \ (1000°C)$$

$$\text{(iii) } CO(g) + H_2O \rightarrow CO_2(g) + H_2(g)$$

These reactions are not very convenient for the preparation of small quantities of hydrogen in the laboratory. Instead, (iv) treatment of an acid with an active metal (such as zinc) or (v) treatment of a metal hydride with water would be suitable. The balanced equations are:

$$\text{(iv) } Zn(s) + 2HCl(aq) \rightarrow Zn^{2+}(aq) + 2Cl^-(aq) + H_2(g)$$

$$\text{(v) } NaH(s) + H_2O \rightarrow Na^+(aq) + OH^-(aq) + H_2(g)$$

10.5 **Properties of hydrides of the elements? (a) Position in the periodic table?** See Figure 10.2.

(b) Trends in $\Delta_f G°$? See Table 10.1.

(c) Different molecular hydrides? Molecular hydrides are found in groups 13/III through 17/VII. Those in group 13/III are electron-deficient, those in group 14/IV are electron-precise, and those in group 15/V through 17/VII are electron-rich.

10.6 What are the physical properties of water without hydrogen bonding? Read Section 10.6 (a) (*iii) Hydrogen bonding*. If water did not have hydrogen bonds, it most likely would be a gas at room temperature like its heavier homologues H_2S, H_2Se, and H_2Te; see Figure 10.6. Also, most pure compounds are denser as a solid than as a liquid. Because of hydrogen bonding, water is actually less dense and has the structure shown in figure 10.7. This is why ice floats. If there were no hydrogen bonds, it would be expected that ice would be denser than water.

10.7 Which molecule has the stronger hydrogen bonds? Hydrogen bonds consist of a hydrogen atom bonded to an atom more electronegative than itself, i.e., F, N, and O (See Table 1.7). This renders polarity between the E–H bond (E = a *p*-block nonmetal) and allows donation of a lone pair from another molecule to form the hydrogen bond. Since S is less electronegative than O, the partial positive charge felt by the H in O–H is stronger than that in S–H, therefore, S–H···O has a weaker hydrogen bond than O–H···S.

10.8 Name and classify the following? (a) BaH_2? This compound is named barium hydride. It is a saline hydride.
(b) SiH_4? This compound is named silane. It is an electron-precise molecular hydride.
(c) NH_3? This familiar compound is known by its common name, ammonia, rather than by the systematic names azane or nitrane. Ammonia is an electron-rich molecular hydride.
(d) AsH_3? This compound is generally known by its common name, arsine, rather than by its systematic name, arsane. It is also an electron-rich molecular hydride.
(e) $PdH_{0.9}$? This compound is named palladium hydride. It is a metallic hydride.
(f) HI? This compound is known by its common name, hydrogen iodide, rather than by its systematic name, iodane. It is an electron-rich molecular hydride.

10.9 Chemical characteristics of hydrides? (a) Hydridic character? Barium hydride is a good example, since it reacts with proton sources such as H_2O to form H_2:

$$BaH_2(s) + 2H_2O(l) \rightarrow 2H_2(g) + Ba(OH)_2(s)$$

$$\text{Net reaction: } 2H^- + 2H^+ \rightarrow 2H_2$$

(b) Brønsted acidity? Hydrogen iodide is a good example, since it transfers its proton to a variety of bases, including pyridine:

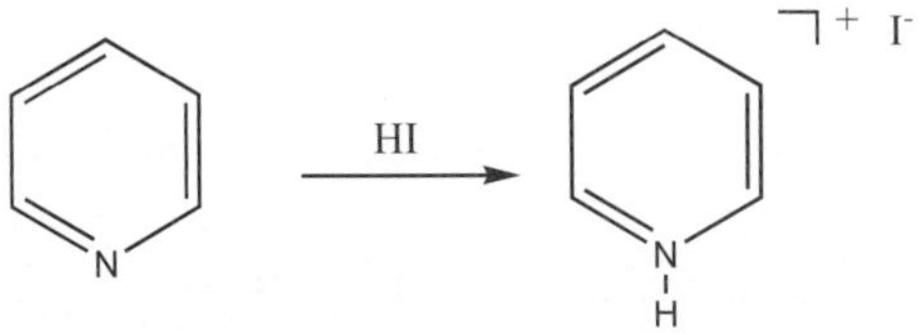

$$C_5H_5N \xrightarrow{HI} [C_5H_5NH]^+ I^-$$

(c) Variable composition? The compound $PdH_{0.9}$ is a good example.
(d) Lewis basicity? Ammonia is a good example, since it forms acid–base complexes with a variety of Lewis acids, including BMe_3:

$$NH_3(g) + BMe_3(g) \rightarrow H_3NBMe_3(g)$$

10.10 Phases of hydrides of the elements? Of the compounds listed in Exercise 10.8, BaH_2 and $PdH_{0.9}$ are solids, none is a liquid, and SiH_4, NH_3, AsH_3, and HI are gases (see Figure 10.6). Only $PdH_{0.9}$ is likely to be a good electrical conductor.

10.11 The structures of H_2Se, P_2H_4, and H_3O^+? The Lewis structures of these three species are:

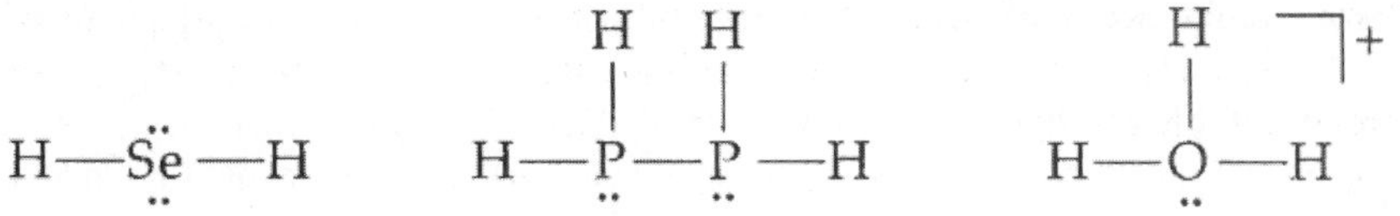

H—S̈e—H (two lone pairs on Se)

H₂P—PH₂: H—P(H)—P(H)—H (one lone pair on each P)

[H—O(H)—H]⁺ (one lone pair on O)

According to the VSEPR model (see Section 2.3), H_2Se should be bent, and so it belongs to the C_{2v} point group; H_3O^+ should be trigonal pyramidal (like NH_3), and so it belongs to the C_{3v} point group; each phosphorus atom of P_2H_4 should have local pyramidal structure. If the molecule adopts the skew conformation (see the Newman diagram below), then it belongs to the C_2 point group (the C_2 axis bisects the P–P bond).
A drawing of the structure of P_2H_4. The P–P and P–H bond distances are 2.22 and 1.42 Å, respectively, and the P–P–H bond angles are all about 94° (cf. PH_3, in which the H–P–H bond angles are 93.8°).

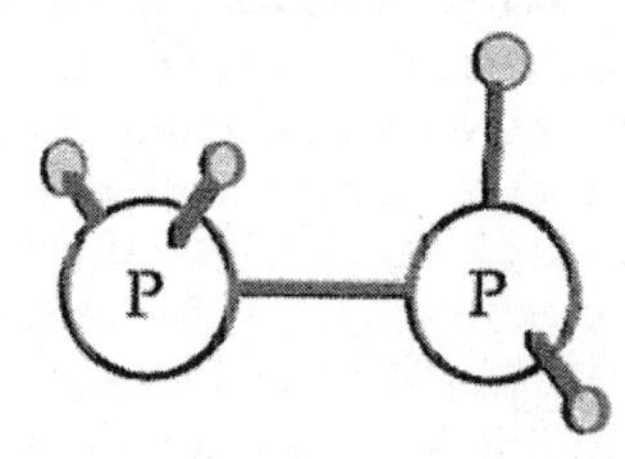

A Newman projection of the skew, or gauche, conformation of P_2H_4. The only element of symmetry that this structure possesses is a C_2 axis that bisects the P–P bond. Therefore, it has C_2 symmetry.

10.12 The reaction that will give the highest proportion of HD? Reactions (a) and (c) both involve the production of both H and D atoms at the surface of a metal. The recombination of these atoms will give a statistical distribution of H_2 (25%), HD (50%), and D_2 (25%). However, reaction (b) involves a source of protons that is 100% $^2H^+$ (i.e., D^+) and a source of hydride ions that is 100% $^1H^-$:

$$D_2O + NaH(s) \rightarrow HD(g) + NaOD(s)$$

$$\text{Net reaction: } D^+ + H^- \rightarrow HD$$

Thus, reaction (b) will produce 100% HD and no H_2 or D_2.

10.13 Most likely to undergo radical reactions? Of the compounds H_2O, NH_3, $(CH_3)_3SiH$, and $(CH_3)_3SnH$, the tin compound is the most likely to undergo radical reactions with alkyl halides. This is because the Sn–H bond in $(CH_3)_3SnH$ is less polar *and* weaker than either O–H, N–H, or Si–H bonds. The formation of radicals involves the homolytic cleavage of the bond between the central element and hydrogen, and a weak nonpolar bond undergoes homolysis most readily.

10.14 Arrange H_2O, H_2S, and H_2Se in order? (a) Increasing acidity? As discussed in Section 4.4, *Periodic trends in aqua acid strength*, acidities of EH_n increase down a group in the *p* block, mostly because the decrease in E–H bond enthalpy lowers the proton affinity of $[EH_{n-1}]^-$ (E is a generic *p*-block element). Therefore, the order of increasing acidity is $H_2O < H_2S < H_2Se$.
(b) Increasing basicity toward a hard acid? In general, soft character increases down a group, so the hardest base of these three compounds is H_2O. The order of increasing basicity toward a hard acid is $H_2Se < H_2S < H_2O$.

10.15 The synthesis of binary hydrogen compounds? The three main methods of synthesis of binary hydrogen compounds are (i) direct combination of the elements, (ii) protonation of a Brønsted base, and (iii) metathesis using a compound such as LiH, $NaBH_4$, or $LiAlH_4$. The first method is limited to those binary hydrogen compounds that are exoergic. An example is:

$$\text{(i)}\quad 2Li(s) + H_2(g) \rightarrow 2LiH(s)$$

The second method can be used for the preparation of EH_n compounds when a source of the E^{n-} anion is available. An example is:

$$\text{(ii)}\quad CaF_2(s) + H_2SO_4(l) \rightarrow 2HF(g) + CaSO_4(s)$$

Almost all of the hydrogen fluoride that is prepared industrially is made this way. The third method can be used to convert the chlorides of many elements E to the corresponding hydrides, as in the following example:

$$PCl_3(l) + 3LiH(s) \rightarrow PH_3(g) + 3LiCl(s)$$

10.16 Compare BH_4^-, AlH_4^-, and GaH_4^-? Since Al has the lowest electronegativity of the three elements B (2.04), Al (1.61), and Ga (1.81, see Table 1.7), the Al–H bonds of AlH_4^- are more hydridic than the B–H bonds of BH_4^- or the Ga–H bonds of GaH_4^-. Therefore, since AlH_4^- is more "hydride-like," it is the strongest reducing agent. The reaction of GaH_4^- with aqueous HCl is as follows:

$$GaH_4^-(aq) + 4HCl(aq) \rightarrow GaCl_4^-(aq) + 4H_2(g)$$

10.17 Compare period 2 and period 3 hydrogen compounds? One important difference between period 2 and period 3 hydrogen compounds is their relative stabilities. The period 2 compounds, except for B_2H_6, are all exoergic (see Table 1.01). Their period 3 homologues either are much less exoergic or are endoergic (cf. HF and HCl, for which $\Delta_f G^o = -273.2$ and -95.3 kJ mol^{-1}, and NH_3 and PH_3, for which $\Delta_f G^o = -16.5$ and $+13.4$ kJ mol^{-1}). Another important chemical difference is that period 2 compounds tend to be weaker Brønsted acids and stronger Brønsted bases than their period 3 homologues. Diborane, B_2H_6, is a gas while AlH_3 is a solid. Methane, CH_4, is inert to oxygen and water, while silane, SiH_4, reacts vigorously with both. The bond angles in period 2 hydrogen compounds reflect a greater degree of sp^3 hybridization than the homologous period 3 compounds (compare the H–O–H and H–N–H bond angles of water and ammonia, which are 104.5° and 106.6°, respectively, with the H–S–H and H–P–H bond angles of hydrogen sulfide and phosphane, which are 92° and 93.8°, respectively). Several period 2 compounds exhibit strong hydrogen bonding, namely, HF, H_2O, and NH_3, while their period 3 homologues do not (see Figure 10.6). As a consequence of hydrogen bonding, the boiling points of HF, H_2O, and NH_3 are all higher than their respective period 3 homologues.

10.18 Suggest a method for the preparation of BiH_3? The current method for the synthesis of bismuthine, BiH_3, is by the redistribution of methylbismuthine, BiH_2Me.

$$3BiH_2Me \rightarrow 2BiH_3 + BiMe_3$$

The starting material BiH_2Me can be prepared by the reduction of $BiCl_2Me$ with $LiAlH_4$. The precursor, BiH_2Me, is unstable as well. The difficulty in synthesizing BiH_3 is it involves a multi-step process, all of which deal with extremely reactive and unstable compounds.

BiH_3 is unstable above -45 °C, yielding hydrogen gas and elemental bismuth. The original preparation was reported in 1961, but, it was not until 2002, that this procedure was repeated and yielded enough BiH_3 to do spectroscopy on!

10.19 Describe the compound formed between water and Kr? This compound is called a clathrate hydrate (see Figure 10.12). It consists of cages of water molecules, all hydrogen bonded together, each surrounding a single krypton atom. Strong dipole–dipole forces hold the cages together, while weaker van der Waals forces hold the krypton atoms in the centers of their respective cages.

10.20 Potential energy surfaces for hydrogen bonds? (See Figure 10.9) There are two important differences between the potential energy surfaces for the hydrogen bond between H_2O and Cl^- ion and for the hydrogen bond in bifluoride ion, HF_2^-. The first difference is that the surface for the H_2O, Cl^- system has a double minimum (as do most hydrogen bonds), since it is a relatively weak hydrogen bond, while the surface for the bifluoride ion has a single minimum (characteristic of only the strongest hydrogen bonds). The second difference is that the surface for the H_2O, Cl^- system is not symmetric, since the proton is bonded to two different atoms (oxygen and chlorine), while the surface for bifluoride ion is symmetric. The two surfaces are shown below.

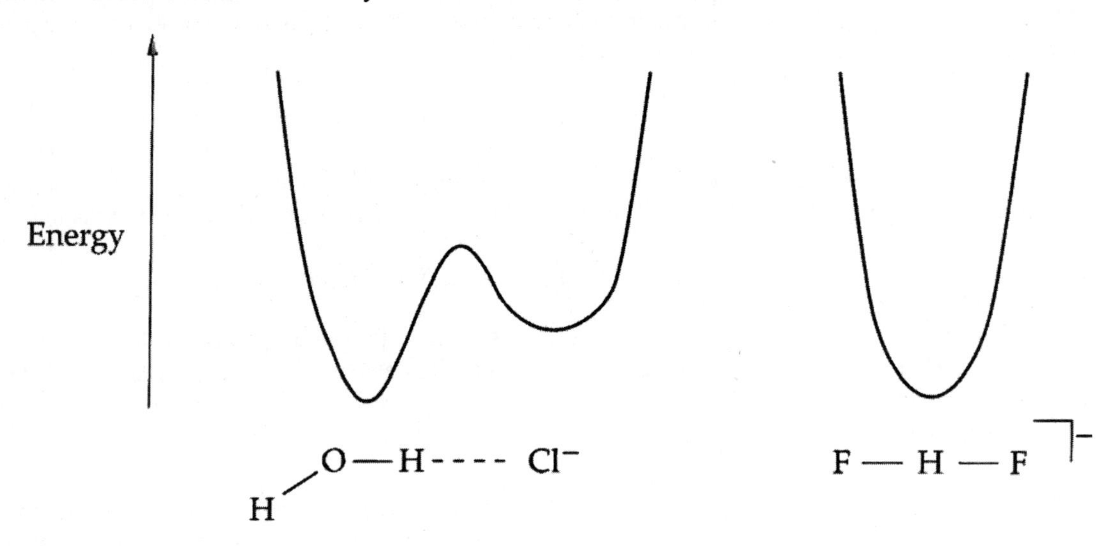

10.21 Dihydrogen as an oxidizing agent? The classic example of H_2 as an oxidizing agent is it's reaction with an active s-block metals such as sodium. In the reaction below, elemental sodium gets oxidized, while elemental hydrodgen gets reduced. Thus H_2 is an oxidizing agent in this case.

$$2Na(s) + H_2(g) \rightarrow 2NaH(s)$$

Chapter 11 The Group 1 Elements

All the Group 1 elements are metallic. Unlike most metals, they have low densities and are very reactive. The Group 1 elements, the **alkali metals,** are lithium, sodium, potassium, rubidium, caesium (cesium), and francium. All the elements form simple, ionic compounds, most of which are soluble in water. Group 1 elements form a limited number of complexes and organometallic compounds. One of the most common type of ligands for group 1 elements are crown ethers, one of which is shown below.

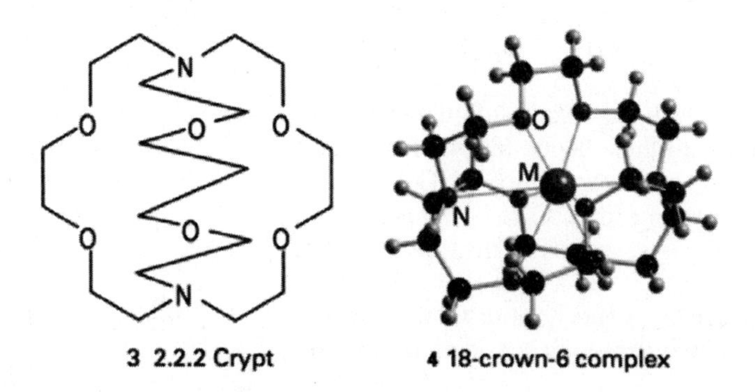

3 2.2.2 Crypt 4 18-crown-6 complex

S11.1 Change in cell parameter for CsCl? We would expect the lattice to expand slightly upon heating before it melts at approximately 650 °C. The lattice type of the CsCl structure type is primitive and the lattice parameter can be calculated by using the ionic radii of Cs and Cl in eightfold coordination (188 and 204 pm, respectively) as 452 pm (from $2(r_+ + r_-)/3^{1/2}$)). At 445 °C the CsCl structure changes to rock-salt and assumes the face centered cubic structure type. The new lattice parameter can be calculated as 784 pm (from $2r_- + 2r_+$). The observed major peak in the Bragg pattern would therefore shift to lower 2 theta values upon heating through the phase transition according to the Bragg relation (Section 8.1).

S11.2 Lattice enthalpies of formation? The first step to find the lattice enthalpy of formation is to use the Kapustinskii equation to calculate the lattice energy for each compound. Then we can use the data from Tables 11.1 and 11.2, combined with equation 11.1, to find that the lattice enthalpy of formation for LiF is 625 kJ mol^{-1} and for NaF is 535 kJ mol^{-1}. Note that the Li^+ cation is smaller than Na^+ and thus leads to the larger lattice enthalpy.

S11.3 Trend is stability of Group 1 ozonides? The ionic radius of the ozonide anion is larger than the radius of the superoxide; as a result the lattice energies are smaller for the ozonide compounds and the Group 1 ozonides are less stable compared to the superoxides. Recall that the lattice energy is inversely proportional to the distance between the center of the two ions. As the group 1 cation size gets larger moving down the group, the lattice energy decreases and the compound is less stable.

S11.4 Sketch the thermodynamic cycle of Group 1 carbonate.

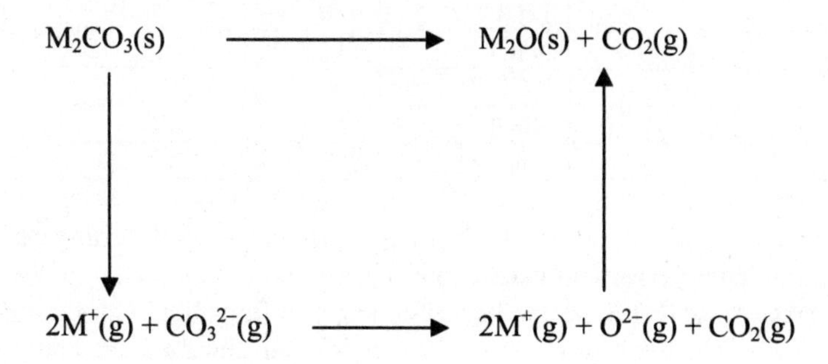

S11.5 Explain the differences in temperature of decomposition of $LiNO_3$ and KNO_3? KNO_3 decomposes in two steps at two different temperatures, 350°C and 450°C, respectfully.

$$KNO_3(s) \rightarrow KNO_2(s) + {}^1/_2O_2(g)$$

$$2KNO_2(s) \rightarrow K_2O(s) + 2NO_2(g) + {}^1/_2O_2(g)$$

While $LiNO_3$ decomposes in one step at the temperature of 600°C.

$$LiNO_3(s) \rightarrow {}^1/_2\,Li_2O(s) + NO_2(g) + {}^1/_4O_2(g)$$

Both end products are the binary oxides of lithium and potassium. Li_2O is more stable, has a higher lattice energy, than K_2O. So the way to think about this problem is to think about the tendency towards the production of Li_2O and K_2O. It is much more favourable for $LiNO_3$ to produce Li_2O than it is KNO_3 to produce K_2O. In other words, it takes a lot more energy, i.e. 600 °C, to decompose $LiNO_3$ to Li_2O because Li_2O is so stable. In addition, lithium nitrite, $LiNO_2$ is significantly less stable than Li_2O. Therefore, $LiNO_3$ goes straight to Li_2O in one step.

On the other hand, KNO_2 is much more stable than $LiNO_2$, and slightly less stable than K_2O. Consequently, KNO_3 decomposes in two steps with considerably less energy.

S11.6 Predicted ^{7}Li NMR of Li_3N? Lithium nitride is the only stable alkali metal nitride, the binry nitrides of the later alkali metals were not successfully made until 2002, when Na_3N was made. Solid Li_3N is a red or purple colored solid and has a high melting point. Li_3N has an unusual crystal structure (Figure 11.11), that consists of two types of layers. The first layer has the composition Li_2N^-, containing 6-coordinate Li ion centers and the other layer consist only of lithium cations. So, the Li ions are in two different chemical and magnetic environments. Therefore, one would expect two peaks in the NMR spectrum. However, the lithium ion is hightly mobile, since there are vacant sites in the structure, the lithium ion can hop from one site to another. So at higher temperatures, all the lithium ions will be fluxional, giving only one resonance in the NMR. At cooler temperatures, you would freeze out the structure, and see two resonances due to the two different environments for the lithium ion in the structure.

11.1 **(a) Why are group 1 metals good reducing agents?** All group 1 metals have one valence electron in the ns^1 subshell. They also have relatively low first ionization energies; therefore loss of one electron to form a closed shell electronic configuration is favorable.
(b) Why are group 1 metals poor complexing agents? Group 1 metals are large, electropositive metals and have little tendency to act as Lewis acids. However, they do complex well with hard Lewis bases, such as oxides, hydroxides, and many other oxygen containing ligands.

11.2 Trends of the fluorides and chlorides of the group 1 metals?

	$-\Delta_f H^\circ$ /(kJ mol^{-1})		$-\Delta_f H^\circ$ /(kJ mol^{-1})
LiF	625	LiCl	470
NaF	535	NaCl	411
KF	564	KCl	466
RbF	548	RbCl	458
CsF	537	CsCl	456

See Figure 11.6, which plots the enthalpy of formation of the group I metals versus the halogens. Fluoride is a hard Lewis base and will form strong complexes with hard Lewis acids. The lithium cation is the hardest Lewis acid of the group I metals, so it makes sense that it has the largest enthalpy of formation with fluoride compared to the rest of the group I metals. However, the trends reverse for the chloride ion; since it is a softer Lewis base than the fluoride ion, it will form stronger bonds with the heavier group I metals.

11.3 **Synthesis of group 1 alkyls?** Most alkyl lithiums are made using elemental lithium with the corresponding alkyl chlorides. The formation of LiCl helps drive the reaction.

$$C_2H_5Cl + 2Li \rightarrow C_2H_5Li + LiCl$$

The sodium analogue can be made the same way.

$$C_2H_5Cl + 2Na \rightarrow C_2H_5Na + NaCl$$

11.4 **Which is more likely to lead to the desired result? (a) Cs^+ or Mg^{2+}, form an acetate complex?** The metal ion with the higher charge, Mg^{2+}, is more likely to form a complex with acetate ion than Cs^+. The reason is that the binding of ligands for these hard metal ions is governed by the electrostatic parameter, z^2/r. As stated in the text, the weakness with which *s*-block metal ions bind ligands explains why until recently very few *s*-metal complexes had been characterized.
(b) Be or Sr, dissolve in liquid ammonia? It is more likely that strontium will dissolve in liquid ammonia than beryllium. There are two ways to arrive at this conclusion. First of all, electropositive metals in low enthalpies of sublimation are prone to dissolve in liquid ammonia. Second, the Be^{2+}/Be reduction potential, −1.97 V, is less negative than E° for the Sr^{2+}/Sr potential, −2.89 V. Even though you were asked to focus on liquid ammonia solutions in this exercise, and E° values refer to redox reactions in *aqueous acid*, the principal steps are the same on going from $M^0 \rightarrow M^{2+}$, namely sublimation, ionization, and solvation. Therefore, you can expect a general agreement between E° values and a tendency to dissolve in liquid ammonia.
(c) Li^+ or K^+, form a complex with C2.2.2? Potassium ion is more likely to form a complex with the cryptate ligand C2.2.2 than Li^+. The difference has to do with the match between the interior size of the cryptate and the ionic radius of the alkali metal ion. According to Figure 11.4, K^+ forms a stronger complex, by about two orders of magnitude, with C2.2.2 than does Li^+.

11.5 **Identify the compounds?**

$$\mathbf{NaOH} \leftarrow H_2O + \textbf{Sodium metal} + O_2 \rightarrow \mathbf{Na_2O_2} + \text{heat} \rightarrow \mathbf{Na_2O}$$

$$\textbf{Sodium metal} \xrightarrow{+NH_3} \mathbf{NaNH_2}$$

11.6 **Trends in solubility?** In LiF and CsI the cation and anion have similar radii. Solubility is lower if there is a large difference between radii of cation and anion, as in CsF and LiI.

11.7 **Thermal stability of hydrides versus carbonates?** Hydrides decompose to elements. Lattice energy decreases down the group as the radius of cation increases. Carbonates decompose to oxides. There is a difference in lattice energy between carbonate and oxide decreases down the group, which results in increased stability.

11.8 **The structures of CsCl and NaCl?** See Figures 3.7 and 3.30; 6-coordinate Na^+, 8-coordinate Cs^+; different r^+/r^-. Cesium is so large that the only way it can pack is in a body-centred cubic lattice.

11.9 **The effect of the alkyl group on the structure of lithium alkyls?** Whether a molecule is monomeric or polymeric is based on the streric size of the alkyl group. Less bulky alkyl groups lead to polymerization; methyl groups are tetrameric or hexameric, while a ^{t}Bu group yields a monomer. The larger aggregates can be broken down into dimmers and monomers using strong Lewis bases such as TMEDA. Below shows how phenyl lithium (which is normally polymeric) forms a dimmer with TMEDA.

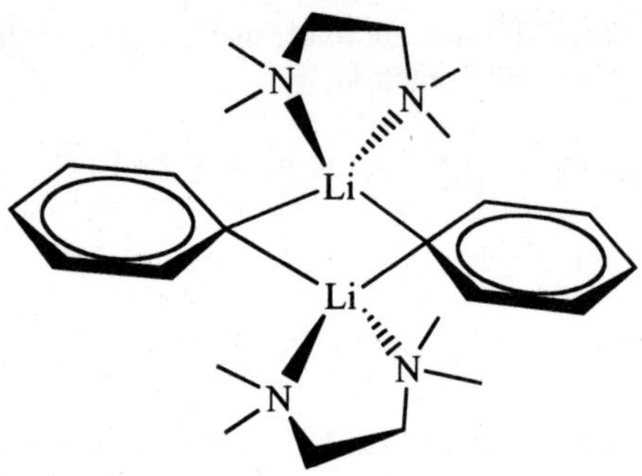

11.10 **Predict the products of the following reactions? (a)** The driving force behind this reaction is the formation of lithium bromide (very large lattice energy). The same reasoning for reaction **(b)**. For reaction **(c)**, the driving force is the loss of ethane gas.

(a) $CH_3Br + Li \rightarrow Li(CH_3) + \mathbf{LiBr}$

(b) $MgCl_2 + LiC_2H_5 \rightarrow Mg(C_2H_5)Br + \mathbf{LiBr}$

(c) $C_2H_5Li + C_6H_6 \rightarrow LiC_6H_5 + \mathbf{C_2H_6}$

Chapter 12 The Group 2 Elements

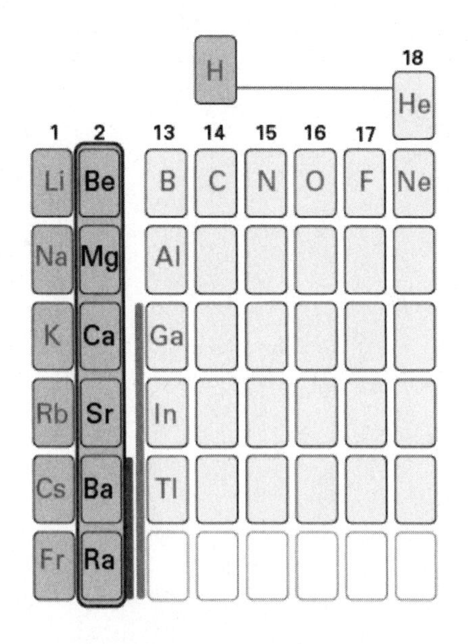

The *Group 2* elements calcium, strontium, barium, and radium are known as the **alkaline earth metals.** All the elements are silvery white metals, but some aspects of the chemical properties of beryllium are more like those of a metalloid. The elements are harder, denser, and less reactive than the elements of Group 1 but are still more reactive than many typical metals. The elements form a limited number of complexes and organometallic compounds. The insolubility of some of the calcium compounds in particular leads to the existence of many inorganic minerals that provide the raw materials for the infrastructure of our built environment and provide the building blocks for the compounds from which many rigid biological structures are formed.

S12.1 Predict whether (a) $BeCl_2$ and (b) $BaCl_2$ are predominantly ionic or covalent?
$BeCl_2$ is covalent because of a smaller radius and high electron density of Be^{2+}, the lightest group 2 metal ion; $BaCl_2$ is ionic because of the larger radius of Ba^{2+}, a heavier and more electropositive group 2 metal ion.

S12.2 Calculate the lattice enthalpies for CaO and CaO_2 and check that the above trend is confirmed?
Use the Kapustinskii equation (equation 3.4), the ionic radii from Table 12.1, and the appropriate constants—which are $K = 1.21 \times 10^5$ MJ Å mol^{-1}, $n = 2$, $z^+_{Ca} = +2$, $z^-_O = -2$—the sum of the thermochemical radii, d_0, 140 pm + 99 pm = 239 pm = 23.9 Å and d = 3.45 Å, to find the lattice enthalpy of CaO. For calcium peroxide use $n = 2$, $z^+_{Ca} = +2$, $z^-_{O_2} = -2$, and d_0 = 180 pm + 99 pm = 279 pm = 27.9 Å. The lattice enthalpies of calcium oxide and calcium peroxide are 3465 kJ mol^{-1} and 3040 kJ mol^{-1}, respectively. The trend that the thermal stability of peroxides decreases down group 2 and that peroxides are less thermally stable than the oxides is confirmed.

S12.3 Use ionic radii to predict a structure type of BeSe?
The ionic radius of Be is 41 pm (coordination number = 4) and the ionic radius of Se is 194 pm. The radius ratio is 41/194 = 0.21. Which according to Table 3.6 should be close to ZnS-like structure. Indeed BeSe crystallizes with hexagonally close packed ZnS structure.

S12.4 Calculate the lattice enthalpy of MgF_2 and comment on how it will affect the solubility compared to $MgCl_2$?
Using the Kapustinskii equation (equation 3.4) and the values of $K = 1.21 \times 10^5$ MJ Å mol^{-1}, $n = 3$, $z^+_{Mg} = +2$, $z^-_{F_2} = -1$, the sum of the thermochemical radii, d_0, 181 pm + 65 pm = 246 pm = 24.6 Å and d = 3.45 Å, the calculated lattice enthalpy of MgF_2 is 2991 kJ mol^{-1} will offset hydration enthalpy and reduce solubility compared to $MgCl_2$.

12.1 Why are compounds of beryllium covalent whereas those of the other group 2 elements are predominantly ionic?
Be has large polarizing power and a high charge density due to its small radius. Descending group 2 elements increase in size, which leads to more electropositive and ionic character. This leads to predominantly ionic compounds with larger group 2 ions.

12.2 Why are the properties of beryllium more similar to aluminium and Zinc than to magnesium?
Because of a diagonal relationship between Be and Al arising from their similar atomic radii.

12.3 Identify the compounds A, B, C, and D of the group 2 element M?

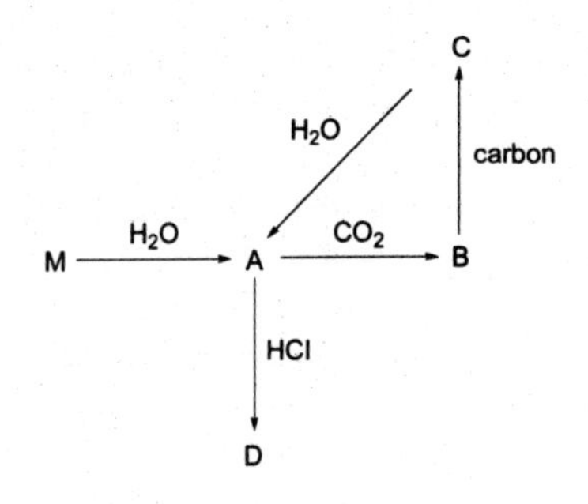

$M + H_2O \rightarrow M(OH)_2$; $A = M(OH)_2$

$M(OH)_2 + CO_2 \rightarrow MCO_3$; $B = MCO_3$

$2MCO_3 + 5C \rightarrow 2MC_2 + 3CO_2$; $C = MC_2$

$MC_2 + 2H_2O \rightarrow M(OH)_2 + C_2H_2$

$M(OH)_2 + 2HCl \rightarrow MCl_2 + 2H_2O$; $D = MCl_2$.

12.4 Why does beryllium fluoride form a glass when cooled from a melt?
Unlike the other alkaline earth metal fluorides MF_2 [M = Mg, Ca, Sr, Ba] which crystallise with CaF_2-like structure, the BeF_2 adopts SiO_2 like arrangement. As with SiO_2 which forms vitreous matter when melted, BeF_2 also forms a glass [a disordered arrangement of molecular fragments] when cooled.

12.5 Why is magnesium hydroxide a much more effective antacid than calcium or barium hydroxide? $Mg(OH)_2$ is sparingly soluble and mildly basic; $Ca(OH)_2$ is more soluble and so moderately basic; $Ba(OH)_2$ is soluble and is so strongly basic, that it also is a poison.

12.6 Explain why Group 1 hydroxides are much more corrosive to metals than Group 2 hydroxides?
Group 1 hydroxides are more soluble than group 2 hydroxides, and therefore have higher OH^- concentrations. The increase in OH^- concentration increases the corrosiveness of group 1 hydroxides.

12.7 Which of the salts $MgSeO_4$ or $BaSeO_4$ would be expected to be more soluble in water?
In general, a larger difference in sizes of the anions and cations of a given salt favors solubility in wter. Assuming that the thermo-chemical radius of selenate ion is similar for both the salts, one would expect a greater difference in the sizes of Mg cation and selenate anion. Accordingly, the MgSeO4 is expected to be more soluble in water.

12.8 Which Group 2 salts are used as drying agents and why?
Anhydrous Mg, and Ca sulphates are preferred as drying agents over the other alkaline earth sulphates. This is because of the higher affinity of Mg and Ca sulphates for water (because of their higher solubility in water) compared to the other alkaline earth sulphates, which are nearly insoluble in water. It may be interesting to note here that BeSO4, although expected to be more hydrophilic, it is not considered as a desiccant because of its higher cost and toxicity.

12.9 How do Group 2 salts give rise to scaling from hard water?
Salts of divalent ions have low solubility. Temporary hardness of water occurs because of the formation of insoluble $CaCO_3$ from $CaHCO_3$ on heating, while permanent hardness is attributed to the presence of $CaSO_4$.

12.10 Predict structures for BeTe and BaTe.

The ionic radius of Be is 41 pm (coordination number = 4) and the ionic radius of Te is 207pm. The radius ratio is 41/207 = 0.19. Which according to Table 3.6 should be close to ZnS-like structure.

The ionic radius of Ba is 152 pm (coordination number = 8) and the ionic radius of Te is 207 pm. The radius ratio is 156/207 = 0.734. Which according to Table 3.6 should be close to CsCl-like structure.

12.11 Use the data in Table 1.7 and the Ketelaar triangle in Fig. 2.38 to predict the nature of the bonding in $BeBr_2$, $MgBr_2$, and $BaBr_2$.

The Pauling electronegativity values of Be, Mg, Ba and Br are 1.57, 1.31, 0.89 and 2.96 respectively.
The average electronegativity of $BeBr_2$ is therefore 2.27 and the difference is 1.39. The values on the Ketelaar triangle indicate that $BeBr_2$ should be covalent.

The average electronegativity of $MgBr_2$ is 2.13 and the difference is 1.65. The values on the Ketelaar triangle indicate that $MgBr_2$ should be ionic.

The average electronegativity of $BaBr_2$ is therefore 1.93 and the difference is 2.07. The values on the Ketelaar triangle indicate that $BaBr_2$ should be ionic

12.12 The two Grignard compounds C_2H_5MgBr and 2,4,6-$(CH_3)_3C_6H_2MgBr$ dissolve in THF. What differences would be expected in the structures of the species formed in these solutions?

In solution, C_2H_5MgBr will be tetrahedral with two molecules of solvent coordinated to the magnesium. The bulky organic group in 2,4,6-$(CH_3)_3C_6$ H_2MgBr leads to a coordination number of two.

12.13 Predict the products of the following reactions:

(a) $MgCl_2 + 2LiC_2H_5 \rightarrow 2LiCl + Mg(C_2H_5)_2$

(b) $Mg + (C_2H_5)_2Hg \rightarrow Mg(C_2H_5)_2 + Hg$

(c) $Mg + C_2H_5HgCl \rightarrow C_2H_5MgCl + Hg$

Chapter 13 The Group 13 Elements

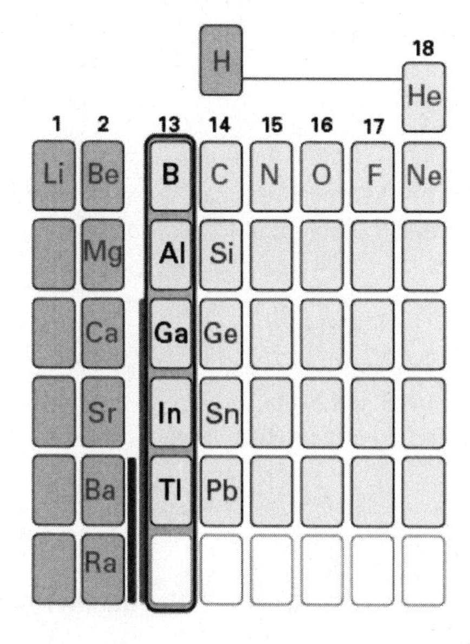

The elements of Groups 13, boron, aluminium, gallium, indium, and thallium, have interesting and diverse physical and chemical properties. The first member of the group, boron, is essentially non-metallic, whereas the properties of the heavier members of the group are distinctly metallic. Aluminium is the most important element commercially and is produced on a massive scale for a wide range of applications. Boron forms a large number of cluster compounds involving hydrogen, metals, and carbon.

S13.1 ^{11}B nuclei have I = 3/2. Predict the number of lines and their relative intensities in the ^1H-NMR spectrum of BH_4^-?

VSEPR theory predicts that a species with four bonding pairs of electrons should be tetrahedral.

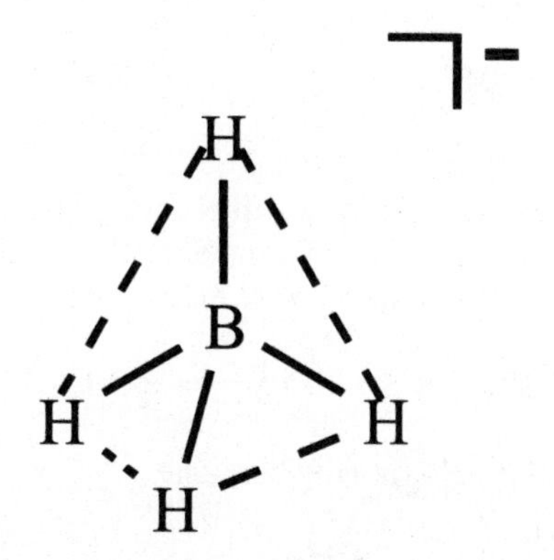

All protons are in equal environment and it gives a quartet by coupling with ^{11}B (I = 3/2). Therefore the number of lines in the ^1H-NMR spectrum of BH_4^- is 4. Their relative intensity ratio is 1:3:3:1.

S13.2 Write an equation for the reaction of $LiBH_4$ with propene in ether solvent and a 1:1 stoichiometry and another equation for its reaction with ammonium chloride in THF with the same stoichiometry?

In General, it is important to note that simple alkenes are inert towards $LiBH_4$. However, by mixing a catalyst such as BH_3, I_2, AcOH, or selected transition metal sals, one can achieve hydroboration of the alkenes with $LiBH_4$. In such cases, the product would either be propyl alcohol or isopropyl alcohol depending on the catalyst used.

The equation for the reaction of LiBH4 with ammonium chloride in THF with the 1:1 stoichiometry:

$$LiBH_4 + NH_4Cl \xrightarrow{THF} BH_3NH_3 + LiCl + H_2$$

S13.3 Write and justify balanced equations for plausible reactions between (a) BCl_3 and ethanol, (b) BCl_3 and pyridine in hydrocarbon solution, (c) BBr_3 and $F_3BN(CH_3)_3$?

(a) BCl_3 and ethanol? As mentioned in the example, boron trichloride is vigorously hydrolyzed by water. Therefore, a good assumption is that it will also react with protic solvents such as alcohols, forming HCl and B–O bonds:

$$BCl_3(g) + 3\ EtOH(l) \rightarrow B(OEt)_3(l) + 3\ HCl(g)$$

(b) BCl_3 and pyridine in hydrocarbon solution? Neither pyridine nor hydrocarbons can cause the protolysis of the B–Cl bonds of boron trichloride, so the only reaction that will occur is a complex formation reaction, such as the one shown below:

$$BCl_3(g) + py(l) \rightarrow Cl_3B - py(s)$$

Note that only a 1:1 complex is formed, even if excess pyridine (py) is used. Boron and the other period 2 atoms cannot become hypervalent, in contrast with the heavier atoms of periods 3 and beyond.

(c) BBr_3 and $F_3BN(CH_3)_3$? Since boron tribromide is a stronger Lewis acid than boron trifluoride, it will displace BF_3 from its complex with $N(CH_3)_3$:

$$BBr_3(l) + F_3BN(CH_3)_3(s) \rightarrow BF_3(g) + Br_3BN(CH_3)_3(s)$$

S13.4 Suggest a reaction or series of reactions for the preparation of N, N', N''-trimethyl-B,B',B''-trimethylborazine starting with methylamine and boron trichloride?

This compound is permethylborazine. The reaction of ammonium chloride with boron trichloride yields *B*-trichlorobocazine, while the reaction of a primary ammonium chloride, with boron trichloride yields *N*-alkyl substituted *B*-trichloroborazine, as shown below:

$$3CH_3NH_3Cl + 3BCl_3 \xrightarrow{\text{Heat}} 9HCl + Cl_3BNCH_3$$

Therefore, if you use methylammonium chloride you will produce $Cl_3BN(CH_3)_3$ (i.e. R = CH_3). This product can be converted to the desired one by treating it with an organometallic methyl compound of a metal that is more electropositive than boron. Either methyllithium or methymagnesium bromide could be used, as shown below:

$$Cl_3BNCH_3 + 3CH_3MgBr \longrightarrow (CH_3)_3B_3N_3(CH_3)_3 + 3Mg(Br, Cl)_2$$

The structure of *N,N',N''*-trimethyl-*B,B',B''*-trimethylborazine.

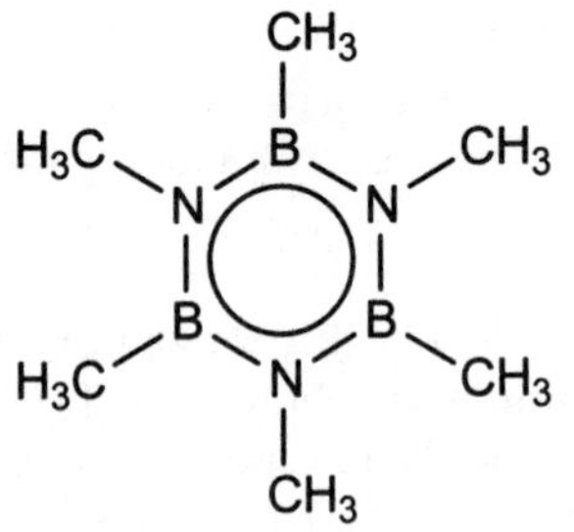

S13.5 How many framework electron pairs are present in B_4H_{10} and to what structural category does it belong? Sketch its structure?

The formula B_4H_{10} belongs to a class of borohydrides having the formula B_nH_{n+6}, which is characteristic of a *arachno* species. Assume one B–H bond per B atom, there are 4 BH units, which contribute 4x2 = 8 electrons, and the six additional H atoms, which contribute a further 6 electrons, giving 14 electrons, or seven electron pairs which is n+3 with n = 4. This is characteristic of *arachno* clusters. The resulting seven pairs are distributed: two are used for the additional terminal B–H bonds, four are used for the four BHB bridges, and one is used for the central B–B bond.

The structure of B_4H_{10}:

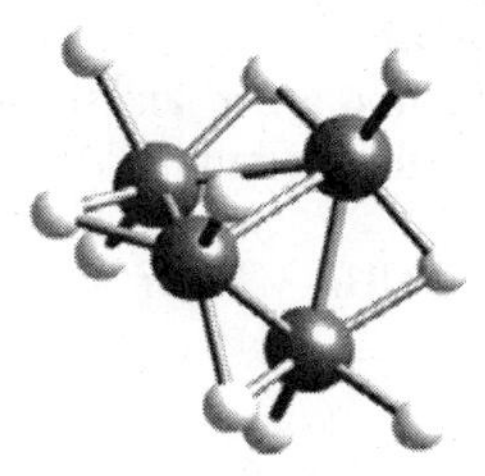

S13.6 Propose a plausible product for the reaction between $Li[B_{10}H_{13}]$ and $Al_2(CH_3)_6$?
By analogy with the reaction of $[B_{11}H_{13}]^{2-}$ with $Al_2(CH_3)_6$, the plausible product would be $[B_{10}H_{11}(AlCH_3)]^-$, which would be formed as follows:

$$2[B_{10}H_{13}]^- + Al_2(CH_3)_6 \rightarrow 2[B_{10}H_{11}(AlCH_3)]^- + 4CH_4$$

S13.7 Propose a synthesis for the polymer precursor 1,7-$B_{10}C_2H_{10}(Si(CH_3)_2Cl)_2$ from 1,2-$B_{10}C_2H_{12}$ and other reagents of your choice?
As in the example, you should consider attaching the $Si(CH_3)_2Cl$ substituents to the carbon atoms of this carborane by using the dilithium derivative 1,7-$B_{10}H_{10}C_2Li_2$. You can first prepare 1,2-$B_{10}C_2H_{12}$ from decaborane as in the example. Then, this compound is thermally converted to a mixture of the 1,7- and 1,12-isomers, which can be separated by chromatography:

$$1{,}2\text{-}B_{10}C_2H_{12} \xrightarrow{\Delta} 1{,}7\text{-}B_{10}C_2H_{12}\ (90\%) + 1{,}12\text{-}B_{10}C_2H_{12}\ (10\%)$$

The pure 1,7-isomer is lithiated with RLi and then treated with $Si(CH_3)_2Cl_2$:

$$1{,}7\text{-}B_{10}C_2H_{10}Li_2 + 2Si(CH_3)_2Cl_2 \rightarrow 1{,}7\text{-}B_{10}C_2H_{10}(Si(CH_3)_2 + 2LiCl$$

S13.8 Propose, with reasons, the chemical equation (or indicate no reaction) for reactions between (a) $(CH_3)_2SAlCl_3$ and $GaBr_3$? $GaBr_3$ is the stronger Lewis acid, so the reaction is $(Me)_2SalCl_3 + GaBr_3 \rightarrow Me_2SGaBr_3 + AlCl_3$.
(b) TlCl and formaldehyde (HCHO) in acidic aqueous solution? Thallium trihalides are very unstable and are easily reduced, as shown below:

$$2\,TlCl_3 + H_2CO + H_2O \rightarrow 2TlCl + CO_2 + 4\,H^+ + 4\,Cl^-$$

13.1 Give a balanced chemical equation and conditions for the recovery of boron?
Boron is recovered from the mineral borax, $Na_2B_4O_5(OH)_4{\cdot}8H_2O$, by formation of B_2O_3 followed by treatment with magnesium:

$$B_2O_3 + 3Mg \rightarrow 2B + 3MgO \quad \Delta H < 0$$

13.2 Describe the bonding in (a) BF_3? Covalent with a strong π component of the B–F bond **(b) $AlCl_3$?** In the solid state, $AlCl_3$ has a layered structure. At melting point, $AlCl_3$ converts to dimers with *2c, 2e* bridging bonds. **(c) B_2H_6?** Structure 1 shows that B_2H_6 is an electron-deficient dimer with *3c, 2e* bridging bonds.

13.3 Arrange the following in order of increasing Lewis acidity: BF_3, BCl_3, $AlCl_3$. In the light of this order, write balanced chemical reactions (or no reaction) for (a) $BF_3N(CH_3)_3 + BCl_3 \rightarrow$, (b) $BH_3CO + BBr_3 \rightarrow$?
For a given halogen, the order of acidity for Group 13 halides toward hard Lewis bases like dimethylether or trimethylamine is $BX_3 > AlX_3 > GaX_3$, while the order toward soft Lewis bases such as dimethylsulfide or trimethylphosphine is $BX_3 < AlX_3 < GaX_3$. For boron halides, the order of acidity is $BF_3 < BCl_3 < BBr_3$, exactly opposite to the order expected from electronegativity trends. This fact establishes the order of Lewis acidity is $BCl_3 > BF_3 > AlCl_3$ toward hard Lewis bases
(a) $BF_3N(CH_3)_3 + BCl_3 \longrightarrow BCl_3N(CH_3)_3 + BF_3$ ($BCl_3 > BF_3$)
(b) $BH_3CO + BBr_3 \longrightarrow$ NR (BH_3 is softer than BBr_3, and CO is a soft base)

13.4 Thallium tribromide (1.11 g) reacts quantitatively with 0.257 g of NaBr to form a product A. Deduce the formula of A. Identify the cation and anion?

For the product in the equation below, the cation is Na^+ and the anion is $TlBr_4^-$. Equal moles (2.5 mmol) of NaBr and $TlBr_3$ react to make $Na[TlBr_4]$, which is the formula of A.

$$TlBr_3 + NaBr \rightarrow NaTlBr_4$$

13.5 Identify compounds A, B, and C?

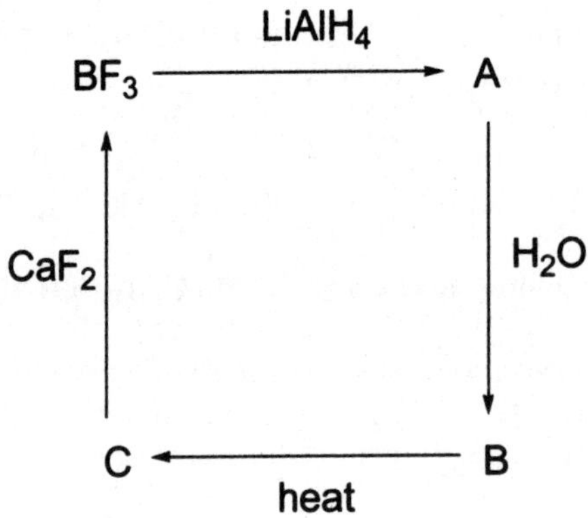

(a) $4\ BF_3 + 3\ LiAlH_4 \rightarrow 2\ B_2H_6 + 3\ LiBF_4$; $A = B_2H_6$
(b) $B_2H_6 + 6\ H_2O \rightarrow 2\ B(OH)_3, + 6\ H_2$; $B = B(OH)_3$
(c) $2\ B(OH)_3$ + heat $\rightarrow B_2O_3 + 3\ H_2O$; $C = B_2O_3$
$B_2O_3 + 3\ CaF_2 \rightarrow 2\ BF_3 + 3\ CaO$

13.6 Does B_2H_6 survive in air? If not, write the equation for the reaction?

No, B_2H_6 does not survive in air; instead it reacts spontaneously with the oxygen in air and forms the hydrated solid oxide, as shown below:

$$B_2H_6 + 3\ O_2 \rightarrow B_2O_3 + 3\ H_2O_2$$

13.7 Predict how many different boron environments would be present in the proton-decoupled ^{11}B-NMR of a) B_5H_{11}, b) B_4H_{10}?

The structure of B_5H_{11}:

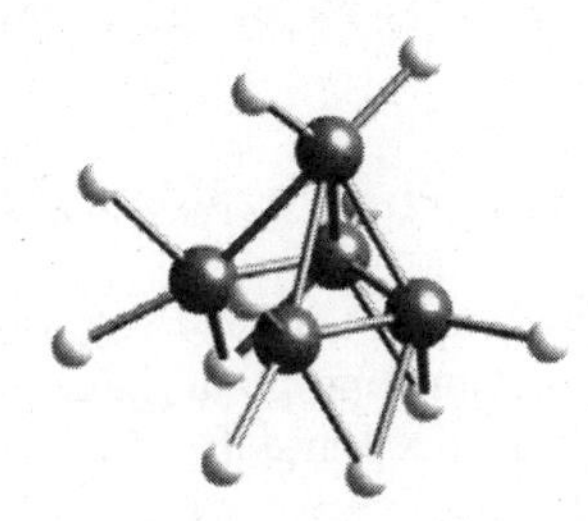

By the observation of the structure of B_5H_{11} we can understand that there are three different boron environments present in the proton-decoupled ^{11}B-NMR..

The structure of B_4H_{10}:

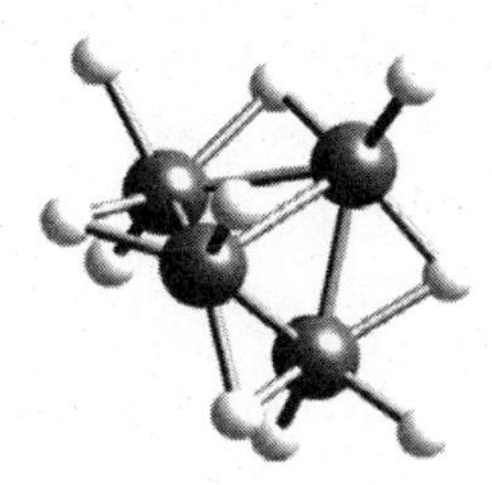

In this molecule two different boron environments are present in the proton-decoupled ^{11}B-NMR B_4H_{10}.

13.8 Predict the products from the hydroboration of (a) $(CH_3)_2C{=}CH_2$, (b) $CH{\equiv}CH$?

(a) $BH_3 + (CH_3)_2C{=}CH_2 \longrightarrow B[CH_2\text{-}CH(CH_3)_2]_3$

(b) $BH_3 + CH{\equiv}CH \longrightarrow B(CH{=}CH_2)_3$

13.9 Diborane has been used as a rocket propellant. Calculate the energy released from 1.00 kg of diborane given the following values of $\Delta_fH°$/kJ mol^{-1}: B_2H_6 = 31, H_2O = -242, B_2O_3 = -1264. The combustion reaction is B_2H_6 (g) + 3 O_2(g) → 3 H_2O (g) + B_2O_3 (s). What would be the problem with diborane as a fuel?

$$B_2H_6\,(g) + 3O_2\,(g) \xrightarrow{\text{Heat}} B_2O_3\,(s) + 3H_2O\,(g)$$

In the first step, we will calculate the ΔH give above reaction by employing equation

$$\Delta H = \Sigma\, \Delta H_{\text{formation of products}} - \Sigma\, \Delta H_{\text{formation of reactants}}$$

Since the heats of formation of B_2H_6, H_2O and B_2O_3 are given and the heat of formation of O_2 is zero as per the definition.

$$\Delta H = \Sigma[\,3(-242\text{ kJ/mol}) + (-1264\text{ kJ/mol})] - [\,31\text{ kJ/mol}]$$
$$= -2021\text{ kJ}$$

To calculate the heat released, when 1 kg of B_2H_6 is used, we follow the following procedure

$$1\text{ kg} \times 10^3\text{g/1 kg} \times 1\text{ ml } B_2H_6/27.62\text{ g } B_2H_6 \times -2021\text{ kJ/ 1 ml } B_2H_6 = -73{,}172\text{ kJ}.$$

Diborane is extremely toxic. It is an extremely reactive gas and hence should be handled in a special apparatus. Moreover, a serious drawback to using diborane is that the boron containing product of combustion is a solid, B_2O_3. If an internal combustion engine is used, the solid will eventually coat the internal surfaces, increasing friction, and will clog the exhaust valves.

13.10 Using BCl_3 as a starting material and other reagents of your choice, devise a synthesis for the Lewis acid chelating agent, $F_2B–C_2H_4–BF_2$?

By analogy with the addition of B_2Cl_4 to ethene (ethylene), you can prepare this compound by adding B_2F_4 to ethene. Starting with BCl_3, you prepare B_2Cl_4 and then convert it to B_2F_4 with a double replacement reagent such as AgF or HgF_2, as follows:

$$2BCl_3 + 2Hg \xrightarrow{\text{Electron Impact}} B_2Cl_4 + 2HgCl_2$$

$$B_2Cl_4 + 4AgF \rightarrow B_2F_4 + 4AgCl$$

$$B_2F_4 + C_2H_4 \rightarrow F_2CH_2CH_2BF_2$$

13.11 Given $NaBH_4$, a hydrocarbon of your choice, and appropriate ancillary reagents and solvents, give formulas and conditions for the synthesis of (a) $B(C_2H_5)_3$, (b) Et_3NBH_3?

$$\text{(a) } BCl_3 + 3C_2H_5MgCl \xrightarrow{\text{Heat, Ether}} B(C_2H_5)_3 + 3MgCl_2$$

(b) We should expect NaCl, with its high lattice enthalpy, to be a likely product. If this is the case we shall be left with BH_4^- and $[HN(C_2H_5)_3]^+$. The interaction of the hydridic BH_4^- ion with the protic $[HN(C_2H_3)_5]^+$ ion will evolve hydrogen to produce triethylamine and BH_3. In the absence of other Lewis bases, the BH_3 molecule would coordinate to THF; however, the stronger Lewis base triethylamine is produced in the initial reactions, so the overall reaction will be

$$[HN(C_2H_5)_3]Cl + NaBH_4 \longrightarrow H_2 + H_3BN(C_2H_5)_3 + NaCl$$

13.12 Draw the B_{12} unit that is a common motif of boron structures; take a viewpoint along a C_2 axis?

C_2-Axis viewpoint of B_{12} is

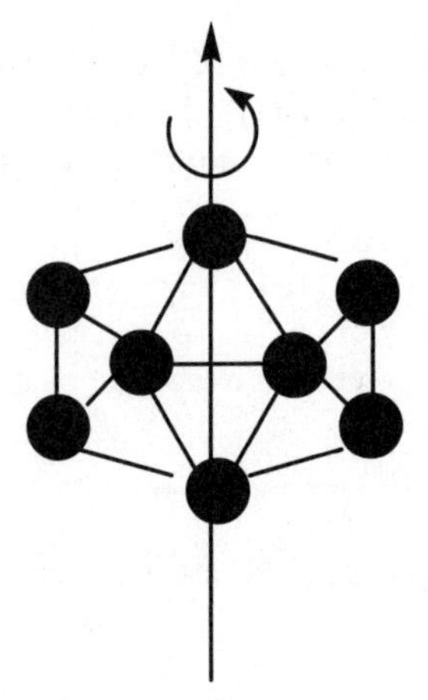

13.13 Which boron hydride would you expect to be more thermally stable, B_6H_{10} or B_6H_{12}? Give a generalization by which the thermal stability of a borane can be judged?

The compounds B_6H_{10} and B_6H_{12} belong to two different classes of boranes, one that has less number of H atoms, B_nH_{n+4} and the other with more hydrogens B_nH_{n+6}. In this case, n happens to be six. In general, boron hydrides with more hydrogen atoms are unstable thermally. Hence we expect B_6H_{10} to be stable relative to B_6H_{12}.

13.14 How many skeletal electrons are present in B_5H_9?

The formula B_5H_9 belongs to a class of borohydrides having the formula B_nH_{n+4}, which is characteristic of a *nido* species. Assume one B–H bond per B atom, there are 5 BH units, which contribute 5x2 = 10 electrons, and the four additional H atoms, which contribute a further 4 electrons, giving 14 electrons, or seven electron pairs which is n+2 with n = 5. This is characteristic of *nido* clusters. Total 14 skeletal electrons are present in B_5H_9.

13.15 (a) Give a balanced chemical equation (including the state of each reactant and product) for the air oxidation of pentaborane(9). (b) Describe the probable disadvantages, other than cost, for the use of pentaborane as a fuel for an internal combustion engine?

The balanced chemical equation for the air oxidation of petaborane (9) is

$$\text{(a) } 2B_5H_9\,(l) + 12O_2\,(g) \xrightarrow{\text{Heat}} 5B_2O_3\,(s) + 9H_2O\,(l)$$

(b) A serious drawback to using pentaborane is that the boron containing product of combustion is a solid, B_2O_3. If an internal combustion engine is used, the solid will eventually coat the internal surfaces, increasing friction, and will clog the exhaust valves. Another disadvantage for the use of pentaborane as a fuel is the incomplete combustion to B_2O_3. Because of this the exhaust nozzles of the rocket became partly blocked with an involatile BO polymer.

13.16 (a) From its formula, classify $B_{10}H_{14}$ as closo, nido, or arachno. (b) Use Wade's rules to determine the number of framework electron pairs for decaborane(14). (c) Verify by detailed accounting of valence electrons that the number of cluster valence electrons of $B_{10}H_{14}$ is the same as that determined in (b)?

(a) The formula $B_{10}H_{14}$ belongs to a class of borohydrides having the formula B_nH_{n+4}, which is characteristic of a *nido* species.

(b) Assume one B–H bond per B atom, there are 10 BH units, which contribute 10x2 = 20 electrons, and the four additional H atoms, which contribute a further 4 electrons, giving 24 electrons, or 12 electron pairs which is n+2 with n = 10. This is characteristic of *nido* clusters.

(c) The total number of valence elections for $B_{10}H_{14}$ is (10x3)+(14x1)=44. Since there are 10 2c2e B-H bonds, which account or 20 of the valence elections, the number of cluster valence is the remainder of 44-20=24.

13.17 Starting with $B_{10}H_{14}$ and other reagents of your choice, give the equations for the synthesis of $[Fe(nido\text{-}B_9C_2H_{11})_2]^{2-}$, and sketch the structure of this species?

(1) $B_{10}H_{14} + 2SEt_2 \longrightarrow B_{10}H_{12}(SEt_2)_2 + H_2$

(2) $B_{10}H_{12}(SEt_2)_2 + C_2H_2 \longrightarrow B_{10}C_2H_{12} + 2SEt_2 + H_2$

(3) $2\,B_{10}C_2H_{12} + 2EtO^- + 4EtOH \longrightarrow 2\,B_9C_2H_{12}^- + 2B(OEt)_3 + 2H_2$

(4) $Na[B_9C_2H_{12}] + NaH \longrightarrow Na_2[B_9C_2H_{11}] + H_2$

(5) $2Na_2[B_9C_2H_{11}] + FeCl_2 \xrightarrow{THF} 2NaCl + Na_2[Fe(B_9C_2H_{11})_2]$

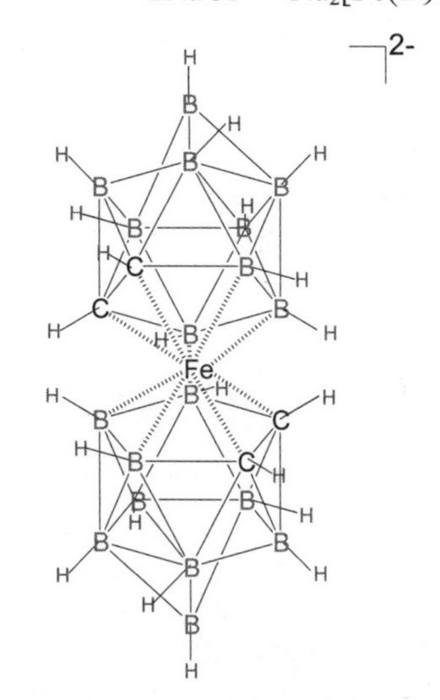

13.18 (a) What are the similarities and differences in structure of layered BN and graphite (Section 13.9)? (b) Contrast their reactivity with Na and Br_2. (c) Suggest a rationalization for the differences in structure and reactivity.

(a) Their structures? Both of these substances have layered structures. The planar sheets in boron nitride and in graphite consist of edge-shared hexagons such that each B or N atom in BN has three nearest neighbors that are the other type of atom and each C atom in graphite has three nearest neighbor C atoms. The structure of graphite is shown in Figure 14.2. The B–N and C–C distances within the sheets, 1.45 Å and 1.42 Å, respectively, are much shorter than the perpendicular interplanar spacing, 3.33 Å and 3.35 Å, respectively. In BN, the B_3N_3 hexagonal rings are stacked directly over one another so that B and N atoms from alternating planes are 3.33 Å apart, while in graphite the C_6 hexagons are staggered (see section 13.9) so that C atoms from alternating planes are either 3.35 Å or 3.64 Å apart (you should determine this yourself using trigonometry).

(b) Their reactivity with Na and Br_2? Graphite reacts with alkali metals and with halogens. In contrast, boron nitride is quite unreactive.

(c) Explain the differences? The large HOMO–LUMO gap in BN, which causes it to be an insulator, suggests an explanation for the lack of reactivity: since the HOMO of BN is a relatively low energy orbital, it is more difficult to remove an electron from it than from the HOMO of graphite, and since the LUMO of BN is a relatively high energy orbital, it is more difficult to add an electron to it than to the LUMO of graphite.

13.19 Devise a synthesis for the borazines (a) $Ph_3N_3B_3Cl_3$ and (b) $Me_3N_3B_3H_3$, starting with BCl_3 and other reagents of your choice. Draw the structures of the products?

(a) $Ph_3N_3B_3Cl_3$? The reaction of a primary ammonium salt with boron trichloride yields *N*-substituted *B*-trichloroborazines:

$$3\ PhNH_3^+Cl^- + 3\ BCl_3 \rightarrow Ph_3N_3B_3Cl_3 + 9\ HCl$$

(b) $Me_3N_3B_3H_3$? You first prepare $Me_3N_3B_3Cl_3$ using $MeNH_3^+Cl^-$ and the method described above, and then perform a Cl^-/H^- metathesis reaction using LiH as the hydride source:

$$3\ MeNH_3^+Cl^- + 3\ BCl_3 \rightarrow Me_3N_3B_3Cl_3 + 9\ HCl$$
$$Me_3N_3B_3Cl_3 + 3\ LiH \rightarrow Me_3N_3B_3H_3 + 3\ LiCl$$

The structures of $Ph_3N_3B_3Cl_3$ and $Me_3N_3B_3H_3$ are shown below:

$Ph_3N_3B_3Cl_3$ $\qquad$ $Me_3N_3B_3H_3$

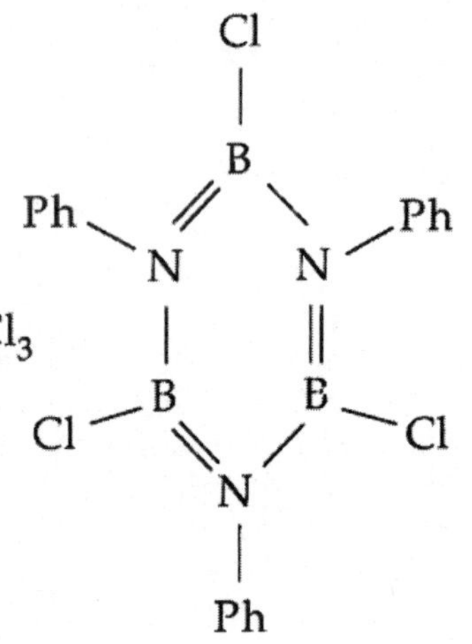

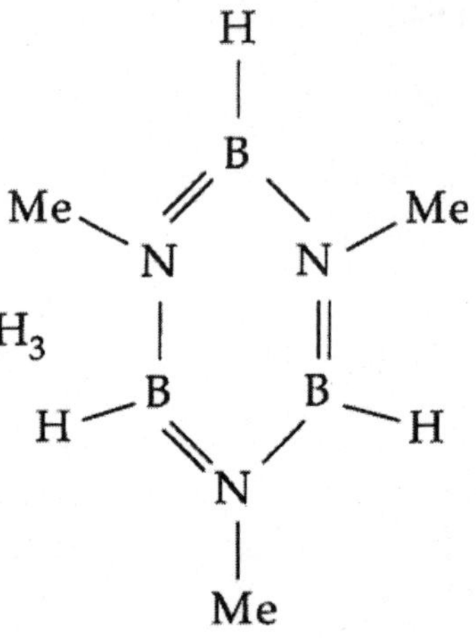

13.20 Give the structural type and describe the structures of B_4H_{10}, B_5H_9, and 1,2-$B_{10}C_2H_{12}$?

Simple polyhedral boranes come in three basic types, $B_nH_n^{2-}$ *closo* structures, B_nH_{n+4} *nido* structures, and B_nH_{n+6} *arachno* structures. The first compound given in this exercise, B_4H_{10}, is an example of a B_nH_{n+6} compound with n = 4, so it is an *arachno* borane. Its name is tetraborane(10) and its structure is shown in Structure 3: two B–H units are joined by a (2c,2e) B–B bond; this B_2H_2 unit is flanked by four hydride bridges to two BH_2 units (the B–H–B bridge bonds are (3c,2e) bonds). The second compound, B_5H_9, is an example of a B_nH_{n+4} compound with $n = 5$, so it is a *nido* borane. Its name is pentaborane(9) and its structure is also shown in Structure 4: four B–H units are joined by four hydride bridges; the resulting B_4H_8 unit, in which the four boron atoms are coplanar, is capped by an apical B–H unit that is bonded to all four of the coplanar boron atoms. The third compound is an example of a carborane in which two C–H units substitute for two $B–H^-$ units in $B_{12}H_{12}^{2-}$, resulting in 1,2-$B_{10}C_2H_{12}$ that retains the *closo* structure of the parent $B_{12}H_{12}^{2-}$ ion. Its name is 1,2-*closo*-dodecaborane(12): a B_5H_5 pentagonal plane is joined to a B_4CH_5 pentagonal plane that is offset from the first plane by 36°; the first plane is capped by a B–H unit, while the second is capped by a C–H unit.

13.21 Arrange the following boron hydrides in order of increasing Brønsted acidity, and draw a structure for the probable structure of the deprotonated form of one of them: B_2H_6, $B_{10}H_{14}$, B_5H_9?

In a series of boranes, the acidity increases as the size of the borane increases. This is because the negative charge, formed upon deprotonation, can be better delocalized over a large anion with many boron atoms than over a small one. Therefore, the Brønsted acidity increases in the order $B_2H_6 < B_5H_9 < B_{10}H_{14}$. See Secton 13.3 for more detail on boron clusters.

Chapter 14 The Group 14 Elements

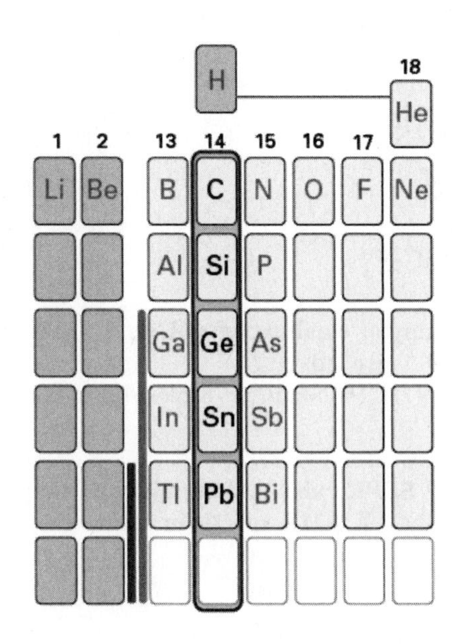

The elements of Group 14, carbon, silicon, germanium, tin, and lead show considerable diversity in their chemical and physical properties. Carbon, of course, is the building block of life and central to organic chemistry. All the elements form binary compounds with other elements. In addition, silicon forms a diverse range of network solids. Many of the organocompounds of the Group 14 elements are commercially important. Silicon is widely distributed in the natural environment and tin and lead find widespread applications in industry and manufacturing.

S14.1 Describe how the electronic structure of graphite is altered when it reacts with (a) potassium, (b) bromine?

(a) With potassium? The extended π system for each of the planes of graphite results in a band of π orbitals. The band is half filled in pure graphite, that is, all of the bonding MOs are filled and all of the antibonding MOs are empty. The HOMO–LUMO gap is ~0 eV, giving rise to the observed electrical conductivity of graphite. Chemical reductants, like potassium, can donate their electrons to the LUMOs (graphite π^* orbitals), resulting in a material with a higher conductivity.

(b) With bromine? Chemical oxidants, like bromine, can remove electrons from the π-symmetry HOMOs of graphite. This also results in a material with a higher conductivity.

S14.2 Use the bond enthalpy data in Table 14.2 and above to calculate the standard enthalpy of formation of CH_4 and SiH_4?

The enthalpy of formation is calculated as the difference between the bonds broken and the bonds formed in the reaction.

Four C–H bonds are formed in the formation of methane from graphite and hydrogen gas.

$C(graphite) + 2H_2(g) \rightarrow CH_4(g)$

$\Delta_fH\ (CH_4,g) = [715+2(436)]\text{kJ mol}^{-1} - [4(412)]\ \text{kJ mol}^{-1}$

$\Delta_fH = -61\text{kJ mol}^{-1}$

Four Si–H bonds are formed in the synthesis of silane from silicon with hydrogen gas.

$Si(s) + 2H_2(g) \rightarrow SiH_4(g)$

$\Delta_fH\ (SiH_4,g) = [439+2(436)]\text{kJ mol}^{-1} - [4(318)]\ \text{kJ mol}^{-1}$

$\Delta_fH = +39\ \text{kJ mol}^{-1}$

S14.3 Propose a synthesis of $D^{13}CO_2^-$ starting from ^{13}CO?

As in the example, you would want to oxidize ^{13}CO to $^{13}CO_2$. However, unlike the example, you would then want to treat the $^{13}CO_2$ with a source of deuteride ion, D^-. A good source would be LiD. The entire synthesis would be:

$$^{13}CO(g) + 2MnO_2(s) \rightarrow {}^{13}CO_2(g) + Mn_2O_3(s)$$

$$2Li(s) + D_2 \rightarrow 2LiD(s)$$

$$^{13}CO_2(g) + LiD(et) \rightarrow Li^+D^{13}CO_2^-(et)$$

14.1 Silicon forms the chlorofluorides $SiCl_3F$, $SiCl_2F_2$, and $SiClF_3$. Sketch the structures of these molecules?

$SiCl_3F$, $SiCl_2F_2$, and $SiClF_3$ have tetrahedral structures as shown below. Si has a conventional oxidation state of +4 and therefore forms a neutrally charged molecule with four bonds for $SiCl_3F$, $SiCl_2F_2$, and $SiClF_3$.

$SiCl_3F$ $SiCl_2F_2$ $SiClF_3$

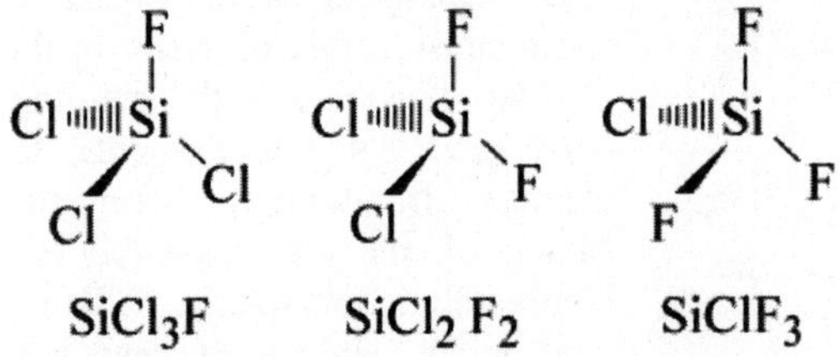

14.2 Explain why CH_4 burns in air whereas CF_4 does not. The enthalpy of combustion of CH_4 is −888 kJ mol^{-1} and the C–H and C–F bond enthalpies are −413 and −489 kJ mol^{-1} respectively?

The bond enthalpy of a C–F bond is higher than the bond enthalpy of a C–H bond; see Table 14.2. Bond enthalpy tends to decrease down a group.

14.3 SiF_4 reacts with $(CH_3)_4NF$ to form $[(CH_3)_4N][SiF_5]$. (a) Use the VSEPR rules to determine the shape of the cation and anion in the product; (b) account for the fact that the ^{19}F NMR spectrum shows two fluorine environments?

The chemical equation for reaction of SiF_4 with $(CH_3)_4NF$:

$$SiF_4 + (CH_3)_4NF \longrightarrow [(CH_3)_4N]^+[SiF_5]^-$$

(a) The cation is $[(CH_3)_4N]^+$

The number of valence electrons present on central N atom = 8

The number of valence electron pairs = 4

Therefore, VSEPR theory predicts that a species with four bonding pairs of electrons should be tetrahedral.

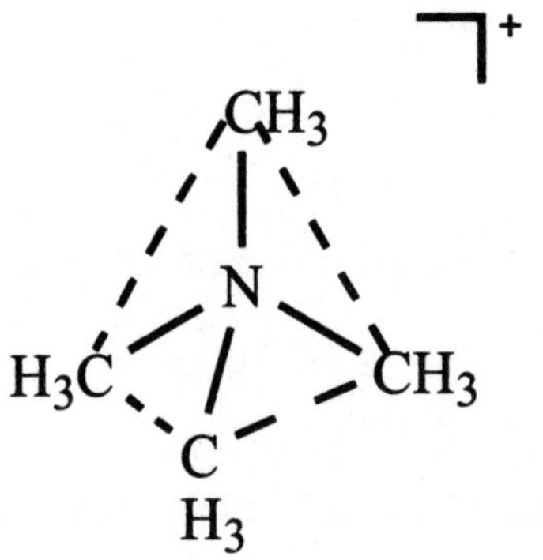

The anion is SiF_5^-

The number of valence electrons present on central Si atom = 4 (valence electrons of central Si atom) + 5 (five electrons from 5 F atom) + 1 (negative charge) = 10

The number of valence electron pairs = 5

Therefore, VSEPR theory predicts that a species with five bonding pairs of electrons should be trigonal bipyramidal.

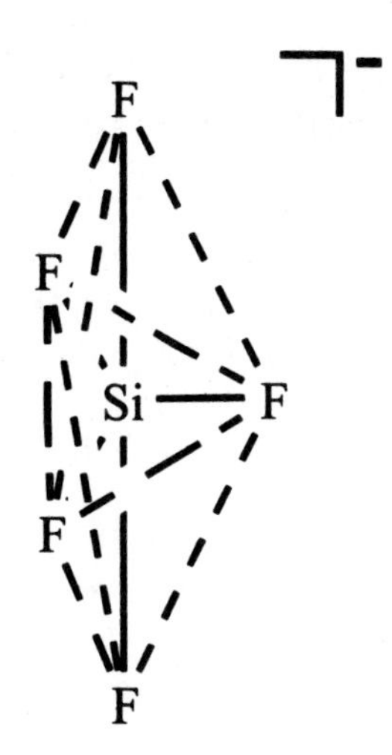

(b) In the structure of anion there are two different fluorine environments (axial and equatorial).

14.4 **Draw the structure and determine the charge on the cyclic anion $[Si_4O_{12}]^{n-}$?**
The structure of cyclic anion $[Si_4O_{12}]^{n-}$:

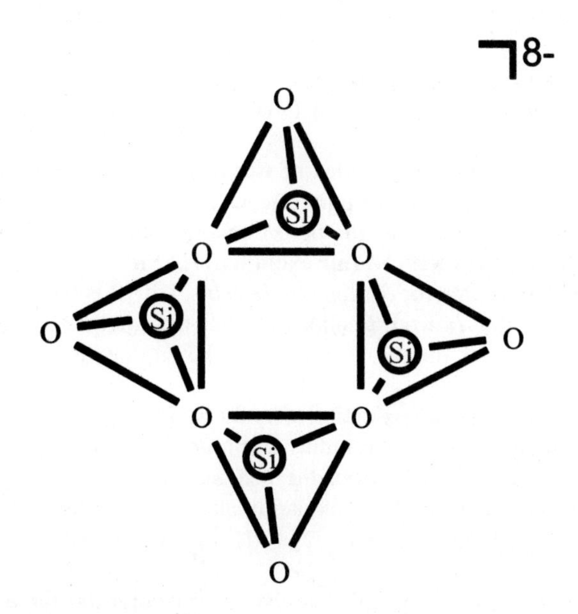

The general formula of silicates is $(SiO_3)_n^{2n-}$

If n = 4 then 2n = 8. Therefore the charge on the cyclic anion $[Si_4O_{12}]^{n-}$ = –8.

14.5 **Predict the appearance of the ^{119}Sn-NMR spectrum of $Sn(CH_3)_4$?**
^{119}Sn-NMR spectrum of $Sn(CH_3)_4$ contains doublet peak for ^{119}Sn (I = ½) nucleus. This doublet is due to coupling of ^{119}Sn nucleus with 12 equivalent protons.

14.6 **Predict the appearance of the ^{1}H-NMR spectrum of $Sn(CH_3)_4$?**
^{1}H-NMR spectrum of $Sn(CH_3)_4$ contains doublet peak for ^{1}H (I = ½) nucleus. This doublet is due to coupling of ^{1}H (I =1/2) with ^{119}Sn (I = 1/2) nucleus.

14.7 **Use the data in Table 14.2 and the additional bond enthalpy data given here to calculate the enthalpy of hydrolysis of CCl_4 and CBr_4. Bond enthalpies/(kJ mol^{-1}): O–H = 463, H–Cl = 431, H–Br = 366?**

The enthalpy of hydrolysis of a compound can be calculated as the difference in energy between the bonds broken and the bonds formed in the hydrolysis reaction. Therefore, the relevant equations for the hydrolysis of CCl_4 and CBr_4.

$$CCl_4\,(l) + 2H_2O\,(l) \longrightarrow CO_2\,(g) + 4HCl\,(aq)$$

$$\begin{aligned}\Delta_h H^\circ_{CCl4} &= [(2 \times 743) + (4 \times 431)] - [(4 \times 322) + 2(2 \times 463)] \\ &= [1526 + 1724] - [1288 + 2 \times 926] \\ &= 3250 - [1288 + 1852] \\ &= 3250 - 3140 \\ &= 110\ \text{kJmol}^{-1}\end{aligned}$$

$$CBr_4\,(l) + 2H_2O\,(l) \longrightarrow CO_2\,(g) + 4HBr\,(aq)$$

$$\begin{aligned}\Delta_h H^\circ_{CBr4} &= [(2 \times 743) + (4 \times 366)] - [(4 \times 288) + 2(2 \times 463)] \\ &= [1526 + 1464] - [1152 + 2 \times 926] \\ &= 3090 - [1152 + 1852] \\ &= 3090 -.\ 3004 \\ &= 86\ \text{kJmol}^{-1}\end{aligned}$$

14.8 **Identify the compounds A to F?** For products **(A)**, **(B)**, **(C)**, **(D)**, **(F)**, refer to Section 14.16. **(A)** $Si + 2\,Cl_2 \rightarrow SiCl_4$ **(B)** $SiCl_4 + RLi \rightarrow SiRCl_3 + LiCl$ **(C)** $SiRCl_3 + 3H_2O \rightarrow RSi(OH)_3 + 3HCl$; if $R = CH_3$, this hydrolysis reaction produces the cross-linked polymer shown in Figure 14.19. **(E)** $2RSi(OH)_3 + \Delta \rightarrow RSiOSiR + H_2O$ **(e)** $SiCl_4 + 4RMgBr \rightarrow SiR_4 + 4MgBrCl$; see Section 14.17 **(F)?** $SiCl_4 + 2H_2O \rightarrow SiO_2 + 4HCl$.

14.9 **(a) Summarize the trends in relative stabilities of the oxidation states of the elements of Group 14, and indicate the elements that display the inert pair effect. (b) With this information in mind, write balanced chemical reactions or NR (for no reaction) for the following combinations, and explain how the answer fits the trends. (i) $Sn^{2+}(aq) + PbO_2(s)$ (excess) → (air excluded) (ii) $Sn^{2+}(aq) + O_2(air) \rightarrow$**

(a) +4 is the most stable oxidation state for the lighter elements, but +2 is the most stable oxidation state of Pb, the heaviest element in group 14. Recall from Section 9.5 that the inert-pair effect is the relative stability of an oxidation state in which the oxidation number is 2 less than the group oxidation number. Pb therefore displays the inert-pair effect.

(b) Table 14.1 shows that the most stable oxidation state for Sn is +4 and the most stable oxidation state for Pb is +2. This is because of an increase in atomic radius and a decrease in ionization energy descending the group. Cations are therefore more readily formed from the heavier elements within the group. Reactions (i) and (ii) illustrate Pb and Sn attaining their most stable oxidation state and are therefore favourable reactions. (i) $Sn^{2+} + PbO_2 + 4\,H^+ \rightarrow Sn^{4+} + Pb^{2+} + 2H_2O$, (ii) $2Sn^{2+} + O_2 + 4H^+ \rightarrow 2Sn^{4+} + 2H_2O$.

14.10 **Use data from Resource section 3 to determine the standard potential for each of the reactions in Exercise 14.5 (b). In each case, comment on the agreement or disagreement with the qualitative assessment you gave for the reactions?**

Reaction (i) is $Sn^{2+} + PbO_2 + 4H^+ \rightarrow Sn^{4+} + Pb^{2+} + 2H_2O$. In this reaction Sn is oxidized from +2 to +4 and Pb is reduced from +4 to +2. The potential for this Sn oxidation is –(0.15) V and is 1.46 V for the Pb reduction. The standard potential is 1.46 V – 0.15 V = +1.31 V. Reaction (ii) is $2Sn^{2+} + O_2 + 4H^+ \rightarrow Sn^{4+} + 2H_2O$. In this reaction Sn is oxidized to +4 from a +2 oxidation state and oxygen is reduced to its conventional –2 oxidation. The standard potential is therefore 1.229 V – 0.15 V= 1.08 V. Reaction (i) is therefore more favorable because it has a higher standard potential. Both reactions agree with the predictions made in Exercise 14.5 because lead and tin are going to their most stable oxidation states.

14.11 **Give balanced chemical equations and conditions for the recovery of silicon and germanium from their ores?**

Silicon is recovered from silica by reduction with carbon:

$$SiO_2(s) + C(s) \rightarrow Si(s) + CO_2(g) \quad \Delta H < 0$$

Germanium is recovered from its oxide by reduction with hydrogen:

$$GeO_2(s) + 2H_2(g) \rightarrow Ge(s) + 2H_2O(g) \qquad \Delta H < 0$$

The recovery of germanium is the most energy efficient, since it is less oxophilic than either silicon or boron. Oxophilicity decreases down a group.

14.12 (a) Describe the trend in band gap energy, *Eg*, for the elements carbon (diamond) to tin (grey). (b) Does the electrical conductivity of silicon increase or decrease when its temperature is changed from 20°C to 40°C?

Elements	Band gap (E_g in eV)
C (diamond)	5.5
Si	1.11
Ge	0.67
Sn	no bond gap

There is a decrease in band gap energy from carbon (diamond) to grey tin. This is because the trend of non-metal to metal character of these elements. Silicon is a semiconductror. This type of material always experiences an increase in conductivity as the temperature is raised. Therefore, the conductivity of Si will be greater at 40°C than at 20°C.

14.13 Preferably without consulting reference material, draw a periodic table and indicate the elements that form saline, metallic, and metalloid carbides?

	Ionic(silane) carbides	Metallic carbides	Metalloid carbides
Group I elements	Li, Na, K, Rb, Cs		
Group II elements	Be, Mg, Ca, Sr, Ba		
Group 13 elements	Al		B
Group 14 elements			Si
3d-Block elements		Sc, Ti, V, Cr, Mn, Fe, Co, Ni	
4d-Block elements		Zr, Nb, Mo, Tc, Ru	
5d-Block elements		La, Hf, Ta, W, Re, Os	
6d-Block elements		Ac	
Lanthanides		Ce, Pr, Nd, Pm, Sm, Eu, Gd, Tb, Dy, Ho, Er, Tm, Yb, Lu	

14.14 Describe the preparation, structure and classification of (a) KC_8, (b) CaC_2, (c) K_3C_{60}?

(a) KC_8? This compound is formed by heating graphite with potassium vapour or by treating graphite with a solution of potassium in liquid ammonia. The potassium atoms are oxidized to K^+ ions; their electrons are added to the LUMO π^* orbitals of graphite. The K^+ ions intercalate between the planes of the reduced graphite, so that there is a layered structure of alternating sp^2 carbon atoms and potassium ions. The structure of KC_8, which is an example of a saline carbide, is shown in Figures 14.13 and 14.4.

(b) CaC_2? There are two ways of preparing calcium carbide, and both require very high temperatures (≥ 2000 °C). The first is the direct reaction of the elements, while the second is the reaction of calcium oxide with carbon:

$$Ca(l) + 2C(s) \rightarrow CaC_2(s)$$
$$CaO(s) + 3C(s) \rightarrow CaC_2(s) + CO(g)$$

The structure is quite different from that of KC_8. Instead of every carbon atom bonded to three other carbon atoms, as in graphite and KC_8, calcium carbide contains discrete C_2^{2-} ions with carbon–carbon triple bonds.

(c) K_3C_{60}? A solution of C_{60}, perhaps in toluene, can be treated with elemental potassium to form this compound. It is ionic, that is, it contains discrete K^+ ions and C_{60}^{3-} ions (see Section 14.6(a), *Carbon clusters*).

14.15 Write balanced chemical equations for the reactions of K_2CO_3 with HCl(aq) and of Na_4SiO_4 with aqueous acid?

Both of these compounds react with acid to produce the oxide, which for carbon is CO_2 and for silicon is SiO_2. The balanced equations are:

$$K_2CO_3(aq) + 2HCl(aq) \rightarrow 2KCl(aq) + CO_2(g) + H_2O(l)$$

$$Na_4SiO_4(aq) + 4HCl(aq) \rightarrow 4NaCl(aq) + SiO_2(s) + 2H_2O(l)$$

The second equation represents one of the ways that silica gel is produced.

14.16 Describe in general terms the nature of the $[SiO_3]_n^{2n-}$ ion in jadeite and the silica-alumina framework in kaolinite?

In contrast with the extended three-dimensional structure of SiO_2, the structures of jadeite and kaolinite consist of extended one- and two-dimensional structures, respectively. The $[SiO_3^{2-}]_n$ ions in jadeite are a linear polymer of SiO_4 tetrahedra, each one sharing a bridging oxygen atom with the tetrahedron before it and the tetrahedron after it in the chain (this is referred to as a chain metasilicate; see Structure 17). Each silicon atom has two bridging oxygen atoms and two terminal oxygen atoms. The two-dimensional aluminosilicate layers in kaolinite represent another way of connecting SiO_4 tetrahedra (see Figure 14.15). Each silicon atom has three oxygen atoms that bridge to other silicon atoms in the plane and one oxygen atom that bridges to an aluminum atom.

14.17 (a) How many bridging O atoms are in the framework of a single sodalite cage? (b) Describe the (supercage) polyhedron at the centre of the Zeolite A structure in Fig. 14.3?

(a) A sodalite cage is based on a truncated octahedron. Each of the heavy lines in the drawing of an octahedron truncated along one of its C_4 axes represents an M–O–M linkage (M = Si or Al). There are eight such lines in the drawing, and since the truncation procedure is carried out six times to produce a sodalite cage, there are $8 \times 6 = 48$ bridging oxygen atoms.

(b) As shown in Figure 14.3, eight sodalite cages are linked together to form the large cage of zeolite A. The polyhedron at the centre has six octagonal faces (the one in the front of the diagram is the most obvious) and eight smaller square faces. If you look at Structure 19, you will see that a truncated octahedron has 8 hexagonal faces and 4 smaller square faces. Thus the fusing together of 8 truncated octahedra produces a central cavity that is different than a truncated octahedron.

Chapter 15 The Group 15 Elements

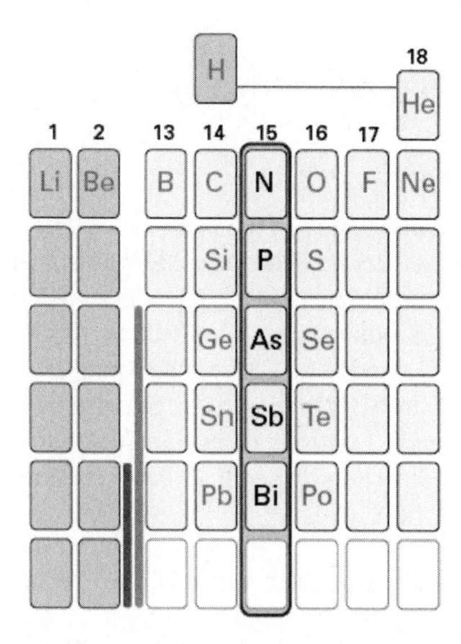

The Group 15 elements—nitrogen, phosphorus, arsenic, antimony, and bismuth—are some of the most important elements for life, geology, and industry. The members of Group 15, the nitrogen group, are sometimes referred to collectively as the **pnictogens**. Group 15 elements exhibit a wide range of oxidation states and form many complex compounds with oxygen. Nitrogen makes up a large proportion of the atmosphere and is widely distributed in the biosphere. Phosphorus is essential for both plant and animal life. In stark contrast, arsenic is a well-known poison.

S15.1 Consider the Lewis structure of a segment of the structure of bismuth shown in Fig. 15.2. Is this puckered structure consistent with the VSEPR model?

Considering the three nearest neighbours only, the structure around each Bi atom is trigonal pyramidal, like NH_3. A side view of the structure of bismuth is shown below. VSEPR theory predicts this geometry for an atom that has a Lewis structure with three bonding pairs and a lone pair.

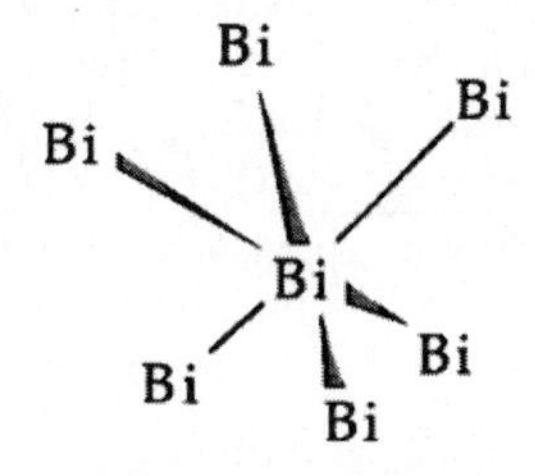

S15.2 Refined hydrocarbons and liquid hydrogen are also used as rocket fuel. What are the advantages of dimethylhydrazine over these fuels?

Dimehylhydrazine ignites spontaneously, it is not a gas and does not need to be liquefied, and it produces less CO_2.

S15.3 From trends in the periodic table, decide whether phosphorus or sulphur is likely to be the stronger oxidizing agent?

The key to estimating the oxidizing strength of a substance is to determine how readily it is reduced. From the table of standard reduction potentials in Resource Section 3 you can obtain the following information:

$$P(s) + 3e^- + 3\,H^+(aq) \rightarrow PH_3(aq) \qquad E^\circ = -0.063\text{ V}$$

$$S(s) + 2e^- + 2\,H^+(aq) \rightarrow H_2S(aq) \qquad E^\circ = 0.144$$

Since $\Delta G = -NFE^\circ$, sulphur is more readily reduced and is therefore the better oxidizing agent. This is in harmony with the higher electronegativity of sulphur ($X = 2.58$) than phosphorus ($X = 2.19$). As you will see in Chapter 17, the most electronegative element, fluorine, is also the most potent elemental oxidizing agent.

S15.4 Summarize the reactions that are used for the synthesis of hydrazine and hydroxylamine. Are these reactions best described as electron-transfer processes or nucleophilic displacements?

The reactions that are employed to synthesize hydrazine are as follows:

$$NH_3 + ClO^- + H^+ \rightarrow [H_3N\text{—}Cl\text{-}O]^- + H^+ \rightarrow H_2NCl + H_2O$$

$$H_2NCl + NH_3 \rightarrow H_2NNH_2 + HCl$$

Both of these can be thought of as redox reactions since the formal oxidation state of the N atoms changes (from –3 to –1 in the first reaction and from –1 and –3 to –2 and –2 in the second reaction). Mechanistically, both reactions appear to involve nucleophilic attack by NH_3 (a Lewis base) on either ClO^- or NH_2Cl (acting as Lewis acids). The reaction employed to synthesize hydroxylamine, which produces the intermediate $N(OH)(SO_3)_2$, probably involves the attack of HSO_3^- (acting as a Lewis base) on NO_2^- (acting as a Lewis acid), although, since the formal oxidation state of the N atom changes from +3 in NO_2^- to +1 in $N(OH)(SO_3)_2$, this can also be seen as a redox reaction. Whether one considers reactions such as these to be redox reactions or nucleophilic substitutions may depend on the context in which the reactions are being discussed. It is easy to see that these reactions do not involve simple electron transfer, such as in $2\,Cu^+ (aq) \rightarrow Cu^0 (s) + Cu^{2+} (aq)$.

S15.5 When titrated against base a sample of polyphosphate gave end points at 30.4 and 45.6 cm³. What is the chain length?

The strongly acidic OH groups are titrated by the first 30.4 cm³ and the two terminal OH groups are titrated by the remaining 45.6 cm³ – 30.4 cm³ = 15.2 cm³. The concentrations of analyte and titrant are such that each OH group requires 15.2 cm³ /2 =7.6 cm³ of NaOH. There are therefore (30.4 cm³)/(7.6 cm³) = 4 strongly acidic OH groups per molecule. A molecule with 2 terminal OH groups and four further OH groups is a tetrapolyphosphate.

15.1 List the elements in Groups 15 and indicate the ones that are (a) diatomic gases, (b) nonmetals, (c) metalloids, (d) true metals. Indicate those elements that display the inert-pair effect?

In the list below, the properties of polonium have been omitted since the chemistry of this element has been little studied owing to its radioactivity.

	Type of element	Diatomic gas?	Achieves maximum oxidation state?	Displays inert pair effect?
N	nonmetal	yes	yes	no
P	nonmetal	no	yes	no
As	nonmetal	no	yes	no
Sb	metalloid	no	yes	no
Bi	metalloid	no	yes	yes
O	nonmetal	yes	no	no
S	nonmetal	no	yes	no
Se	nonmetal	no	yes	no
Te	nonmetal	no	yes	no

15.2 (a) Give complete and balanced chemical equations for each step in the synthesis of H_3PO_4 from hydroxyapatite to yield (a) high-purity phosphoric acid and (b) fertilizer-grade phosphoric acid. (c) Account for the large difference in costs between these two methods?

(a) high-purity phosphoric acid? The starting point is hydroxyapatite, $Ca_5(PO_4)_3OH$, which is converted to crude $Ca_3(PO_4)_2$. This compound is treated with (i) carbon to reduce phosphorus from P(V) in PO_4^{3-} to P(0) in P_4 and with (ii) silica, SiO_2, to keep the calcium-containing products molten for easy removal from the furnace. The impure P_4 is purified by sublimation, then oxidized with O_2 to form P_4O_{10}, which is hydrated to form pure H_3PO_4.

$$2Ca_3(PO_4)_2 + 10C + 6SiO_2 \rightarrow P_4^- + 10CO + 6CaSiO_3$$

$$P_4\,(\text{pure}) + 5O_2 \rightarrow P_4O_{10}$$

$$P_4O_{10} + 6H_2O \rightarrow 4H_3PO_4 \text{ (pure)}$$

(b) Fertilizer grade H_3PO_4? In this case, hydroxyapatite is treated with sulfuric acid, producing phosphoric acid that contains small amounts of many impurities.

$$Ca_5(PO_4)_3OH + 5H_2SO_4 \rightarrow 3H_3PO_4 \text{ (impure)} + 5CaSO_4 + H_2O$$

(c) Account for the difference in cost? The synthesis of fertilizer-grade (i.e., impure) phosphoric acid involves a single step and gives a product that requires little or no purification. In contrast, the synthesis of pure phosphoric acid involves several synthetic steps and a time-consuming and expensive purification step, the sublimation of white phosphorus, P_4.

15.3 Ammonia can be prepared by (a) the hydrolysis of Li_3N or (b) the high-temperature, high-pressure reduction of N_2 by H_2. Give balanced chemical equations for each method starting with N_2, Li, and H_2, as appropriate. (c) Account for the lower cost of the second method?

(a) Hydrolysis of Li_3N? Lithium nitride is one of two binary metal nitrides that can be prepared directly from the elements (the other is Mg_3N_2). Since it contains N^{3-} ions, which are extremely basic, it can be hydrolyzed with water to produce ammonia.

$$6Li + N_2 \rightarrow 2Li_3N$$

$$2Li_3N + 3H_2O \rightarrow 2NH_3 + 3Li_2O$$

(b) Reduction of N_2 by H_2? This reaction requires high temperatures and pressures to proceed at an appreciable rate.

$$N_2 + 3H_2 \rightarrow 2NH_3$$

(c) Account for the difference in cost? The second process is considerably cheaper than the first, even though the second process must be carried out at high temperature and pressure. This is because lithium, like many very electropositive metals (including aluminum), is very expensive.

15.4 Show with an equation why aqueous solutions of NH_4NO_3 are acidic?

$$NH_4NO_3(s) + H_2O \rightarrow NH_4^+ + NO_3^-(aq)$$

The ammonium ion is the conjugate acid of the weak base ammonia and hence expected to be moderately acidic (relative to water). Accordingly the following equation explains the observed acidity of ammonium ions. Note that NO_3^- is the conjugate base of the strong acid HNO_3 and therefore an extremely weak base.

$$NH_4^+(aq) + H_2O(l) \rightarrow NH_3(aq) + H_3O^+(aq)$$

15.5 Carbon monoxide is a good ligand and is toxic. Why is the isoelectronic N_2 molecule not toxic?

CO is toxic because it forms a complex with haemoglobin in the blood, and this complex is more stable than oxy-haemoglobin. This prevents the haemoglobin in the red blood corpuscles from carrying oxygen round the body. This causes an oxygen deficiency, leading to unconsciousness and then death.

N_2 itself, with a triple bond between the two atoms, is strikingly unreactive. The unreactivity of N_2 appears to be the result of several factors. One is the strength of the N–N triple bond. Second one is the relatively large size of the HOMO – LUMO gap in N_2. A third factor is the low polarizability of N_2. So that N_2 can not form complex with haemoglobin in the blood. Therefore, N_2 molecule not toxic.

15.6 Compare and contrast the formulas and stabilities of the oxidation states of the common nitrogen chlorides with the phosphorus chlorides?

The only isolable nitrogen chloride is NCl_3, and it is thermodynamically unstable with respect to its constituent elements (i.e., it is endoergic). The compound NCl_5 is unknown. In contrast, both PCl_3 and PCl_5 are stable can be prepared directly from phosphorus and chlorine.

15.7 Use the VSEPR model to predict the probable shapes of (a) PCl_4^+, (b) PCl_4^-, (c) $AsCl_5$?

(a) PCl_4^+? The Lewis structure is shown below. With four bonding pairs of electrons around the central phosphorus atom, the structure is a tetrahedron.

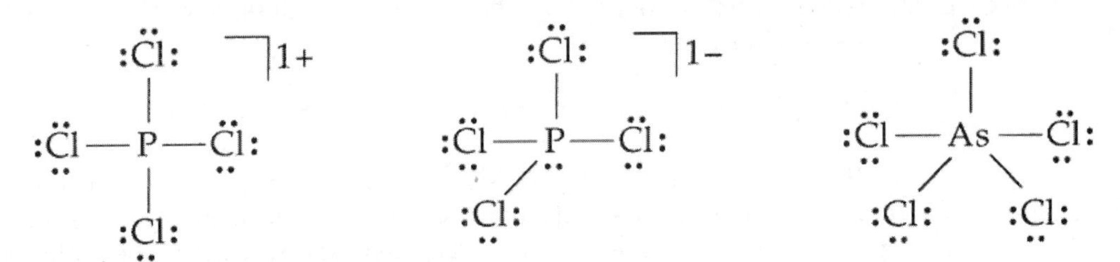

(b) PCl_4^-? The Lewis structure is shown above. With four bonding pairs of electrons and one lone pair of electrons around the central phosphorus atom, the structure is a see-saw (trigonal bipyramidal array of electron pairs with the lone pair in the equatorial plane).

(c) $AsCl_5$? The Lewis structure is shown above. With five bonding pairs of electrons around the central arsenic atom, the structure is a trigonal bipyramid.

15.8 Give balanced chemical equations for each of the following reactions: (a) oxidation of P_4 with excess oxygen, (b) reaction of the product from part (a) with excess water, (c) reaction of the product from part (b) with a solution of $CaCl_2$ and name the product?

(a) Oxidation of P_4 with excess O_2? The balanced equation is:

$$P_4 + 5O_2 \rightarrow P_4O_{10}$$

(b) Reaction of the product from part (a) with excess H_2O? The balanced equation is:

$$P_4O_{10} + 6H_2O \rightarrow 4H_3PO_4$$

(c) Reaction of the product from part (b) with $CaCl_2$? The products of this reaction would be calcium phosphate and a solution of hydrochloric acid. The balanced equation is:

$$2H_3PO_4(l) + 3CaCl_2(aq) \rightarrow Ca_3(PO_4)_2(s) + 6HCl(aq)$$

15.9 Starting with NH3(g) and other reagents of your choice, give the chemical equations and conditions for the synthesis of (a) HNO_3, (b) NO_2^-, (c) NH_2OH, (d) N_3^-?

(a) HNO_3? The synthesis of nitric acid, an example of the most oxidized form of nitrogen, starts with ammonia, the most reduced form:

$$4NH_3(aq) + 7O_2(g) \rightarrow 6H_2O(g) + 4NO_2(g)$$

High temperatures are obviously necessary for this reaction to proceed at a reasonable rate: ammonia is a flammable gas, but at room temperature it does not react rapidly with air. The second step in the formation of nitric acid is the high-temperature disproportionation of NO_2 in water:

$$3NO_2(aq) + H_2O(l) \rightarrow 2HNO_3(aq) + NO(g)$$

(b) NO_2^-? Whereas the disproportionation of NO_2 in acidic solution yields NO_3^- and NO (see part **(a)**), in basic solution nitrite ion is formed:

$$2NO_2(aq) + 2OH^-(aq) \rightarrow NO_2^-(aq) + NO_3^-(aq) + H_2O(l)$$

(c) NH_2OH? The protonated form of hydroxylamine is formed in a very unusual reaction between nitrite ion and bisulfite ion in cold aqueous acidic solution (the NH_3OH^+ ion can be deprotonated with base):

$$NO_2^-(aq) + 2HSO_3^-(aq) + H_2O(l) \rightarrow NH_3OH^+(aq) + 2SO_4^{2-}(aq)$$

(d) N_3^-? The azide ion can be prepared from anhydrous molten sodium amide (m.p. ~ 200°C) and either nitrate ion or nitrous oxide at elevated temperatures:

$$3NaNH_2(l) + NaNO_3 \rightarrow NaN_3 + 3NaOH + NH_3(g)$$

$$2NaNH_2(l) + N_2O \rightarrow NaN_3 + NaOH + NH_3$$

15.10 **Write the balanced chemical equation corresponding to the standard enthalpy of formation of $P_4O_{10}(s)$. Specify the structure, physical state (s, l, or g), and allotrope of the reactants. Does either of the reactants differ from the usual practice of taking as reference state the most stable form of an element?**
This compound is formed by the complete combustion of white phosphorus, as shown below:

$$P_4(s) + 5O_2(g) \rightarrow P_4O_{10}(s)$$

The structure of P_4 is a tetrahedron of phosphorus atoms with six P–P σ bonds. As discussed in Section 15.1, white phosphorus (P_4) is adopted as the reference phase for thermodynamic calculations even though it is not the most stable phase of elemental phosphorus.

15.11 **Without reference to the text, sketch the general form of the Frost diagrams for phosphorus (oxidation states 0 to +5) and bismuth (0 to +5) in acidic solution and discuss the relative stabilities of the +3 and +5 oxidation states of both elements?**
The main points you will want to remember are (i) Bi(III) is much more stable than Bi(V), and (ii) P(III) and P(V) are both about equally stable (i.e., Bi(V) is a strong oxidant but P(V) is not). This suggests that the point for Bi(III) on the Frost diagram lies *below* the line connecting Bi(0) and Bi(V), while the point for P(III) lies very close to the line connecting P(0) and P(V). The essential parts of the Frost diagrams for these two elements are shown below (see Figure 14.6 for the complete Frost diagrams).

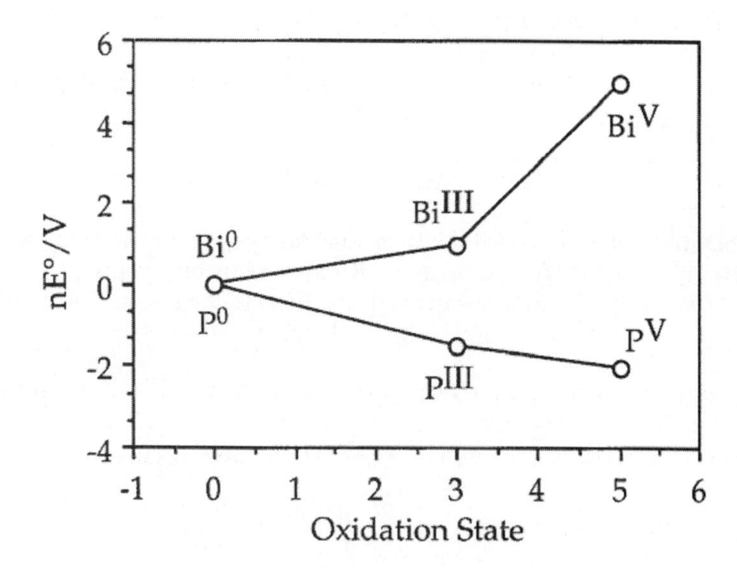

15.12 **Are reactions of NO_2^- as an oxidizing agent generally faster or slower when pH is lowered? Give a mechanistic explanation for the pH dependence of NO_2^- oxidations?**
The rates of reactions in which nitrite ion is reduced (i.e., in which it acts as an oxidizing agent) are increased as the pH is lowered. That is, acid enhances the rate of oxidations by NO_2^-. The reason is that NO_2^- is converted to the nitrosonium ion, NO^+, in strong acid:

$$HNO_2(aq) + H^+(aq) \rightarrow NO^+(aq) + H_2O(l)$$

This cationic Lewis acid can form complexes with the Lewis bases undergoing oxidation (species that are oxidizable are frequently electron rich and hence are basic). Therefore, at low pH the oxidant (NO^+) is a different chemical species than at higher pH (NO_2^-).

15.13 When equal volumes of nitric oxide (NO) and air are mixed at atmospheric pressure a rapid reaction occurs, to form NO_2 and N_2O_4. However, nitric oxide from an automobile exhaust, which is present in the parts per million concentration range, reacts slowly with air. Give an explanation for this observation in terms of the rate law and the probable mechanism?

The given observations suggest that more than one NO molecule is involved in the activated complex (i.e., in the rate-determining step), since the rate of reaction is dependent on the concentration of NO. Therefore, the rate law must be more than first order in NO concentration. It turns out to be second order: an equilibrium between NO and its dimer N_2O_2 precedes the rate-determining reaction with oxygen. At high concentrations of NO, the concentration of the dimer is higher.

15.14 Give balanced chemical equations for the reactions of the following reagents with PCl_5 and indicate the structures of the products: (a) water (1:1), (b) water in excess, (c) $AlCl_3$, (d) NH_4Cl?

(a) H_2O? In this case, water will form a strong P=O double bond, releasing two equivalents of HCl and forming tetrahedral $POCl_3$ (which actually has C_{3v} symmetry).

$$PCl_5 + H_2O \rightarrow POCl_3 + 2HCl$$

(b) H_2O in excess? When water is in excess, all of the P–Cl bonds will be hydrolyzed, and phosphoric acid will result:

$$2PCl_5(g) + 8H_2O(l) \rightarrow 2H_3PO_4(aq) + 10HCl(aq)$$

(c) $AlCl_3$? By analogy to the reaction with another group 13/III Lewis acid, BCl_3, PCl_5 will donate a chloride ion and a salt will result. Both the cation and the anion are tetrahedral.

$$PCl_5 + AlCl_3 \rightarrow [PCl_4]^+[AlCl_4]^-$$

(d) NH_4Cl? A phosphazene, containing cyclic molecules or linear chain polymers of $(=P(Cl)_2^-N=P(Cl)_2^-N=P(Cl)_2^-N=)$ bonds will be formed:

$$nPCl_5 + nNH_4Cl \rightarrow -[(N=P(Cl)_2)_n]^- + 4nHCl$$

15.15 Use standard potentials (Resource section 3) to calculate the standard potential of the reaction of H_3PO_2 with Cu^{2+}. Are HPO_2^{2-} and $H_2PO_2^{2-}$ useful as oxidizing or reducing agents?

From the standard potentials for the following two half-reactions, the standard potential for the net reaction can be calculated:

$$H_3PO_2(aq) + H_2O(l) \rightarrow H_3PO_3(aq) + 2e^- + 2H^+(aq) \qquad E^o = 0.499V$$

$$Cu^{2+}(aq) + 2e^- \rightarrow Cu^0(s) \qquad E^o = 0.340V$$

Therefore the net reaction is:

$$H_3PO_2(aq) + H_2O(l) + Cu^{2+}(aq) \rightarrow H_3PO_3(aq) + Cu^0(s) + 2H^+(aq)$$

and the net standard potential is $E^o = 0.839$ V. To determine whether HPO_3^{2-} and $H_2PO_2^-$ are useful as oxidizing or reducing agents, you must compare E values for their oxidations and reductions at pH 14 (these potentials are listed in Resource Section 3):

$$HPO_3^{2-}(aq) + 3OH^-(aq) \rightarrow PO_4^{3-}(aq) + 2e^- + 2H_2O(l) \qquad E(pH14) = 1.12V$$

$$HPO_3^{2-}(aq) + 2e^- + 2H_2O(l) \rightarrow H_2PO_2^-(aq) + 3OH^-(aq) \quad E(pH14) = -1.57V$$

$$H_2PO_2^-(aq) + e^- \rightarrow P(s) + 2OH^-(aq) \quad E(pH14) = -2.05V$$

Since a positive potential will give a negative value of ΔG, the oxidations of HPO_3^{2-} and $H_2PO_2^-$ are much more favorable than their reductions, so these ions will be much better reducing agents than oxidizing agents.

15.16 Identify the compounds A, B, C, and D?

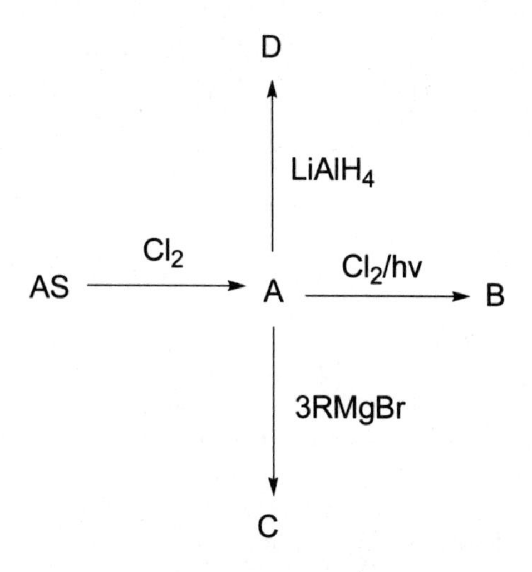

(a) $As + {}^3/2\ Cl_2 \rightarrow AsCl_3$; $A = AsCl_3$
(b) $AsCl_3 + Cl_2/h\nu \rightarrow AsCl_5$; $B = AsCl_5$
(c) $AsCl_3 + 3RMgBr \rightarrow AsR_3 + 3MgClBr$; see Section 14.12; $C = AsR_3$
(d) $AsCl_3 + LiAlH_4 \rightarrow AsH_3$; $D = AsH_3$

15.17 Identify the nitrogen compounds A, B, C, D, and E?

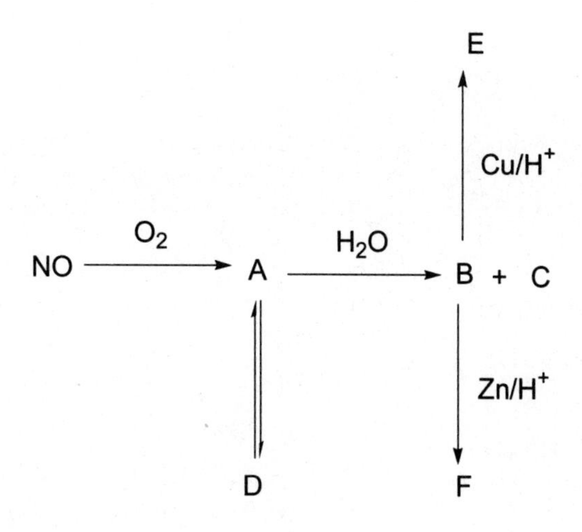

(a) $NO + O_2 \rightarrow NO_2$; $A = NO_2$
(b) $H_2O + 3NO_2 \rightarrow 2HNO_3 + NO$; $B = HNO_3$; $C = NO$
(c) $2NO_2 \rightarrow N_2O_4$; $D = N_2O_4$
(d) $2HNO_3 + Cu + 2H^+ \rightarrow 2NO_2 + Cu^{2+} + 2H_2O$; $E = NO_2$
(e) $HNO_3 + 4Zn + 9H^+ \rightarrow NH_4^+ + 3H_2O + 4Zn^{2+}$; $F = NH_4^+$

15.18 Sketch the two possible geometric isomers of the octahedral $[AsF_4Cl_2]^-$ and explain how they could be distinguished by ^{19}F–NMR?

The *cis* isomer gives two ^{19}F signals and the *trans* isomer gives one signal. The two possible isomers of $[AsF_4Cl_2]^-$ are shown below for A=F and B=Cl. Recall Structures 53 and 54 from Chapter 8. The chlorines are adjacent in the *cis* form and are diagonal in the *trans* form, leading to two fluorine environments in the *cis* form and one in the *trans*.

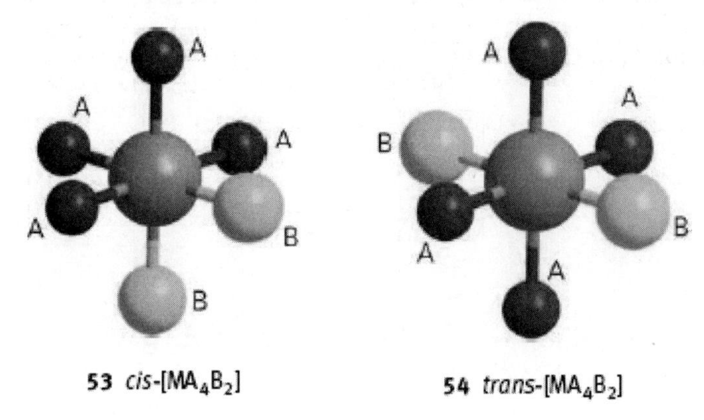

53 *cis*-$[MA_4B_2]$ **54** *trans*-$[MA_4B_2]$

15.19 Use the Latimer diagrams in Resource section 3 to determine which species of N and P disproportionate in acid conditions?

Recall that if the potential to the right is more positive than the potential going left, the species is unstable and likely to disproportionate. However, no kinetic information can be deduced. The species of N and P that disproportionate are N_2O_4, NO, N_2O, NH_3OH^+, $H_4P_2O_6$, and P.

Chapter 16 The Group 16 Elements

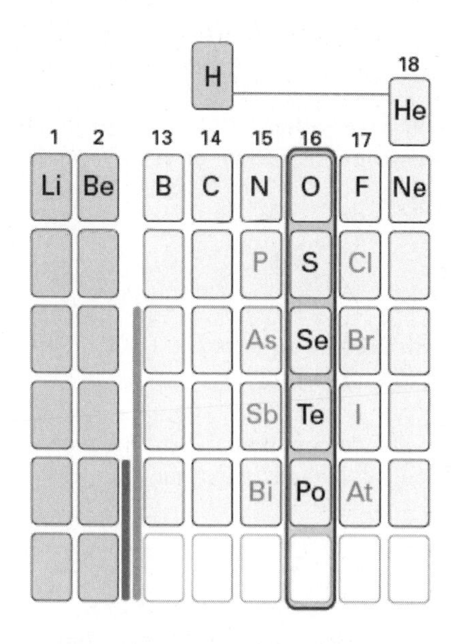

The Group 16 elements, oxygen, sulfur, selenium, tellurium and polonium, are often called the **chalcogens.** The name derives from the Greek word for bronze, and refers to the association of sulfur and its congeners with copper. Elemental oxygen is essential for respiration of all living organisms and for the existence of water. In stark contrast, other members of the group form some compounds that are highly toxic.

S16.1 Determine whether the decomposition of H_2O_2 is spontaneous in the presence of either Br_2 or Cl_2?

1). Br_2 (aq) + H_2O_2 (aq) $\longrightarrow$ 2Br– (aq) + O_2 (g) + $2H^+$ (aq)
2). $2Br^-$ (aq) + H_2O_2 (aq) + $2H^+$ (aq) $\longrightarrow$ Br_2 + $2H_2O$ (l)
So the net reaction is simply the decomposition of H_2O_2.
3). $2H_2O_2$ $\longrightarrow$ O_2 (g) + $2H_2O$ (l)
The first reaction is the difference between the half reactions
$Br_2 + 2e^-$ $\longrightarrow$ $2Br^-$ $E^\circ = +1.06$ V
O_2 (g) + $2H^+ + 2e^-$ $\longrightarrow$ H_2O_2 $E^\circ = +0.70$ V
And therefore E°_{cell} + + 0.36 V. This reaction is therefore spontaneous.
The second reaction is the difference between the half reactions
H_2O_2 (aq) + $2H^+$ (aq) + $2e^-$ $\longrightarrow$ $2H_2O$ (l) $E^\circ = +1.77$ V
$Br_2 + 2e^-$ $\longrightarrow$ $2Br^-$ $E^\circ = +1.06$ V
And therefore E°_{cell} = + 0.71 V. This reaction is therefore spontaneous. Because of both reactions are spontaneous, the decomposition of H_2O_2 is thermodynamically favoured in presence of Br^-.

In Presence of Cl^-

1). Cl_2 (aq) + H_2O_2 (aq) $\longrightarrow$ $2Cl^-$ (aq) + O_2 (g) + $2H^+$ (aq)
2). $2Cl^-$ (aq) + H_2O_2 (aq) + $2H^+$ (aq) $\longrightarrow$ Cl_2 + $2H_2O$ (l)
So the net reaction is simply the decomposition of H_2O_2.
3). $2H_2O_2$ $\longrightarrow$ O_2 (g) + $2H_2O$ (l)
The first reaction is the difference between the half reactions
$Cl_2 + 2e^-$ $\longrightarrow$ $2Cl^-$ $E^\circ = +1.36$ V
O_2 (g) + $2H^+ + 2e^-$ $\longrightarrow$ H_2O_2 $E^\circ = +0.70$ V
And therrefore E°_{cell} + + 0.66 V. This reaction is therefore spontaneous.
The second reaction is the difference between the half reactions
H_2O_2 (aq) + $2H^+$ (aq) + $2e^-$ $\longrightarrow$ $2H_2O$ (l) $E^\circ = +1.77$ V
$Cl_2 + 2e^-$ $\longrightarrow$ $2Cl^-$ $E^\circ = +1.36$ V

And therrefore E°_{cell} = + 0.41 V. This reaction is therefore spontaneous. Because of both reactions are spontaneous, the decomposition of H_2O_2 is thermodynamically favoured in presence of Cl^-.

16.1 State whether the following oxides are acidic, basic, neutral, or amphoteric: CO_2, P_2O_5, SO_3, MgO, K_2O, Al_2O_3, CO?

Refer to Section 4.6; CO_2, SO_3, P_2O_5, and Al_2O_3 can react with either acids or bases, so they are considered amphoteric. CO is neutral; MgO and K_2O are acidic.

16.2 (a) Use standard potentials (Resource section 3) to calculate the standard potential of the disproportionation of H_2O_2 in acid solution. (b) Is Cr^{2+} a likely catalyst for the disproportionation of H_2O_2? (c) Given the Latimer diagram

$$O_2 \xrightarrow{-0.13} HO_2^- \xrightarrow{1.51} H_2O_2$$

in acidic solution, calculate Δ_rG^o for the disproportionation of hydrogen superoxide (HO^{2-}) into O_2 and H_2O_2, and compare the result with its value for the disproportionation of H_2O_2?

(a) The disproportionation of H_2O_2 and HO_2? To calculate the standard potential for a disproportionation reaction, you must sum the potentials for the oxidation and reduction of the species in question. For hydrogen peroxide in acid solution, the oxidation and reduction are:

$$H_2O_2(aq) \rightarrow O_2(g) + 2\,e^- + 2\,H^+(aq) \qquad E^o = -0.695\text{ V}$$

$$H_2O_2(aq) + 2\,e^- + 2\,H^+(aq) \rightarrow 2\,H_2O(l) \quad E^o = 1.763\text{ V}$$

Therefore, the standard potential for the net reaction $2\,H_2O_2(aq) \rightarrow O_2(g) + 2\,H_2O(1)$ is (–0.695 V) + (1.763 V) = 1.068 V.

(b) Catalysis by Cr^{2+}? Cr^{2+} can act as a catalyst for the decomposition of hydrogen peroxide if the Cr^{3+}/Cr^{2+} reduction potential falls between the values for the reduction of O_2 to H_2O_2(0.695 V) and the reduction of H_2O_2 to H_2O(1.76 V). Reference to Resource Section 3 reveals that the Cr^{3+}/Cr^{2+} reduction potential is –0.424 V, so Cr^{2+} is *not* capable of decomposing H_2O_2.

(c) The disproportionation of HO_2? The oxidation and reduction of superoxide ion (O_2^-) in acid solution are:

$$HO_2(aq) \rightarrow O_2(g) + e^- + H^+(aq) \qquad E^o = 0.125\text{ V}$$

$$HO_2(aq) + e^- + H^+(aq) \rightarrow H_2O_2(aq) \qquad E^o = 1.51\text{ V}$$

Therefore, the standard potential for the net reaction $2\,HO_2(aq) \rightarrow O_2(g) + H_2O_2(aq)$ is (0.125 V) + (1.51 V) = 1.63 V. Since 1 V is equivalent to 96.5 kJ, Δ_rG° = 157 kJ. For the disproportionation of H_2O_2 (part **(a)**), Δ_rG° = 103 kJ.

16.3 Which hydrogen bond would be stronger: S—H . . . O or O—H . . . S?

Hydrogen bonds are stronger in molecules where the hydrogen is bonded directly to highly electronegative atom such as O, N, or F. Since S has smaller electronegativity compared to oxygen, we expect O–H hydrogen bonds to be stronger.

16.4 Which of the solvents ethylenediamine (which is basic and reducing) or SO_2 (which is acidic and oxidizing) might not react with (a) Na_2S_4, (b) K_2Te_3?

Anionic species such as S_4^{2-} and Te_3^{2-} are intrinsically basic species that cannot be studied in solvents that are Lewis acids because complex formation will destroy the independent identity of the anion. Therefore, since the basic solvent ethylenediamine will not react with Na_2S_4 or with K_2Te_3, it is a better solvent for them than sulfur dioxide.

16.5 Rank the following species from the strongest reducing agent to the strongest oxidizing agent: SO_4^{2-}, SO_3^{2-}, $O_3SO_2SO_3^{2-}$?

The three values of $E°$ for the $S_2O_8^{2-}/SO_4^{2-}$, the SO_4^{2-}/SO_3^{2-}, and the $SO_3^{2-}/S_2O_3^{2-}$ couples are 1.96 V, 0.158 V, and 0.400 V, respectively. Peroxydisulfate, $S_2O_8^{2-}$, is very easily reduced, so it is the strongest oxidizing agent. Sulfate dianion, SO_4^{2-}, is neither strongly oxidizing nor strongly reducing. Sulfite dianion, SO_3^{2-}, on the other hand, is relatively readily oxidized to sulfate, and hence is a moderate reducing agent. Two plausible redox reactions are:

$$S_2O_8^{2-}(aq) + NO_2^-(aq) + H_2O(l) \rightarrow 2\,HSO_4^-(aq) + NO_3^-(aq) \qquad E° = 1.02\text{ V}$$

$$SO_3^{2-}(aq) + Cu^{2+}(aq) + H_2O(l) \rightarrow SO_4^{2-}(aq) + Cu(s) + 2\,H^+(aq) \quad E° = 0.182\text{ V}$$

16.6 Predict which oxidation states of Mn will be reduced by sulfite ions in basic conditions?

The reduction potential of sulfite ions in basic solution = –0.576 V

Different Oxidation states of Mn	Standard Reduction potential of Mn	$E^0 = E^0_{cathode} - E^0_{anode}$
+VII	+VII to +VI = +0.56	+ 1.136 V
+VI	+VI to +V = +0.27	+ 0.847 V
+V	+V to +IV = +0.93	+ 1.506 V
+IV	+IV to +III = +0.15	+ 0.726 V
+III	+III to +II = -0.25	+ 0.326 V
+II	+II to 0 = -1.56	– 0.984

Except the +II oxidation state of Mn, all other oxidation states of Mn (+VII, +VI, +V, +IV, +III) will be reduced by sulfite ions in basic solution.

16.7 (a) Give the formula for Te(VI) in acidic aqueous solution and contrast it with the formula for S(VI). (b) Offer a plausible explanation for this difference?

(a) Formulas? The formulas for these ions in acidic solution are $H_5TeO_6^-$ and HSO_4^- (the parent acids are H_6TeO_6, or $Te(OH)_6$, and H_2SO_4).

(b) An explanation? Tellurium is a much larger element than sulfur and readily can increase its coordination number. This trend is found in other groups in the *p* block. An example involving perchlorate and periodate will be discussed in Chapter 17.

16.8 Use the standard potential data in Resource section 3 to predict which oxoanions of sulfur will disproportionate in acidic conditions?

If the potential to the right is more positive than the potential going left, the species is unstable and likely to disproportionate. $S_2O_6^{2-}$ and $S_2O_3^{2-}$ are unstable.

16.9 Use the standard potential data in Resource section 3 to predict whether SeO_3^{2-} is more stable in acidic or basic solution?

In general, from Latimer diagrams the stability of a given species against disproportionation depends on whether the right hand side potential is greater or smaller than the left hand side potential. If the right hand side potential greater than the left hand side potential the species is inherently unstable against disproportionation. Now looking at the Latimer diagram of SeO_3^{2-} in both acidic and basic solutions, the right hand side potential is smaller than the left hand side potential indicating that SeO_3^{2-} is stable both in acidic as well as basic solution. However, if one wants to estimate the relative stability then it would be instructive to calculate the overall potentials for disproportionation reaction.

Standard Electrode potential (E^0) = Cathodic reduction potential (E^0_c) – Anodic reduction potential (E^0_a)

For $P^H > 7$,
The cathodic reduction potential of $SeO_3^{2-} = -0.36$
The anodic reduction potential of $SeO_3^{2-} = +0.03$

$$E^0_{\text{selenate ion}} = -0.36-(+0.03)$$
$$= -0.39\ \text{V}$$

For $P^H < 7$,
The cathodic reduction potential of $SeO_3^{2-} = 0.74$
The cathodic reduction potential of $SeO_3^{2-} = +1.15$

$$E^0_{\text{selenate ion}} = 0.74 - (+1.15)$$
$$= -0.41$$

From this it can be concluded that the SeO_3^{2-} is marginally more stable in acid solutions.

16.10 Predict whether any of the following will be reduced by thiosulfate ions, $S_2O_3^{2-}$, in acidic conditions: VO^{2+}, Fe^{3+}, Cu^{+}, Co^{3+}?
The half potentials for the reduction of Fe^{3+} and Co^{3+} are 0.771V and 1.92 V, respectively. Fe^{3+} and Co^{3+} will be reduced as their potentials are greater than the potential of thiosulfate.

16.11 SF_4 reacts with BF_3 to form $[SF_3][BF_4]$. Use VSEPR theory to predict the shapes of the cation and anion?
SF_4^+ trigonal pyramidal, BF_4^- tetrahedral.

16.12 Tetramethylammonium fluoride (0.70 g) reacts with SF_4 (0.81 g) to form an ionic product. (a) Write a balanced equation for the reaction and (b) sketch the structure of the anion. (c) How many lines would be observed in the ^{19}F-NMR spectrum of the anion?
(a) $SF_4 + (CH_3)_4NF \rightarrow [(CH_3)_4N]^+[SF_5]^-$; **(b)** square pyramidal structure; **(c)** two F environments.

16.13 Identify the sulfur-containing compounds A, B, C, D, E, and F?

F ← (O_2) — S — (Cl_2) → A — (excess NH_3) → B — (Ag/heat) → C
S — (K_2SO_3) → D
D — (I_2) → E
E — (H_2S) → F

A = S_2Cl_2, B = S_4N_4, C = S_2N_2, D = $K_2S_2O_3$, E = $S_2O_6^{2-}$, F = SO_2.

Chapter 17 The Group 17 Elements

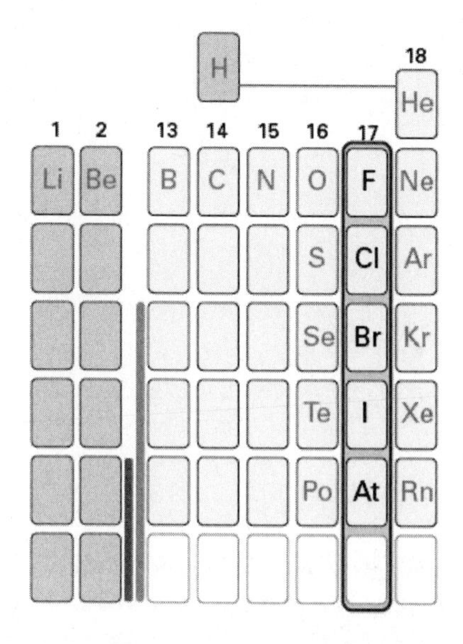

The Group 17 elements, fluorine, chlorine, bromine, iodine, and astatine, are known as the **halogens** from the Greek for "salt giver." Fluorine and chlorine are poisonous gases, bromine is a toxic, volatile liquid, and iodine is a sublimable solid. They are among the most reactive nonmetallic elements. The chemical properties of the halogens are extensive and their compounds have been mentioned in many of the previous chapters.

S17.1 One source of iodine is sodium iodate, $NaIO_3$. Which of the reducing agents SO_2(aq) or Sn^{2+}(aq) would seem practical from the standpoints of thermodynamic feasibility and plausible judgements about cost? Standard potentials are given in Resource section 3?

Since the IO_3^-/I_2 standard reduction potential is 1.19 V (see Resource Section 3), many species can be used as reducing agents, including SO_2(aq) and Sn^{2+}(aq) (the HSO_4^-/H_2SO_3 and Sn^{4+}/Sn^{2+} reduction potentials are 0.158 V and 0.15 V, respectively). The balanced equations would be as follows:

$$2\,IO_3^- + 5\,H_2SO_3 \rightarrow I_2 + 5\,HSO_4^- + H_2O + 3\,H^+$$

$$2\,IO_3^- + 5\,Sn^{2+} + 12\,H^+ \rightarrow I_2 + 5Sn^{4+} + 6\,H_2O$$

The reduction of iodate with aqueous sulfur dioxide would be far cheaper than reduction with Sn^{2+}, because sulfur dioxide is much cheaper than tin. One reason this is so is because sulfuric acid, for which SO_2 is an intermediate, is prepared worldwide on an enormous scale.

S17.2 Predict the ^{19}F-NMR pattern for IF_7?

Iodine is the central atom in the molecule. Seven unpaired electrons form bonds with seven fluorine atoms, giving a total of seven electron pairs structure is pentagonal bipyramid.

The structure of IF_7:

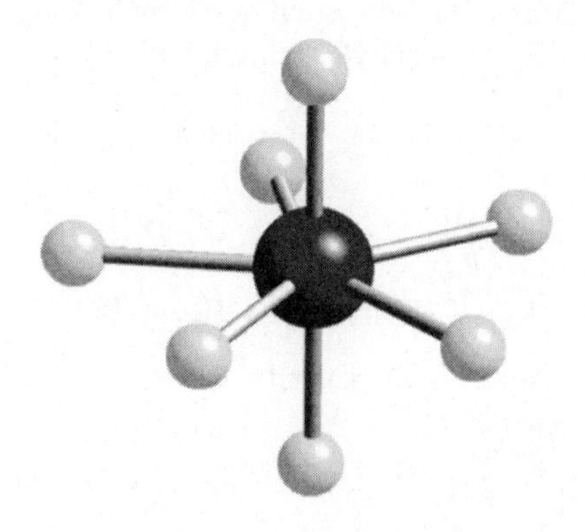

The ^{19}F-NMR spectrum of IF_7 contains two resonances to integrate seven fluorine atoms. One is sextet for two apical F atoms and another one is a triplet arising from five equatorial F atoms.

S17.3 From the perspective of structure and bonding, indicate several polyhalides that are analogous to [py–I–py]$^+$, and describe their bonding?

Examples include I_3^-, IBr_2^-, ICl_2^-, and IF_2^-. In all cases, consider that the central I^+ species has a vacant orbital that interacts with two electron pair donors, py in the case of [py–I–py]$^+$ and I^- in the case of I_3^-. The three centres contribute four electrons to three molecular orbitals, one bonding, one nonbonding, and one antibonding.

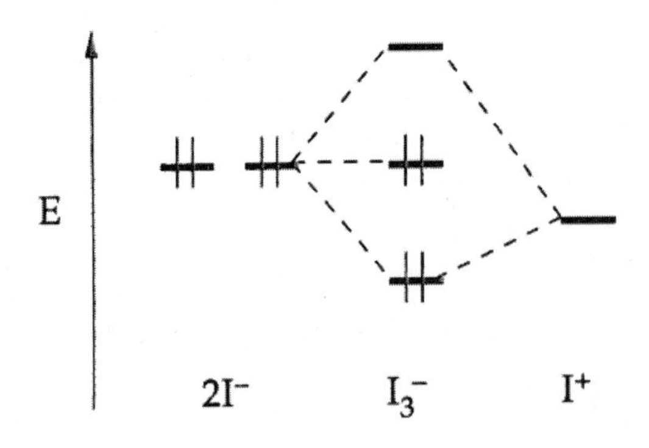

17.1 Preferably without consulting reference material, write out the halogens and noble gases as they appear in the periodic table, and indicate the trends in (a) physical state (s, l, or g) at room temperature and pressure, (b) electronegativity, (c) hardness of the halide ion, (d) colour?

	Physical state	Electronegativity	Hardness of halide ion	Color
F_2	gas	highest (4.0)	hardest	light yellow
Cl_2	gas	lower	softer	yellow-green
Br_2	liquid	lower	softer	dark red-brown
I_2	solid	lowest	softest	dark violet
He	gas			colorless
Ne	gas			colorless
Ar	gas			colorless
Kr	gas			colorless
Xe	gas			colorless

17.2 Describe how the halogens are recovered from their naturally occurring halides and rationalize the approach in terms of standard potentials. Give balanced chemical equations and conditions where appropriate?

The principal source of fluorine is CaF_2. It is converted to HF by treating it with a strong acid such as sulfuric acid. Liquid HF is electrolyzed to H_2 and F_2 in an anhydrous cell, with KF as the electrolyte:

$$CaF_2 + H_2SO_4 \rightarrow CaSO_4 + 2\,HF$$

$$2\,HF + 2\,KF \rightarrow 2\,K^+HF_2^-$$

$$2\,K^+HF_2^- + \text{electricity} \rightarrow F_2 + H_2 + 2\,KF$$

The principal source of the other halides is seawater (natural brines). Chlorine is liberated by the electrolysis of aqueous NaCl (the chloralkali process):

$$2\,Cl^- + 2\,H_2O + \text{electricity} \rightarrow Cl_2 + H_2 + 2\,OH^-$$

Bromine and iodine are prepared by treating aqueous solutions of the halides with chlorine:

$$2\,X^- + Cl_2 \rightarrow X_2 + 2\,Cl^- \qquad (X^- = Br^-, I^-)$$

17.3 Sketch a choralkali cell. Show the half-cell reactions and indicate the direction of diffusion of the ions. Give the chemical equation for the unwanted reaction that would occur if OH^- migrated through the membrane and into the anode compartment?

A drawing of the cell is shown in Figure 17.3. Note that Cl_2 is liberated at the anode and H_2 is liberated at the cathode, according to the following half-reactions:

$$\text{anode: } 2\ Cl^-(aq) \rightarrow Cl_2(g) + 2\ e^-$$

$$\text{cathode: } 2\ H_2O(l) + 2\ e^- \rightarrow 2\ OH^-(aq) + H_2(g)$$

To maintain electroneutrality, Na^+ ions diffuse through the polymeric membrane. Because of the chemical properties of the membrane, anions such as Cl^- and OH^- cannot diffuse through it. If OH^- did diffuse through the membrane, it would react with Cl_2 and spoil the yield of the electrolysis, as shown in the following equation:

$$2\ OH^-(aq) + Cl_2(aq) \rightarrow ClO^-(aq) + Cl^-(aq) + H_2O(l)$$

17.4 Sketch the form of the vacant s* orbital of a dihalogen molecule and describe its role in the Lewis acidity of the dihalogens?

According to Figure 17.5, the vacant antibonding orbital of a halogen molecule is $2\sigma_u{}^*$ and is composed primarily of halogen atomic *p* orbitals; recall that the *ns–np* gap is larger toward the right-hand side of the periodic table, and the larger the gap, the smaller the amount of *s–p* mixing. Sketches of the filled $2\sigma_g$ bonding orbital and the empty $2\sigma_u{}^*$ antibonding orbital are shown below:

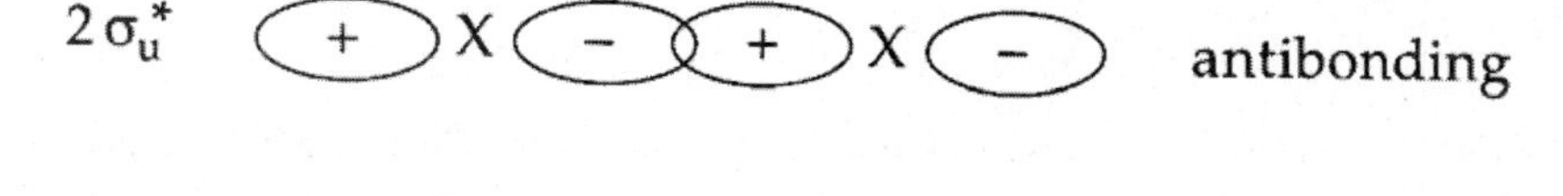

Since the $2\sigma_u{}^*$ antibonding orbital is the LUMO for a X_2 molecule, it is the orbital that accepts the pair of electrons from a Lewis base B when a dative $B \rightarrow X_2$ bond is formed. From the shape of the LUMO, the B–X–X unit should be linear.

17.5 Which dihalogens are thermodynamically capable of oxidizing H_2O to O_2?

The oxidation potential of different dihalogens and H_2O:

$$2I^- \longrightarrow I_2 + 2e^- \qquad E^{\circ}_{I-/I2} = -0.53\ V$$
$$2Br^- \longrightarrow Br_2 + 2e^- \qquad E^{\circ}_{Br-/Br2} = -1.06\ V$$
$$2Cl^- \longrightarrow Cl_2 + 2e^- \qquad E^{\circ}_{Cl-/Cl2} = -1.36\ V$$
$$2F^- \longrightarrow F_2 + 2e^- \qquad E^{\circ}_{F-/F2} = -2.85\ V$$
$$2H_2O \longrightarrow O_2\ (g) + 2H^+ + 4e^- \qquad E^{\circ}_{H2O/O2} = -1.23\ V$$

To oxidise H_2O to O_2, the oxdation potential of dihalogens should be lesser than the oxidation potential of H_2O. From the above given data, the oxidation potential of I_2 and Br_2 are higher than the oxidation potential of H_2O whereas the oxidation potential of Cl_2 and F_2 are lower than the oxidation potential of H_2O. Therefore Cl_2 and F_2 are thermodynamically capable of oxidizing H_2O to O_2

17.6 Nitrogen trifluoride, NF_3, boils at –129°C and is devoid of Lewis basicity. By contrast, the lower molar mass compound NH_3 boils at –33°C and is well known as a Lewis base. (a) Describe the origins of this very large difference in volatility. (b) Describe the probable origins of the difference in basicity?

(a) Difference in volatility? Ammonia is one of a number of substances that exhibit very strong hydrogen bonding, as described in Sections 4.3 and 9.4(b). The very strong intermolecular interactions lead to a relatively large enthalpy of vaporization and a relatively high boiling point for NH_3 (–33°C). In contrast, the intermolecular forces in liquid NF_3 are relatively weaker dipole–dipole forces, resulting in a smaller enthalpy of vaporization and a low boiling point (–129°C).

(b) Explain the difference in basicity? The strong electron-withdrawing effect of the three fluorine atoms in NF_3 lowers the energy of the nitrogen atom lone pair. This lowering of energy has the effect of reducing the electron-donating ability of the nitrogen atom in NF_3, reducing the basicity.

17.7 Based on the analogy between halogens and pseudohalogens write: (a) the balanced equation for the probable reaction of cyanogen, $(CN)_2$, with aqueous sodium hydroxide, (b) the equation for the probable reaction of excess thiocyanate with the oxidizing agent $MnO_2(s)$ in acidic aqueous solution, (c) a plausible structure for trimethylsilyl cyanide?

(a) The reaction of NCCN with NaOH? When a halogen such as chlorine is treated with aqueous base, it undergoes disproportionation to yield chloride ion and hypochlorite ion, as follows:

$$Cl_2(aq) + 2\,OH^-(aq) \rightarrow Cl^-(aq) + ClO^-(aq) + H_2O(l)$$

The analogous reaction of cyanogen with base is:

$$NCCN(aq) + 2\,OH^-(aq) \rightarrow CN^-(aq) + NCO^-(aq) + H_2O(l)$$

The linear NCO^- ion is named cyanate.

(b) The reaction of SCN^- with MnO_2 in aqueous acid? If MnO_2 is an oxidizing agent, then the probable reaction is oxidation of thiocyanate ion to thiocyanogen, $(SCN)_2$, coupled with reduction of MnO_2 to Mn^{2+} (if you do not recall that Mn^{3+} is unstable to disproportionation, you will discover it when you refer to the Latimer diagram for manganese in Resource Section 3). The balanced equation is:

$$2SCN^-(aq) + MnO_2(s) + 4\,H^+(aq) \rightarrow (SCN)_2(aq) + Mn^{2+}(aq) + 2\,H_2O(l)$$

(c) The structure of trimethylsilyl cyanide? Just as halides form Si–X single bonds, trimethylsilyl cyanide contains an Si–CN single bond. Its structure is shown in Structure 6.

17.8 Given that 1.84 g of IF_3 reacts with 0.93 g of $[(CH_3)_4N]F$ to form a product X, (a) identify X, (b) use the VSEPR model to predict the shapes of IF_3 and the cation and anion in X, (c) predict how many ^{19}F-NMR signals would be observed in IF_3 and X?

(a)

IF_3	+	$[(CH_3)_4N]F$	$\longrightarrow$	$IF_4N(CH_3)_4$
(1.84 g, 10 mmol)		(0.93 g, 10 mmol)		(10 mmol)

Therefore the product X = $IF_4N(CH_3)_4$

(b) A reliable way to predict the structure of IF_3 is to draw its Lewis structure and then apply VSEPR theory. The Lewis structure for IF_3 is shown below. In this molecule, iodine is the central atom. Three bonding pairs and two lone pairs on the iodine yield trigonal bipyramid geometry, as shown below. Lone pairs occupy two of the corners, and F atoms occupy the other three corners. Three different arrangements are theoretically possible, as shown below. Groups at 90° to each other repel each other strongly, whilst groups 120°apart repel each other much less.
Structure 1 is the most symmetrical, but has six 90° repulsions between lone pairs and atoms.
Structure 2 has one 90° repulsion between two lone pairs, plus three 90° repulsions between lone pairs and atoms.
Structure 3 has four 90° repulsions between lone pairs and atoms.
This confirms that the correct structure is 3. As a general rule, if lone pairs occur in a trigonal bipyramid they will be located in the equatorial position rather than the apical positions (top and bottom), since this arrangement minimizes repulsive forces.

Lewis structure of IF_3:

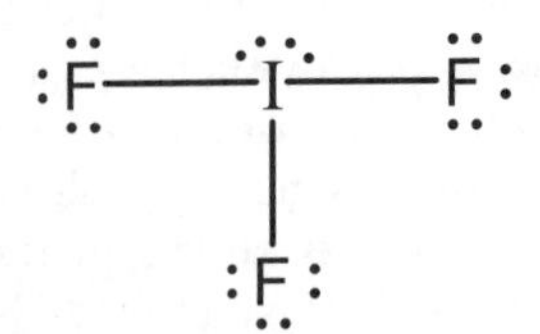

Different possible arrangements of IF_3:

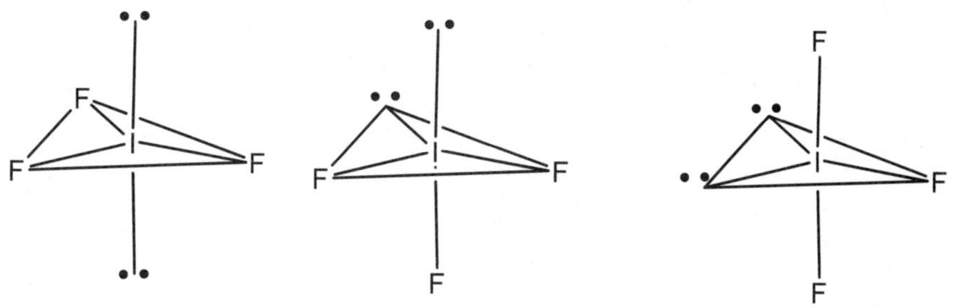

Structure-1 Structure-2 Structure-3

The shape of anion IF_4^- is Square planar.
The iodine atom is at the centre of the molecule and determines its shape. Iodine has seven outer electrons. Four electrons forms bonds to F, and three electrons do not take part in bonding. Thus in IF_4^- the I atom has six electron pairs including negative charge in the outer shell: hence the structure is octahedral with two positions occupied by a lone pairs (alternatively described as square planar).

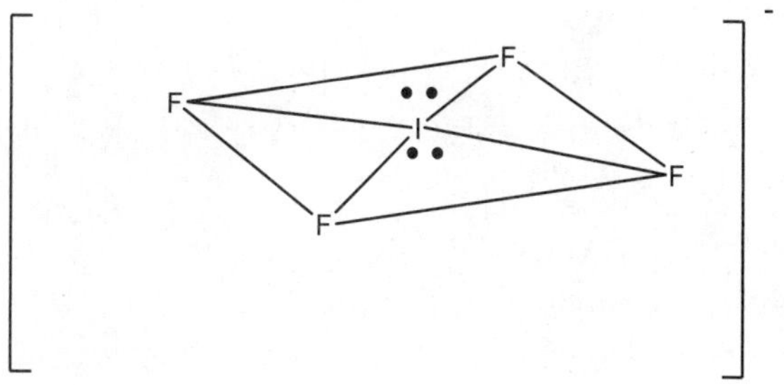

The shape of cation $(CH_3)_4N^+$ is Tetrahedral.
The Nitrogen atom is at the centre of the molecule and determines its shape. Nitrogen has five outer electrons. Three electrons form bonds to CH_3, plus one lone pair forms coordinate covalent bond with CH_3^+. Thus in $(CH_3)_4N^+$ the N atom has four electron pairs in the outer shell: hence the structure is tetrahedral.

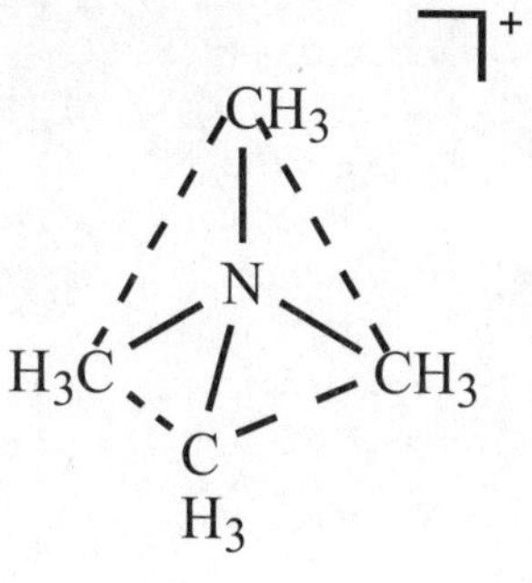

c).The ^{19}F-NMR spectrum of IF_3 contains two resonances to integrate three F atoms. One doublet is for axial 2F atoms and another triplet is for 1equatorial F atom.
The ^{19}F-NMR spectrum of IF_4^- contains one resonance to integrate four fluorine atoms. One singlet is assigned for four equivalent F atoms.

17.9 Use the VSEPR model to predict the shapes of $SbCl_5$, $FClO_3$, and $[ClF_6]^+$?

$SbCl_5$ is trigonal bipyramidal, $FClO_3$ is pyramidal, and ClF_6 is octahedral.

$SbCl_5$ $FClO_3$ $[ClF_6]^+$

17.10 Indicate the product of the reaction between ClF_5 and SbF_5?

SbF5 is a strong Lewis acid and extracts a fluoride ion from the interhalogen compound ClF_5, leading to the formation of $[ClF_4]^+[SbF_6]^-$.

17.11 Sketch all the isomers of the complexes MCl_4F_2 and MCl_3F_3. Indicate how many fluorine environments would be indicated in the ^{19}F-NMR spectrum of each isomer?

For MCl_4F_2, the *cis* isomer has one resonance and the *trans* isomer has one resonance. For MCl_3F_3, the *fac* isomer has one resonance and the *mer* isomer has two resonances of intensity. The structures below, taken from Chapter 8, show the differences in the isomers for A = Cl and B = F.

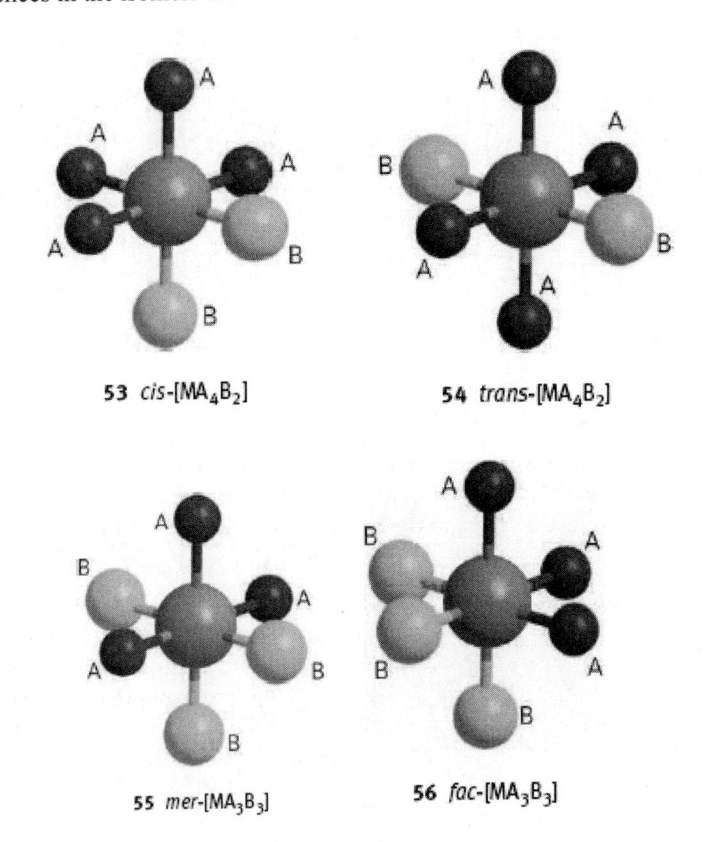

53 *cis*-$[MA_4B_2]$ 54 *trans*-$[MA_4B_2]$

55 *mer*-$[MA_3B_3]$ 56 *fac*-$[MA_3B_3]$

17.12 (a) Use the VSEPR model to predict the probable shapes of $[IF_6]^+$ and IF_7. (b) Give a plausible chemical equation for the preparation of $[IF_6][SbF_6]$?

(a) The structures of $[IF_6]^+$ and IF_7? The structures are shown below. The structures actually represent the Lewis structures as well, except that the three lone pairs of electrons on each fluorine atom have been omitted. VSEPR theory predicts that a species with six bonding pairs of electrons, like IF_6^+, should be octahedral. Possible structures of species with seven bonding pairs of electrons, like IF_7, were not covered in Section 2.3, but a reasonable and symmetrical structure would be a pentagonal bipyramid.

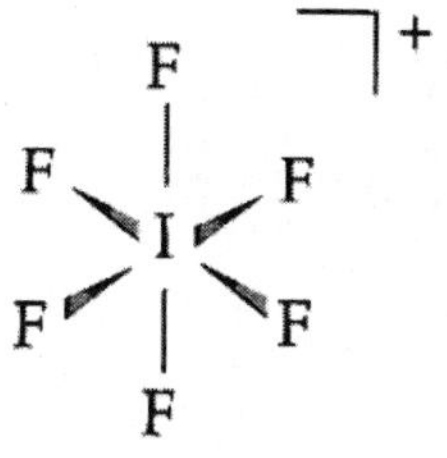

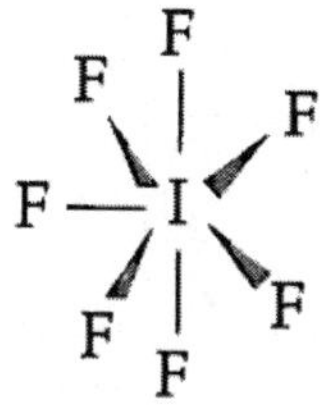

IF_6^+ IF_7

(b) The preparation of $[IF_6]$ $[SbF_6]$? Judging from the structures shown above, it should be possible to abstract an F^- ion from IF_7 to produce IF_6^+. The strong Lewis acid SbF_5 should be used for the fluoride abstraction so that the salt $[IF_6]$ $[SbF_6]$ will result:

$$IF_7 + SbF_5 \rightarrow [IF_6][SbF_6]$$

17.13 Predict the shape of the doubly chlorine–bridged I_2Cl_6 molecule by using the VSEPR model and assign the point group?

I_2Cl_6 is the dimer of ICl_3 and in this molecule 2Cl are bridged atoms. Therefore each I atom contains 12 valence electrons, or 4 bonding pairs and 2 non-bonding pairs. VSEPR theory predicts that the shape around each I atom with 6 pairs of electrons should be octahedral. In this structure 4 Cl atoms occupy the corners of square plane in octahedral geometry. Therefore in the solid state it is present as a planar dimer (I_2Cl_6) with bridging Cl atoms.

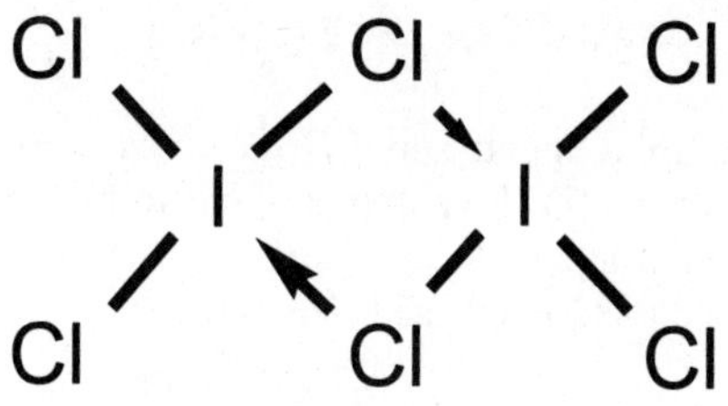

The point group of I_2Cl_6 is D_{2h}.

17.14 Predict the structure and identify the point group of ClO_2F?

The Lewis structure for ClO_2F is shown below. (Note that chlorine is the central atom, not oxygen or fluorine: the heavier, less electronegative element is always the central atom in interhalogen compounds and in compounds containing two different halogens and oxygen). Three bonding pairs and one lone pair yield a trigonal pyramidal geometry, also shown below. This structure possesses only a single symmetry element, a mirror plane that bisects the O–Cl–O angle and contains the Cl and F atoms. Therefore, ClO_2F has C_s symmetry.

The Lewis structure of ClO_2F:

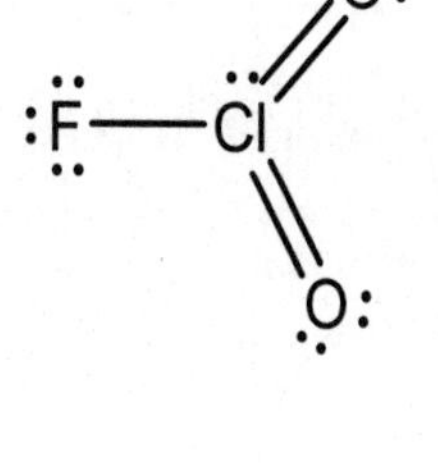

The trigonal pyramidal structure of ClO_2F:

17.15 Predict whether each of the following solutes is likely to make liquid BrF_3 a stronger Lewis acid or a stronger Lewis base: (a) SbF_5, (b) SF_6, (c) CsF?

(a) SbF_5? The acid/base properties of this inorganic solvent result from the following autoionization reaction:

$$2\,BrF_3 \rightarrow BrF_2^+ + BrF_4^-$$

The acidic species is the cation BrF_2^+ and the basic species is the anion BrF_4^-. Adding the powerful Lewis acid SbF_5 will increase the concentration of BrF_2^+, thus increasing the acidity of BrF_3, by the following reaction:

$$BrF_3 + SbF_5 \rightarrow BrF_2^+ + SbF_6^-$$

(b) SF_6? Adding SF_6 will have no effect on the acidity or basicity of BrF_3 because SF_6 is neither Lewis acidic nor Lewis basic.

(c) CsF? Adding CsF, which contains the strong Lewis base F^-, will increase the concentration of BrF_4^-, thus increasing the basicity of BrF_3, by the following reaction:

$$BrF_3 + F^- \rightarrow BrF_4^-$$

17.16 Predict the appearance of the ^{19}F-NMR spectrum of IF_5^+?

Iodine is the central atom in the molecule IF_5. Five unpaired electrons form bonds with five fluorine atoms, plus one lone pair giving a total of six electron pairs. Therefore the structure is octahedral with one position occupied by a lone pair (alternatively described as square based pyramid).

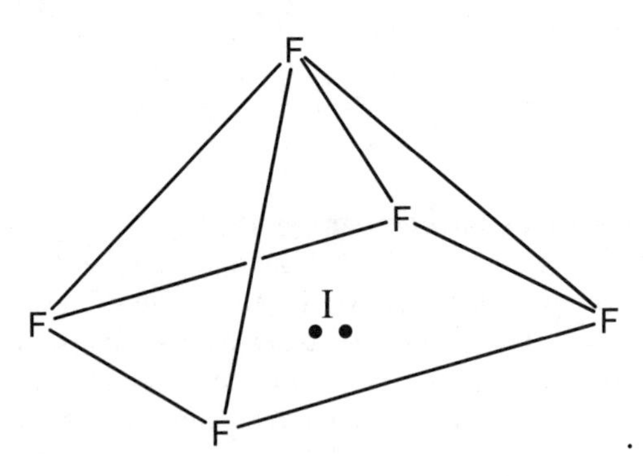

The ^{19}F-NMR spectrum of IF_5^+ contains two resonances to integrate five F atoms. One is doublet for four equatorial F atoms and another one is quintate for axial F atom.

17.17 Predict whether each of the following compounds is likely to be dangerously explosive in contact with BrF_3 and explain your answer: (a) SbF_5, (b) CH_3OH, (c) F_2, (d) S_2Cl_2?

(a) SbF_5? All interhalogens, including BrF_3, are strong oxidizing agents. The antimony atom in SbF_5 is already in its highest oxidation state, and the fluorine atom ligands cannot be chemically oxidized (fluorine is the most electronegative element). Since SbF_5 cannot be oxidized, it will not form an explosive mixture with BrF_3.

(b) CH_3OH? Methanol, being an organic compound, is readily oxidized by strong oxidants. Therefore, you should expect it to form an explosive mixture with BrF_3.
(c) F_2? As stated in part **(a)**, fluorine atoms cannot be oxidized by a simple chemical oxidant like an interhalogen compound. For this reason, it will not form an explosive mixture with BrF_3.
(d) S_2Cl_2? In this compound, sulfur is in the +1 oxidation state. Recall that SF_4 and SF_6 are stable compounds. Therefore, you should expect S_2Cl_2 to be oxidized to higher valent sulfur fluorides. The mixture $S_2Cl_2 + BrF_3$ will be an explosion hazard.

17.18 The formation of Br_3^- from a tetraalkylammonium bromide and Br_2 is only slightly exoergic. Write an equation (or NR for no reaction) for the interaction of $[NR_4][Br_3]$ with I_2 in CH_2Cl_2 solution and give your reasoning?
Since Br_3^- has only a moderate formation constant, you can think of it as being chemically equivalent to $Br_2 + Br^-$. Since IBr is fairly stable, the probable reaction will be:

$$Br_3^- + I_2 \rightarrow 2\,IBr + Br^-$$

The IBr may associate to some extent with Br^- to produce IBr_2^-.

17.19 Explain why $CsI_3(s)$ is stable with respect to the elements but $NaI_3(s)$ is not?
This is an example of an important principle of inorganic chemistry: large cations stabilize large, unstable anions. This trend appears to have a simple electrostatic origin. A detailed analysis of this particular situation is as follows. The enthalpy change for the reaction:

$$NaI_3(s) \rightarrow NaI(s) + I_2(s)$$

is negative, which is just another way of stating that NaI_3 is not stable. This enthalpy change is composed of four terms, as shown here:

$$\begin{aligned} \Delta H &= \text{lattice enthalpy of } NaI_3 \\ &+ I-I_2^- \text{ bond enthalpy} \\ &- \text{lattice enthalpy of } I_2 \\ &- \text{lattice enthalpy of NaI} \end{aligned}$$

The fourth term is larger than the first term, since I^- is smaller than I_3^-. This difference provides the driving force for the reaction to occur as written. If you substitute Cs^+ for Na^+, the two middle terms will remain constant. Now, the fourth term is still larger than the first term, *but by a significantly smaller amount*. The net result is that ΔH for the cesium system is positive. In the limit where the cation becomes infinitely large, the difference between the first and fourth terms becomes negligible.

17.20 Write plausible Lewis structures for (a) ClO_2 and (b) I_2O_5 and predict their shapes and the associated point group?
The Lewis structure and the predicted shape of ClO_2 are shown below. The angular shape is a consequence of repulsions between the bonding and nonbonding electrons. With three nonbonding electrons, the O–Cl–O angle in ClO_2 is 118°. With four nonbonding electrons, as in ClO_2^-, the repulsions are greater and the O–Cl–O angle is only 111°.

O — Cl — O

Cl
O O

(b) I_2O_5? The Lewis structure and the predicted shape of I_2O_5 are shown below. The central O atom has two bonding pairs of electrons and two lone pairs, like H_2O, so it should be no surprise that the I–O–I bond angle is less than 180° (it is 139°). Each I atom is trigonal pyramidal, since it has three bonding pairs and one lone pair. If you had difficulty working this exercise, you should review VSEPR theory, covered in Section 2.3.

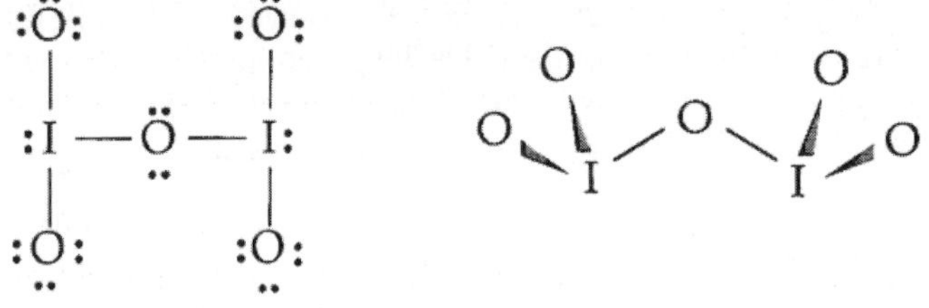

17.21 (a) Give the formulas and the probable relative acidities of perbromic acid and periodic acid. (b) Which is the more stable?

The formulas are $HBrO_4$ and H_5IO_6. The difference lies in iodine's ability to expand its coordination shell, a direct consequence of its large size. Recall from Section 4.4 that the relative strength of an oxoacid can be estimated from the q/p ratio (p is the number of *oxo* groups and q is the number of OH groups attached to the central atom). A high value of p/q correlates with strong acidity, while a low value correlates with weak acidity. For $HBrO_4$, $p/q = 3/1$, so it is a strong acid. For H_5IO_6, $p/q = 1/5$, so it is a weak acid.

(b) Relative stabilities? Periodic acid is thermodynamically more stable with respect to reduction than perbromic acid. Bromine, like its period 4 neighbors arsenic and selenium, is more oxidizing in its highest oxidation state than the members of the group immediately above and below.

17.22 (a) Describe the expected trend in the standard potential of an oxoanion in a solution with decreasing pH. (b) Demonstrate this phenomenon by calculating the reduction potential of ClO_4^- at pH = 7 and comparing it with the tabulated value at pH = 0?

(a) The expected trend? The trend is that E decreases as the pH increases.

(b) E at pH 0 and pH 7 for ClO_4^-? The balanced equation for the reduction of ClO_4^- is:

$$ClO_4^-(aq) + 2\,H^+(aq) + 2\,e^- \rightarrow ClO_3^-(aq) + H_2O(l)$$

The value of $E°$ for this reaction is 1.201 V (see Appendix 2). The potential at any $[H^+]$, given by the Nernst equation, is

$$E = E° - (0.059\text{ V}/2)(\log([ClO_3^-]/[ClO_4^-][H^+]^2))$$

At pH 7, $[H^+] = 10^{-7}$ M. Assuming that both perchlorate and chlorate ions are present at unit activity, at pH 7 the reduction potential is

$$E = 1.201\text{ V} - (0.0295\text{ V})(\log 10^{14}) = 1.201\text{ V} - 0.413\text{ V} = 0.788\text{ V}$$

17.23 With regard to the general influence of pH on the standard potentials of oxoanions, explain why the disproportionation of an oxoanion is often promoted by low pH?

Let's take ClO_3^- as an example. Its disproportionation can be broken down into a reduction and an oxidation, as follows:

$$ClO_3^-(aq) + 6\,H^+(aq) + 6\,e^- \rightarrow Cl^-(aq) + 3\,H_2O(l)$$

$$3\,ClO_3^-(aq) + 3H_2O(l) \rightarrow 3\,ClO_4^-(aq) + 6\,H^+(aq) + 6\,e^-$$

Any effect that changing the pH has on the potential for the reduction reaction will be counteracted by an equal but opposite change on the potential for the oxidation reaction. In other words, the net reaction does not include H^+, so the net potential for the disproportionation cannot be pH dependent. Therefore, the promotion of disproportionation reactions of some oxoanions at low pH *cannot* be a thermodynamic promotion. Low pH results in a kinetic promotion: protonation of an *oxo* group aids oxygen–halogen bond scission (see Section 17.11). The disproportionation reactions have the same driving force at high pH and at low pH, but they are much faster at low pH.

17.24 Which oxidizing agent reacts more readily in dilute aqueous solution, perchloric acid or periodic acid? Give a mechanistic explanation for the difference?

Periodic acid is by far the quicker oxidant. Recall that it exists in two forms in aqueous solution, H_5IO_6 (the predominant form) and HIO_4. Even though the concentration of HIO_4 is low, this four-coordinate species can form a complex with a potential reducing agent, providing for an efficient and rapid mechanism by which the redox reaction can occur (see Section 17.13).

17.25 (a) For which of the following anions is disproportionation thermodynamically favourable in acidic solution: OCl_2, ClO_2^-, ClO_2^-, and ClO_4^-? (If you do not know the properties of these ions, determine them from a table of standard potentials.) (b) For which of the favourable cases is the reaction very slow at room temperature?

The Frost diagram for chlorine in aqueous acid solution is shown below. It is worth your while to construct this diagram yourself from the potential data in Resource Section 2. Then, if you connect the points for Cl^- and ClO_4^- with a line, you will see that the intermediate oxidation state species HClO, $HClO_2$, and ClO_3^- lie above that line. Therefore, they are unstable with respect to disproportionation. As discussed in Section 17.13, the redox reactions of halogen *oxo* anions become progressively faster as the oxidation number of the halogen *decreases*. Therefore, the rates of disproportionation are probably HClO > $HClO_2$ > ClO_3^-. Note that ClO_4^- cannot undergo disproportionation, since there are no species with a higher oxidation number.

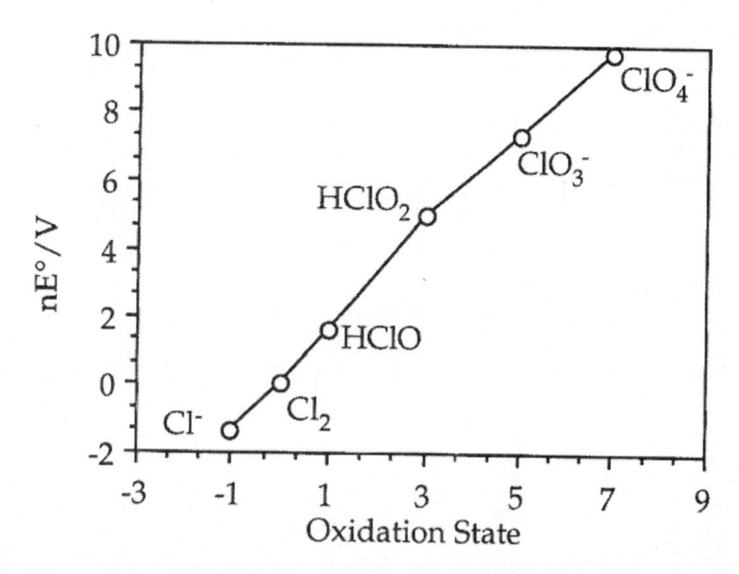

17.26 Which of the following compounds present an explosion hazard? (a) NH_4ClO_4, (b) $Mg(ClO_4)_2$, (c) $NaClO_4$, (d) $[Fe(H_2O)_6][ClO_4]_2$. Explain your reasoning?

(a) NH_4ClO_4? The key to answering this question is to decide if the perchlorate ion, which is a very strong oxidant, is present along with a species that can be oxidized. If so, the compound *does* represent an explosion hazard. Ammonium perchlorate is a dangerous compound, since the N atom of the NH_4^+ ion is in its lowest oxidation state and can be oxidized.

(b) $Mg(ClO_4)_2$? Since Mg(II) cannot be oxidized to a higher oxidation state, magnesium perchlorate is a stable compound and is not an explosion hazard.

(c) $NaClO_4$? The same answer applies here as above for magnesium perchlorate. Sodium has only one common oxidation state. Even the strongest oxidants, such as F_2, FOOF, and ClF_3, cannot oxidize Na(I) to Na(II).

(d) $[Fe(H_2O)_6]$ $[ClO_4]_2$? Although the H_2O ligands cannot be oxidized, the metal ion can. This compound presents an explosion hazard, since Fe(II) can be oxidized to Fe(III) by a strong oxidant such as perchlorate ion.

17.27 Use standard potentials to predict which of the following will be oxidized by ClO^- ions in acidic conditions: (a) Cr^{3+}, (b) V^{3+}, (c) Fe^{2+}, (d) Co^{2+}?

(a) Cr^{3+}? No. **(b) V^{3+}?** Yes. **(c) Fe^{2+}?** Yes. **(d) Co^{2+}?** No. The reduction potential of $Co^{3+} \rightarrow Co^{2+}$ is 1.92 V, which is higher than the 1.630 V that corresponds with a ClO^- reduction to Cl_2.

Chapter 18 The Group 18 Elements

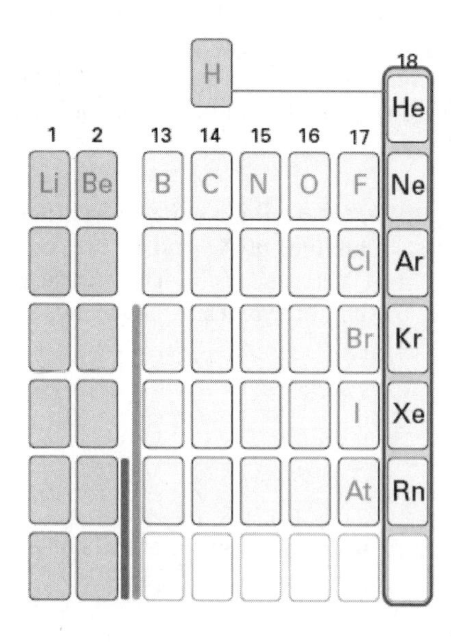

The Group 18 elements, helium, neon, argon, krypton, xenon, and radon, are all monatomic gases. They are the least reactive elements. The existence of the Group 18 elements was not suspected until late in the nineteenth century, and their discovery led to the restructuring of the periodic table and played a key role in the development of important bonding theories. They have been called the rare gases and the inert gases, and are currently called the **noble gases.** This name is now accepted because it gives the sense of low but significant reactivity.

S18.1 Write a balanced equation for the decomposition of xenate ions in basic solution for the production of perxenate ions, xenon, and oxygen? You are told that xenate ($HXeO_4^-$) decomposes to perxenate (XeO_6^{4-}), xenon, and oxygen, so the equation you must balance is:

$$HXeO_4^- \rightarrow XeO_6^{4-} + Xe + O_2 \text{ (not balanced)}$$

Since the reaction occurs in basic solution, we can use OH^- and H_2O to balance the equation. You notice immediately that there is no species containing hydrogen on the right hand-side of the equation, so obviously H_2O will go on the right and OH^- will go on the left. The balanced equation is:

$$2\,HXeO_4^-(aq) + 2\,OH^-(aq) \rightarrow XeO_6^{4-}(aq) + Xe(g) + O_2(g) + 2\,H_2O(l)$$

Since the products are perxenate (an Xe(VIII) species) and elemental xenon, this reaction is a disproportionation of the Xe(VI) species $HXeO_4^-$. Oxygen is produced from a thermodynamically unstable intermediate of Xe(IV), possibly as follows:

$$HXeO_3(aq) \rightarrow Xe(g) + O_2(g) + OH^-(aq)$$

18.1 Explain why helium is present in low concentration in the atmosphere even though it is the second most abundant element in the universe? All of the original He and H_2 that was present in the earth's atmosphere when the earth originally formed has been lost. Earth's gravitational field is not strong enough to hold these light gases, and they eventually diffuse away into space. The small amount of helium that is present in today's atmosphere is the product of ongoing radioactive decay.

18.2 Which of the noble gases would you choose as (a) The lowest-temperature refrigerant? The noble gas with the lowest boiling point would best serve as the lowest-temperature refrigerant. Helium, with a boiling point of 4.2 K, is the refrigerant of choice for very low-temperature applications, such as cooling superconducting magnets in modern NMR spectrometers. The boiling points of the noble gases are directly related to their size, or more correctly, to their number of electrons. The larger atoms Kr and Xe are more polarizable than He or Ne, so Kr and Xe experience larger van der Waals attractions and have higher boiling points.

(b) An electric discharge light source requiring a safe gas with the lowest ionization energy? Of the "safe" noble gases, i.e., He–Xe, the largest one, Xe, has the lowest ionization potential (see Table 18.1). Radon is even larger than Xe and has a lower ionization potential, but since it is radioactive it is not safe.
(c) The least expensive inert atmosphere? Geographic location might play a role in your answer to this question. Since Ar is so plentiful in the atmosphere relative to the other noble gases, it is generally cheaper than helium, which is rare in the atmosphere. However, a great deal of helium is collected as a byproduct of natural gas production. In places where large amounts of natural gas are produced, such as the United States, helium may be marginally less expensive than argon.

18.3 **By means of balanced chemical equations and a statement of conditions, describe a suitable synthesis of (a) XeF_2?** This compound can be prepared in two different ways. First, a mixture of Xe and F_2 that contains an excess of Xe is heated to 400°C. The excess Xe prevents the formation of XeF_4 and XeF_6. The second way is to photolyze a mixture of Xe and F_2 in a glass reaction vessel. For either method of synthesis, the balanced equation is:

$$Xe(g) + F_2(g) \rightarrow XeF_2(s)$$

At 400°C the product is a gas, but at room temperature it is a solid.

(b) XeF_6? For the synthesis of this compound you would want also to use a high temperature, but unlike the synthesis of XeF_2, you want to have a *large* excess of F_2:

$$Xe(g) + 3\ F_2(g) \rightarrow XeF_6(s)$$

As with XeF_2, xenon hexafluoride is a solid at room temperature.

(c) XeO_3? This compound is endoergic, so it cannot be prepared directly from the elements. However, a sample of XeF_6 can be carefully hydrolyzed to form the product:

$$XeF_6(s) + 3\ H_2O(l) \rightarrow XeO_3(s) + 6\ HF(g)$$

If a large excess of water is used, an aqueous solution of XeO_3 is formed instead.

18.4 **Draw the Lewis structures of (a) $XeOF_4$? (b) XeO_2F_2? (c) XeO_6^{4-}?**

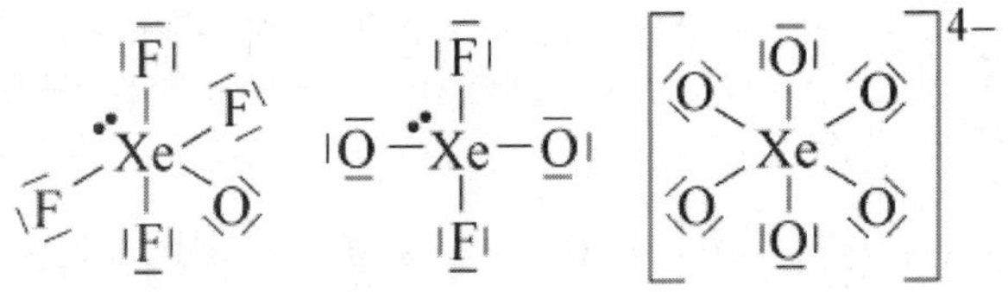

$XeOF_4$ XeO_2F_2 $[XeO_6]^{4-}$

18.5 **Give the formula and describe the structure of a noble gas species that is isostructural with (a) ICl_4^-?** The Lewis structure of this anion has four bonding pairs and two lone pairs of electrons around the central iodine atom. Therefore, the anion has a square-planar geometry. The noble gas compound that is isostructural is XeF_4, as shown below.
(b) IBr_2^-? The Lewis structure of this anion has two bonding pairs and three lone pairs of electrons around the central iodine atom. Therefore, the anion has a linear geometry. The noble gas compound that is isostructural is XeF_2, as shown below.

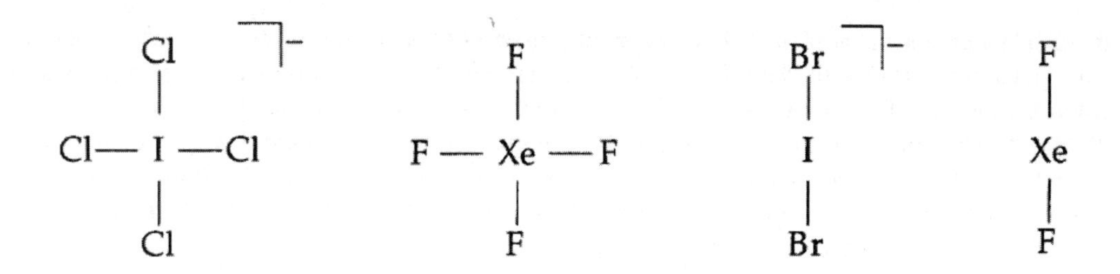

(c) BrO_3^-? The Lewis structure of this anion has three bonding pairs and one lone pair of electrons around the central bromine atom. Therefore, the anion has a trigonal pyramidal geometry. The noble gas compound that is isostructural is XeO_3, as shown below.

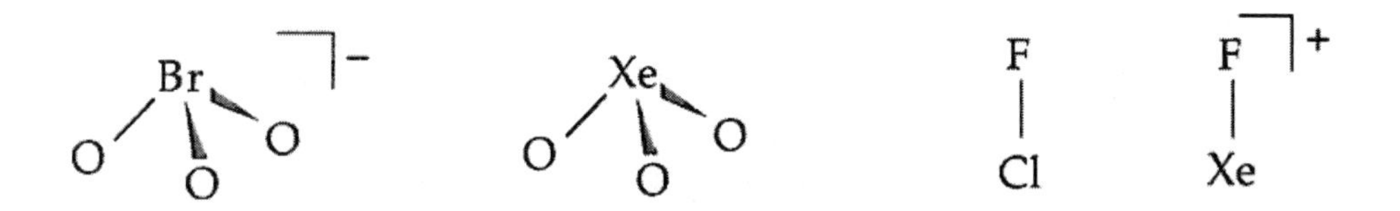

(d) ClF? This diatomic molecule does not have a *molecular* noble gas counterpart (i.e., all noble gas compounds have three or more atoms). However, it is isostructural with the cation XeF^+, as shown above.

18.6 **(a) Give a Lewis structure for XeF_7^-?** The Lewis structure is shown below.

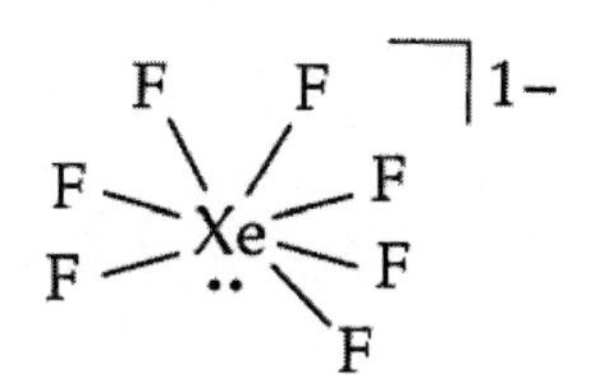

(b) Speculate on its possible structures by using the VSEPR model and analogy with other xenon fluoride anions? The structure of XeF_5^-, with seven pairs of electrons, is based on a pentagonal bipyramidal array of electron pairs with two stereochemically active lone pairs (see Structure 9). The ion XeF_8^{2-}, on the other hand, has nine pairs of electrons, but only eight are stereochemically active (see Structure 10). It is possible that the structure of XeF_7^-, with eight pairs, would be a pentagonal bipyramid, with one stereochemically inactive lone pair.

18.7 **Use molecular orbital theory to calculate the bond order of the diatomic species E_2^+ with E=He and Ne?** According to molecular orbital theory, the electronic configuration of $He_2^{2+}$is $1\sigma_g^2 1\sigma_u^1$. There are two electrons in bonding σ orbitals and one electron in an antibonding σ orbital, therefore the bond order is ½(2 – 1) = 0.5. The electronic configuration of $Ne_2^{2+}$is $1\sigma_g^2 1\sigma_u^2 2\sigma_g^2 1\pi_u^4 1\pi_g^4 2\sigma_u^1$. Recall that the bonding orbitals are $1\sigma_g$, $2\sigma_g$, $1\pi_u$, and $2\pi_u$. The bond order is therefore ½(8 – 7) = 0.5.

18.8 **Identify the xenon compounds A, B, C, D, and E?**
$Xe + F_2 \rightarrow XeF_2$; A = XeF_2(g); see Section 18.5
XeF_2(g) + $MeBF_2 \rightarrow [XeF]^+[MeBF_3]^-$; B = $[XeF]^+[MeBF_3]^-$; see Section 18.6
$Xe + F_2$ (excess) $\rightarrow XeF_6$; C = XeF_6; see Section 18.5
$XeF_6 + H_2O \rightarrow XeO_3$; D = XeO_3; see Section 18.6
$Xe + 2F_2 \rightarrow XeF_4$(g); E = XeF_4 (g); see Section 18.5.

18.9 **Predict the appearance of the ^{129}Xe-NMR spectrum of $XeOF_3^+$.**
A 1:3:3:1 quartet (both are spin ½ nuclei)

18.10 Predict the appearance of the ^{19}F-NMR spectrum of $XeOF_4$.

The molecular structure of XeOF4 is a square pyramid with the O-atom residing at the apical position. The four fluorine atoms are located at the equatorial positions in a basal plane. Because of the equivalency of the Fluorine atoms, only one strong line is expected. Because of the interactions between Fluorine and the isotopes of Xe [^{129}Xe (spin = ½) and ^{131}Xe (spin = 3/2)] we do expect two additional weak features. So the predicted spectra will consist of a strong central line and two other lines symmetrically distributed around this central line.

Chapter 19 The *d*-block Metals

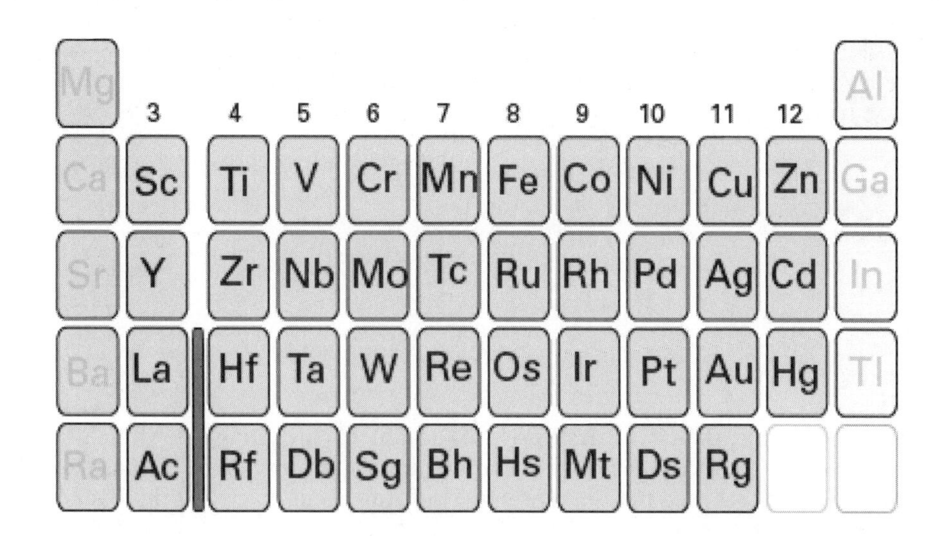

The thirty ***d*-block elements** have chemical properties that are central to both industry and contemporary research. The two terms "*d*-block metal" and "transition metal" are often used interchangeably; however, they do not mean the same thing. The IUPAC definition of a **transition element** is that it is an element that has an incomplete *d* subshell in either the neutral atom or its ions. Thus the Group 12 elements (Zn, Cd, Hg) are members of the *d* block but are not transition elements. Elements toward the left of the *d* block are often referred to as early and those toward the right are referred to as late. Simple binary compounds, such as metal halides and oxides, often follow systematic trends; however, when the metal is in a low oxidation state, interesting variations such as metal–metal bonding may be encountered.

S19.1 **Refer to the appropriate Latimer diagram in *Resource section* 3 and identify the oxidation state and formula of the species that is thermodynamically favoured when an acidic aqueous solution of V^{2+} is exposed to oxygen?** The Latimer diagram for vanadium in aqueous acid is shown below (see Resource Section 3; reduction potentials are given in volts).

$$VO_2^{+} \xrightarrow{1.000} VO^{2+} \xrightarrow{0.337} V^{3+} \xrightarrow{-0.255} V^{2+}$$

$$VO_2^{+} \xrightarrow{0.361} V^{2+}$$

Since the reduction potential for the O_2/H_2O couple is 1.229 V, V^{2+} will be oxidized all the way to VO_2^{+} as long as sufficient O_2 is present. The net reaction is:

$$V^{2+} + O_2 + 2\,H^{+} \rightarrow VO_2^{+} + H_2O \quad E^{\circ} = 1.229\text{V} - 0.361\text{ V} = 0.868\text{ V}$$

S19.2 **Suggest a use for molybdenum(IV) sulfide that makes use of its solid-state structure. Rationalize your suggestion?** MoS_2 used as an effective lubricant. The layered structure of molybdenum(IV) sulfide is the same as that shown in Figure 19.14 for CdI_2. Although the Mo–S bonds are undoubtedly quite strong, the S…S interactions between adjacent S–Mo–S layers are weak and easily disrupted (these are best thought of as van der Waals interactions). The slipperiness of MoS_2 is because of the ease with which one layer can glide over another.

S19.3 **Describe the probable structure of the compound formed when Re_3Cl_9 is dissolved in a solvent containing PPh_3?** Re_3Cl_9 molecules are linked in the solid state by weak intercluster chlorine bridges. When ligands are added, say Cl^- or PPh_3, these bridges are broken and discrete molecular species such as $Re_3Cl_{12}^{3-}$ or $Re_3Cl_9(PPh_3)_3$ are formed. Sterically, the most favourable place for each bulky triphenylphosphine ligand to go is in the terminal position in the Re_3 plane, as shown below.

Here is a portion of the structure of $Re_3Cl_9(PPh_3)_3$, showing the six ligands in the same plane as the Re_3 triangle. Six other Cl^- ligands are part of the molecule. Three terminal Re–Cl bonds are above the plane shown and three are below the plane shown.

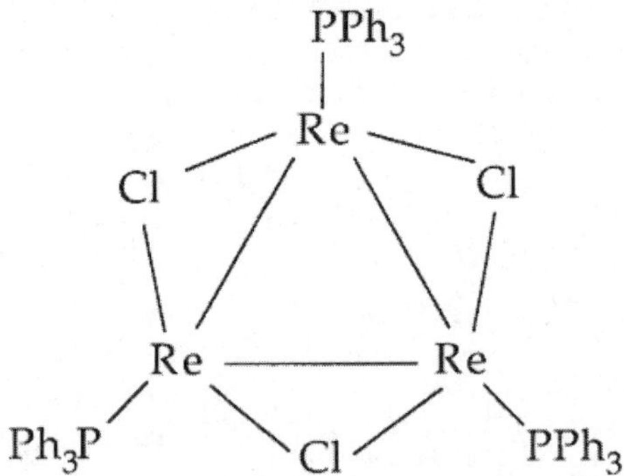

19.1 Without reference to a periodic table, sketch the first series of the *d* block, including the symbols of the elements. Indicate those elements for which the group oxidation number is common by C, those for which the group oxidation number can be reached but is a powerful oxidizing agent by O, and those for which the group oxidation number is not achieved by N?

The metals from scandium through zinc are shown below, with the group numbers written above. Each metal has a notation indicating whether its group oxidation number is common and relatively unreactive (C), a strong oxidant (O), or not achieved (N). Examples of the first three are $ScCl_3$, TiO_2, and the VO_2^+ ion. Examples of the next three are CrO_3, MnO_4^-, and FeO_4^{2-}.

3	4	5	6	7	8	9	10	11	12
Sc	Ti	V	Cr	Mn	Fe	Co	Ni	Cu	Zn
C	C	C	O	O	N	N	N	N	N

19.2 Explain why the enthalpy of sublimation of Re(s) is significantly greater than that of Mn(s)?

The enthalpy of sublimation of Re(s) is 704 kJ mol^{-1}, whereas that of Mn(s) is 221 kJ mol^{-1}. Recall that metal–metal bonding becomes stronger on descending a group because the the *d* orbitals become larger. As the atoms become heavier, more energy is needed to varporize them.

19.3 State the trend in the stability of the group oxidation state on descending a group of metallic elements in the *d* block. Illustrate the trend using standard potentials in acidic solution for Groups 5 and 6.

In the *d* block of the periodic table, the group oxidation number *increases* in stability as you descend a group. For example, CrO_3 is a strong oxidizing agent but WO_3 is not. In the *p* block, the group oxidation number *decreases* in stability as you descend a group. For example, Tl^{3+} is a strong oxidizing agent but Al^{3+} is not. Values for $E°$ are shown below. Remember that the less positive (or the more negative) a reduction potential, the greater the stability of the oxidized form of the redox couple in question.

VO_2^+/V – 0.236 V	**$Cr_2O_7^{2-}/Cr$**	**–0.320 V**	**Al^{3+}/Al**	**–1.676 V**
Nb_2O_5/Nb – 0.65 V	**H_2MoO_4/MO**	**0.114 V**	**Ga^{3+}/Ga**	**– 0.529 V**
Ta_2O_5/Ta – 0.81 V	**WO_3/W**	**– 0.090 V**	**In^{3+}/In**	**– 0.338 V**
			Tl^{3+}/Tl	**0.72 V**

19.4 For each part, give balanced chemical equations or NR (for no reaction) and rationalize your answer in terms of trends in oxidation states.

(a) Cr^{2+} (aq) + Fe^{3+} (aq) →? As you move from left to right across the *d* block, stable oxidation states in aqueous solution tend to get lower (for example, from TiO^{2+} and VO_2^+ on the left to Co^{2+}, Ni^{2+}, Cu^{2+}, and Zn^{2+} on the right). Based on this trend, you should expect Cr^{2+} to be oxidized to Cr^{3+} at the expense of Fe^{3+}, which will be reduced to Fe^{2+}. That is, iron is to the right of chromium, and the trend is left higher, right lower. In fact, the net potential for this redox reaction in aqueous acid is 1.195 V (see Resource Section 3).

$$Cr^{2+}(aq) + Fe^{3+}(aq) \rightarrow Cr^{3+}(aq) + Fe^{2+}(aq) \qquad E^\circ = 1.195\ V$$

(b) CrO_4^{2-} (aq) + MoO_2 (s) →? Recall the trend that higher oxidation numbers become increasingly more stable as you descend a group in the *d* block. Since you are given two compounds containing Cr(VI) and Mo(IV), you should expect a redox reaction to occur, with the oxidation of MoO_2 and reduction of CrO_4^{2-}. The products should be Cr^{3+} and H_2MoO_4 (see Resource Section 3). Note that Cr(V) and Cr(IV) are unstable with respect to disproportionation.

$$2\ CrO_4^{2-} + 3\ MoO_2(s) + 10\ H^+ \rightarrow 2\ Cr^{3+} + 3\ H_2MoO_4 + 2\ H_2O \quad E^\circ = 0.73\ V$$

(c) MnO_4^- (aq) + Cr^{3+} (aq) →? This is another case where you should apply the trend left higher, right lower, predicting a reaction between these two species in which Cr^{3+} is oxidized and MnO_4^- is reduced. However, since these two elements are immediate neighbors, you should not expect the net E° to be very large. In fact, the products are Mn^{2+} and $Cr_2O_7^{2-}$ and E° = 0.12 V.

$$6\ MnO_4^- + 10Cr^{3+} + 11H_2O \rightarrow 6Mn^{2+} + 5Cr_2O_7^{2-} + 22\ H^+ \qquad E^\circ = 0.12\ V$$

19.5 (a) Which ion, Ni^{2+}(aq) or Mn^{2+}(aq), is more likely to form a sulfide in the presence of H_2S? (b) Rationalize your answer with the trends in hard and soft character across Period 4. (c) Give a balanced chemical equation for the reaction. In general, hardness decreases and softness increases from left to right in the *d* block, especially when comparing elements in the same oxidation state. Therefore, it is more likely for Ni^{2+} to form a sulfide than it is for Mn^{2+} to form a sulfide:

$$Ni^{2+}(aq) + H_2S(aq) \rightarrow NiS(s) + 2\ H^+(aq)$$

19.6 Preferably without reference to the text (a) write out the *d* block of the periodic table, (b) indicate the metals that form difluorides with the rutile or fluorite structures, and (c) indicate the region of the periodic table in which metal-metal bonded halide compounds are formed, giving one example?

The *d* block of the periodic table is shown below. Those metals that form difluorides having the rutile structure are indicated. The region in which metal–metal bonded halide compounds are found is indicated with a bold border. Examples of metal–metal bonded halides are Sc_5Cl_6, ZrCl, and Re_3Cl_9.

Sc	Ti rutile	V rutile	Cr rutile	Mn rutile	Fe rutile	Co rutile	Ni rutile	Cu	Zn
Y	Zr	Nb	Mo	Tc	Ru	Rh	Pd	Ag	Cd
La/Lu	Hf	Ta	W	Re	Os	Ir	Pt	Au	Hg

19.7 Write a balanced chemical equation for the reaction that occurs when *cis*-$[RuLCl(OH_2)]^+$ (see Fig. 19.9) in acidic solution at +0.2 V is made strongly basic at the same potential. Write a balanced equation for each of the successive reactions when this same complex at pH = 6 and +0.2 V is exposed to progressively more

oxidizing environments up to +1.0 V. Give other examples and a reason for the redox state of the metal centre affecting the extent of protonation of coordinated oxygen. According to Figure 19.9, the Ru(II) complex *cis*-$[RuLCl(OH_2)]^+$ is stable at 0.2 V below about pH 8. What this means is that if there is a redox couple (Red and Ox) with $E^o = 0.2$ V in solution with the ruthenium complex, no reaction will occur. Assume that E^o for Red/Ox does not change with pH. Now, if you raise the pH past 8, the ruthenium complex will be oxidized and Ox will be reduced:

$$cis\text{-}[Ru^{II}LCl(OH_2)]^+ + Ox + OH^- \rightarrow cis\text{-}[Ru^{III}LCl(OH)]^+ + Red + H_2O$$

Note that the charge on the two ruthenium complexes is the same, since the H_2O ligand in the Ru(II) complex has become a OH^- ligand in the Ru(III) complex. If the pH is held at 6 and the potential is raised (by adding stronger and stronger oxidants to the solution), the original Ru(II) complex will be oxidized to a Ru(III) complex and then to a Ru(IV) complex, with a single deprotonation occurring at each oxidation:

$$cis\text{-}[Ru^{II}LCl(OH_2)]^+ + H_2O \rightarrow cis\text{-}[Ru^{III}LCl(OH)]^+ + H_3O^+ + e^-$$

$$cis\text{-}[Ru^{III}LCl(OH_2)]^+ + H_2O \rightarrow cis\text{-}[Ru^{IV}LCl(OH)]^+ + H_3O^+ + e^-$$

As the oxidation state of the metal increases, its ability to accept electron density from an OH^- or O^{2-} ligand through π-bonding also increases, and deprotonation of the original H_2O ligand occurs.

19.8 **Give plausible balanced chemical reactions (or NR for no reaction) for the following combinations, and state the basis for your answer: (a) MoO_4^{2-} (aq) plus Fe^{2+} (aq) in acidic solution?** The only possible reaction that might occur here is a redox reaction. However, since the Fe^{3+}/Fe^{2+} reduction potential is 0.771 V and since there is no redox couple of molybdenum involving MoO_4^{2-} that is more positive than 0.771 V, no reaction will occur.

(b) The preparation of $[Mo_6O_{19}]^{2-}$(aq) from K_2MoO_4(s)? Acidification of solutions of MoO_4^{2-} will produce polyoxometallates such as $[Mo_6O_{19}]^{2-}$ (see Section 19.8(c), *Polyoxometallates*). The balanced equation is:

$$6\ MoO_4^{2-}(aq) + 10\ H^+(aq) \rightarrow [Mo_6O_{19}]^{2-}(aq) + 5\ H_2O(l)$$

(c) $ReCl_5$ (s) plus $KMnO_4$(aq)? In this case you have a compound of Re(V) and Mn(VII). Recall that the group oxidation number is more readily achieved by the heavier metals in a *d*- block group. Therefore, the rhenium compound will be oxidized and the managanese compound will be reduced. One plausible reaction is:

$$5\ ReCl_5(s) + 2\ MnO_4^-(aq) + 12\ H_2O(l) \rightarrow$$
$$5\ ReO_4^-(aq) + 2\ Mn^{2+}(aq) + 25\ Cl^-(aq) + 24\ H^+(aq)$$

(d) $MoCl_2$(s) plus warm HBr(aq)? Aqueous Br^- is a reducing agent in that it can form Br_2 (although E^o, at ~ 1 V, is high). Therefore, a plausible reaction is the following cluster formation:

$$6\ MoCl_2(s) + 2\ Br^-(aq) \rightarrow [Mo_6Cl_{12}]^{2-}(aq) + Br_2(aq)$$

(e) TiO(s) with HCl(aq) under an inert atmosphere?
Titanium in the 2+ oxidation state in aqueous acid is a strong reducing agent, strong enough even to reduce protons to H2. The balanced equation is:

$$2TiO\ (s) + 6H^+\ (aq) \rightarrow 2Ti^{3+} + H_2(g) + 2H_2O(l)\quad E^\circ = +0.37\ V.$$

(f) Cd(s) added to Hg^{2+}(aq)?
Of the three metals in group 12, mercury is by far the most noble. The consequence of this is that either zinc or cadmium metal will reduce Hg2+ to metallic mercury according the following equation:

$$Hg^{2+}\ (aq) + Cd(s) \rightarrow Hg(l) + Cd^{2+}\ (aq)$$

19.9 Speculate on the structures of the following species and present bonding models to justify your answers: (a) $[Re(O)_2(py)_4]^+$, (b) $[V(O)_2(ox)_2]^{3-}$, (c) $[Mo(O)_2(CN)_4]^{4-}$, (d) $[VOCl_4]^{2-}$? The first three of these complexes are *dioxo* complexes and will have *cis*-dioxometal structural units, as reviewed in Section 19.8 (c), *Mononuclear oxo complexes*. The reason that a *cis* geometry is observed rather than a *trans* geometry is that in the *cis* geometry the two *oxo* ligands only have to share a single *d* orbital for O→M π-bonding. In the *trans* geometry, on the other hand, the two *oxo* ligands must share the same two *d* orbitals. The structures of the first three complexes are shown below.

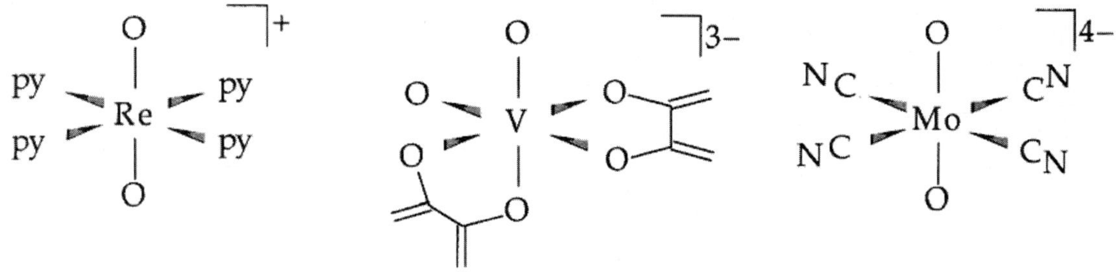

d^2 *trans* $\quad$ d^0 *cis* $\quad$ d^2 *trans*

The complex $VOCl_4^{2-}$, like $VO(acac)_2$, has a square-pyramidal structure with an apical *oxo* ligand. Its structure is shown below.

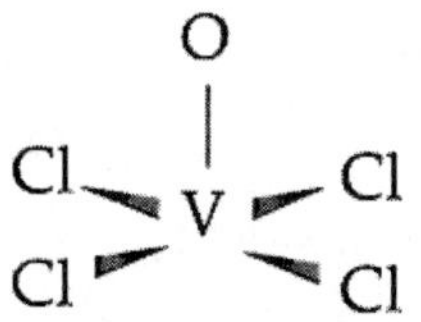

19.10 Which of the following are likely to have structures that are typical of (a) predominantly ionic, (b) significantly covalent, (c) metal-metal bonded compounds: NiI_2, $NbCl_4$, FeF_2, PtS, and WCl_2? Rationalize the differences and speculate on the structures? (a) NiI_2? This is a typical example of a divalent metal diiodide. It should be an ionic compound with a significant degree of covalent character, since I^- is such a soft and polarizable anion. You should expect NiI_2 to have a layered structure as opposed to a more ionic structure such as the rutile structure.

(b) $NbCl_4$? There are very few ionic compounds with metal ions in the 4+ oxidation state. You should expect $NbCl_4$ to be significantly covalent. You should not expect it to contain metal–metal bonds, since the oxidation number is too high. In addition, you should not expect it to be molecular, since a relatively large *d*-block metal would naturally be more than four-coordinate. Therefore, a structure with bridging halides is likely.

(c) FeF_2? A compound containing a divalent metal ion and hard fluoride ions is going to be ionic. Expect the rutile structure.

(d) PtS? With a soft metal ion like Pt^{2+} and a soft anion like S^{2-}, you should expect a significant amount of covalent character. Some metal monosulfides exhibit the nickel-arsenide structure, so that would be a good choice here. In fact, PtS exhibits a different type of covalent solid-state structure, with four-coordinate square-planar Pt^{2+} ions. Most of the d^8 ions, including Pd^{2+} and Pt^{2+}, tend to form structures containing four-coordinate square-planar metal ions, as discussed in Section 19.6, *Noble character*.

(e) WCl_2? This is a compound of a low valent period 4 or 5 early *d*-block metal. This is precisely the kind of situation in which metal–metal bonding occurs.

19.11 Indicate the probable occupancy of s, p, and d bonding and antibonding orbitals, and the bond order for the following tetragonal prismatic complexes? (a) $[Mo_2(O_2CCH_3)_4]$? This is a neutral molecule with two molybdenum ions and four acetate ions. Therefore, each of the two molybdenum ions has a 2+ charge and a d^4

configuration. Consequently, the Mo_2 fragment has eight electrons and a $\sigma^2\pi^4\delta^2$ configuration. This molecule has a molybdenum-molybdenum quadruple bond.

(b) $[Cr_2(O_2CC_2H_5)_4]$? This molecule is very similar to the one in part (a). Instead of Mo^{2+} ions, it contains another divalent ion from group 6, Cr^{2+}. Instead of acetate ions, it contains propionate ions. Consequently, it has the same electron configuration as far as the metal–metal bonding orbitals are concerned, $\sigma^2\pi^4\delta^2$. This molecule has a chromium–chromium quadruple bond.

(c) $[Cu_2(O_2CCH_3)_4]$? With two copper ions and four acetate ions, this molecule contains two d^9 Cu^{2+} ions. With 18 electrons between the two Cu^{2+} ions, all of the metal–metal bonding and antibonding orbitals are filled. Sixteen of the electrons from the $\sigma^2\pi^4\delta^2\delta^{*2}\pi^{*4}\sigma^{*2}$ configuration. Each copper ion also has an additional unpaired electron in its $d_{x^2-y^2}$ orbital (the one that points directly at the acetate oxygen ligands). There is no metal–metal bond in this molecule.

19.12 Explain the differences in the following redox couples, measured at 25°C? Higher oxidation states become more stable on descending a group. The given trends in the redox potentials corroborate this observation because the ReO_4^-/ReO_2 redox couple has the lowest voltage. The order of stability of the higher oxidation states for group 8 transition metals given here is $Re^{+7} > Tc^{+7} > Mn^{+7}$.

19.13 Addition of sodium ethanoate to aqueous solutions of Cr(II) gives a red diamagnetic product. Draw the structure of the product, noting any features of interest? The structure of the complex (shown below) can be described with a quadruple Cr to Cr bond and four bidentate bridging acetate ligands. The formula of the complex would be $[Cr_2(\mu\text{-}CH_3CO_2)_4]$.

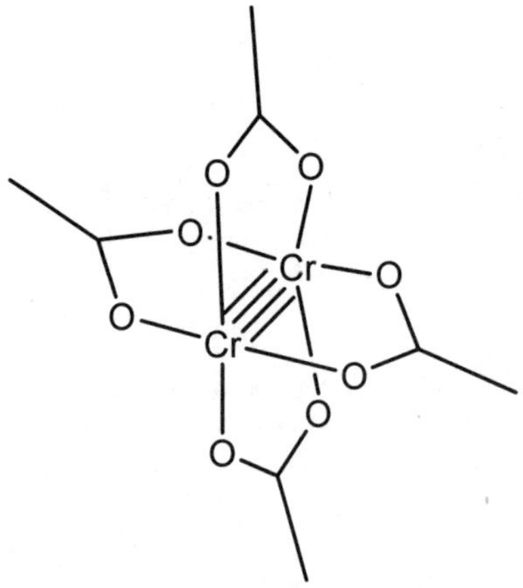

19.14 Consider the two ruthenium complexes in Table 19.8. Using the bonding scheme depicted in Figure 19.19, confirm the bonding orders and electron configurations given in the table.

$[Ru_2Cl_2(ClCO_2)_4]^-$: Assuming that the chloro and chlorformato ligands contribute one electron each the average oxidation state of Ru should be 2.5 +. Clearly this indicates the presence of mixed valence at the Ru center. Since Ru has a total of 8 valence electrons, we will have 11 electrons ($2 \times 8 - 2 \times 2.5$) to fill various bonding and anti-bonding levels given in the Fig. 19.19. Thus the presence of 8 electrons in bonding and 3 in anti-bonding levels is consistent with this picture. Also the total bond order of 2.5 is also in line with the energy level scheme [0.5 (8-3)].

$[Ru_2(CH_3COCH_3)_2(ClCO_2)_4]$: In this complex, only chlorformato ligands contribute one electron each to this neutral complex [note that the acetonato ligands bind with the metal using lone pairs of the oxygen atoms with no charge transfer]. Therefore the average formal oxidation state of Ru in this complex is expected to be 2+. Of the 16 valence electrons of two ruthenium atoms, only 12 electrons ($16 - 2 \times 2$) need to be accommodated in the energy level diagram of Fig. 19.19. The presence of 8 electrons in bonding levels and remaining 4 in the antibonding levels is consistent with this model. Further, the bond order of 2.0 should also be expected [$0.5 \times (8-4)$].

Chapter 20 *d*-Metal Complexes: Electronic Structure and Spectra

This is a correlation diagram between a free d^2-metal ion (left side) and the octahedral strong-field terms of a d^2-metal ion (right side). Only the spin triplet terms are shown on the left. There are three spin-allowed transitions between the $^3T_{1g}$ ground state and higher-energy spin triplet states, $^3T_{2g} \leftarrow {}^3T_{1g}$, $^3T_{1g} \leftarrow {}^3T_{1g}$, and $^3A_{2g} \leftarrow {}^3T_{1g}$. Many properties of *d*-metal complexes can be understood on the basis of the relative energies of the five *d* orbitals that are part of the valence shells of *d*-metal atoms or ions. These include their structure, their spectra, their magnetic behavior, and aspects of their thermodynamic and kinetic reactivity. Ligands in an octahedral complex are frequently placed on the *x, y*, and *z* axes. This results in the metal $d_{x^2-y^2}$ orbital, shown at the right, becoming a metal-ligand σ antibonding MO.

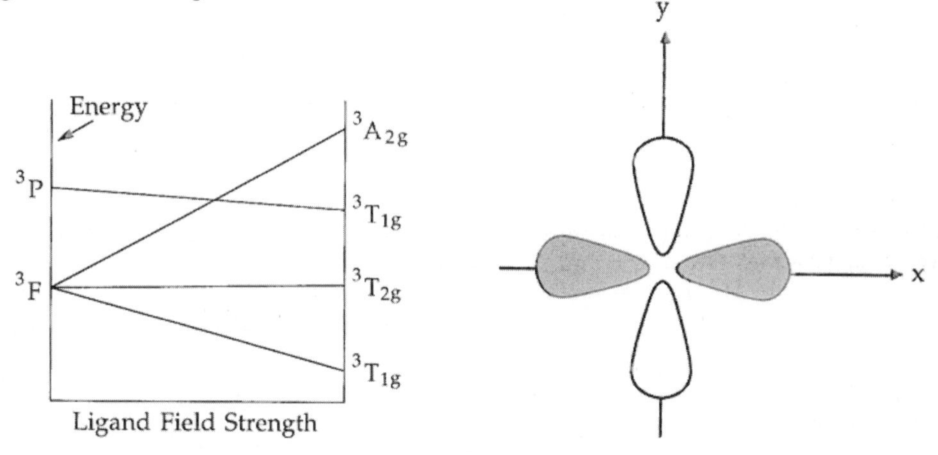

S20.1 **What is the LFSE for both high- and low-spin d^7 configurations?** A high-spin d^7 configuration is $t_{2g}^5e_g^2$. To calculate the LFSE we should note that each electron in the t_{2g} stabilizes the CFSE by 0.4 units and destabilizes by 0.6 units when placed in the e_g level. This is because of the bonding and anti-bonding nature of the t_{2g} and e_g orbitals respectively. Accordingly, for the high spin configuration for the d^7, we should expect [5x0.4 – 2x 0.6] Δ_o or 0.8 Δ_o. The LFSE of a high-spin d^7 with a configuration of $t_{2g}^6e_g^1$, is expected to have 1.8 Δ_o [6x0.4 – 1x 0.6 = 1.8].

S20.2 **The magnetic moment of the complex $[Mn(NCS)_6]^{4-}$ is 6.06μ_B. What is its electron configuration?** Since each isothiocyanate ligand has a single negative charge, the oxidation state of the manganese ion is II. Since Mn(II) is d^5, there are two possibilities for an octahedral complex, low-spin (t_{2g}^5), with one unpaired electron, or high-spin ($t_{2g}^3e_g^2$), with five unpaired electrons. The observed magnetic moment of 6.06 μ_B is close to the spin-only value for five unpaired electrons, $(5 \times 7)^{1/2} = 5.92\ \mu_B$. Therefore, this complex is high spin and has a $t_{2g}^3e_g^2$ configuration.

S20.3 **Account for the variation in lattice enthalpy of the solid fluorides in which each metal ion is surrounded by an octahedral array of F^- ions: MnF^- (2780 kJ mol^{-1}), FeF_2 (2926 kJ mol^{-1}), CoF_2 (2976 kJ mol^{-1}), NiF_2 (3060 kJ mol^{-1}), and ZnF_2 (2985 kJ mol^{-1}).**
If it were not for ligand field stabilization energy (LFSE), MF_2 lattice enthalpies would increase from Mn(II) to Zn(II). This is because the decreasing ionic radius, which is due to the increasing Z_{eff} as you cross through the *d* block from left to right, leads to decreasing M–F separations. Therefore, you expect that ΔH_L for MnF_2 (2780 kJ mol^{-1}) will be smaller than ΔH_L for ZnF_2 (2985 kJ mol^{-1}). In addition, as discussed for aqua ions and oxides, we expect additional LFSE for these compounds. From Table 20.2, you see that LFSE = 0 for Mn(II), 0.4Δ_0 for Fe(II), 0.8Δ_0 for Co(II), 1.2Δ_0 for Ni(II), and 0 for Zn(II). The deviations of the observed values from the straight line connecting Mn(II) and Zn(II) are not quite in the ratio 0.4:0.8:1.2, but note that the deviation for Fe(II) is smaller than that for Ni(II).

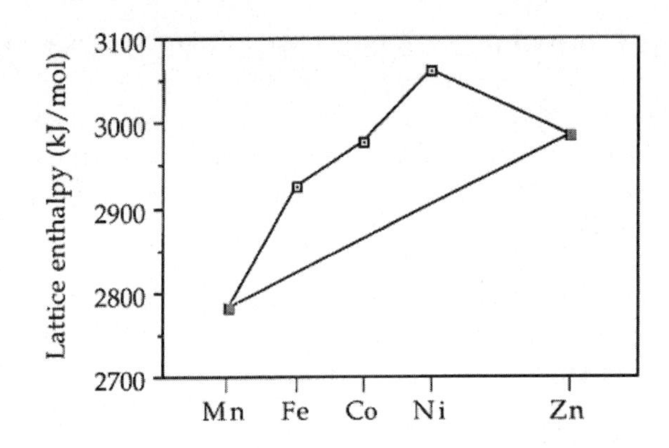

S20.4 Suggest an interpretation of the photoelectron spectra of $[Fe(C_5H_5)_2]$ and $[Mg(C_5H_5)_2]$ shown in Fig. 20.18?
In the spectrum of $[Mo(CO)_6]$, the ionization energy around 8 eV was attributed to the t_{2g} electrons that are largely metal-based. The biggest difference between the photoelectron spectra of ferrocene and magnesocene is in the 6–8 eV region. Although Fe(II) has six *d* electrons (like Mo(0)), Mg(II) has no *d* electrons. Therefore, the differences in the 6–8 eV region can be attributed to the lack of *d* electrons for Mg(II).

S20.5 What terms arise from a p^1d^1 configuration? An atom with a p^1d^1 configuration will have $L = 1 + 2 = 3$ (an electron in a *p* orbital has $l = 1$, while an electron in a *d* orbital has $l = 2$). This value of L corresponds to a F term. Since the electrons may be paired ($S = 0$, multiplicity $2S + 1 = 1$) or parallel ($S = 1$, multiplicity $2S + 1 = 3$), both ^{1}F and ^{3}F terms are possible.

S20.6 Identifying ground terms (Hint: Because d^9 is one electron short of a closed shell with $L = 0$ and $S = 0$, treat it on the same footing as a d^1 configuration.) (a) $2p^2$? Two electrons in a *p* subshell can occupy separate *p* orbitals and have parallel spins ($S = 1$), so the maximum multiplicity $2S + 1 = 3$. The maximum value of M_L is $1 + 0 = 1$, which corresponds to a P term (according to the Pauli principle, m_l cannot be +1 for both electrons if the spins are parallel). Thus, the ground term is ^{3}P (called a "triplet P" term).

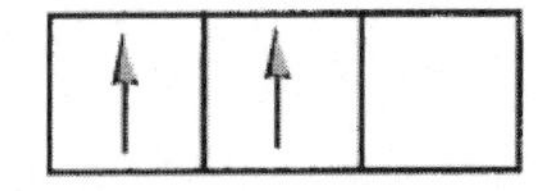

(b) $3d^9$? The largest value of M_S for nine electrons in a *d* subshell is 1/2 (eight electrons are paired in four of the five orbitals, while the ninth electron has $m_s = \pm 1/2$). Thus $S = 1/2$ and the multiplicity $2S + 1 = 2$. Notice that the largest value of M_S is the same for one electron in a *d* subshell. Similarly, the largest value of M_L is 2, which results from the following nine values of m_l: +2, +2, +1, +1, 0, 0, −1, −1, −2. Notice that $M_L = 2$ also corresponds to one electron in a *d* subshell. Thus the ground term for a d^9 or a d^1 configuration is ^{2}D (called a "doublet D" term).

S20.7 What terms in a d^2 complex of O_h symmetry correlate with the ^{3}F and ^{1}D terms of the free atom? An F term arising from a d^n configuration correlates with $T_{1g} + T_{2g} + A_{2g}$ terms in an O_h complex. The multiplicity is unchanged by the correlation, so the terms in O_h symmetry are $^3T_{1g}$, $^3T_{2g}$, and $^3A_{2g}$. Similarly, a D term arising from a d^n configuration correlates with $T_{2g} + E_g$ terms in an O_h complex. The O_h terms retain the singlet character of the ^{1}D free ion term, and so are $^1T_{2g}$ and ^1E_g.

S20.8 Use the same Tanabe-Sugano diagram to predict the energy of the first two spin-allowed quartet bands in the spectrum of $[Cr(OH_2)_6]^{3+}$ for which $\Delta_o = 17\ 600\ cm^{-1}$ and $B = 700\ cm^{-1}$.
The two transitions in question are $^4T_2 \leftarrow {}^4A_2$ and $^4T_1 \leftarrow {}^4A_2$ (the first one has a lower energy and hence a lower wave number). Since $\Delta_O = 17600\ cm^{-1}$ and $B = 700\ cm^{-1}$, you must find the point 17600/700 = 25.1 on the *x*-axis in Figure 20.27. Then, the ratios *E/B* for the two bands are the *y* values of the points on the $^4T_2$ and $^4T_1$ lines, 25

and 32, respectively, as shown below. Therefore, the two lowest-energy spin-allowed bands in the spectrum of $[Cr(H_2O)_6]^{3+}$ will be found at (700 cm^{-1})(25) = 17500 cm^{-1} and (700 cm^{-1})(32) = 22400 cm^{-1}.

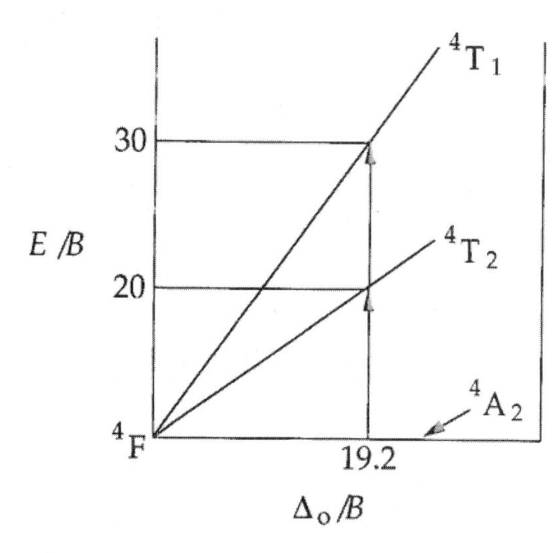

A simplified version of the Tanabe–Sugano diagram for the d^3 configuration of $[Cr(H_2O)_6]^{3+}$.

S20.9 The spectrum of $[Cr(NCS)_6]^{3-}$ has a very weak band near 16 000 cm^{-1}, a band at 17 700 cm^{-1} with ε_{max}= 5160 dm^3 mol^{-1} cm^{-1}, a band at 23 800 cm^{-1} with ε_{max}= 5130 dm^3 mol^{-1} cm^{-1}, and a very strong band at 32 400 cm^{-1}. Assign these transitions using the d^3 Tanabe-Sugano diagram and selection rule considerations. (*Hint:* NCS^- has low-lying π* orbitals.)

This six-coordinate d^3 complex undoubtedly has O_h symmetry, so the general features of its spectrum will resemble the spectrum of $[Cr(NH_3)_6]^{3+}$, shown in Figure 20.22. The very low intensity of the band at 16,000 cm^{-1} is a clue that it is a spin-forbidden transition, probably $^2E_g \leftarrow {}^4A_{2g}$. Spin-allowed but Laporte-forbidden bands typically have ε ~ 100 M^{-1} cm^{-1}, so it is likely that the bands at 17,700 cm^{-1} and 23,800 cm^{-1} are of this type (they correspond to the $^4T_{2g} \leftarrow {}^4A_{2g}$ and $^4T_{1g} \leftarrow {}^4A_{2g}$ transitions, respectively). The band at 32,400 cm^{-1} is probably a charge transfer band, since its intensity is too high to be a ligand field (*d*–*d*) band. Since you are provided with the hint that the NCS^- ligands have low-lying π* orbitals, it is reasonable to conclude that this band corresponds to a MLCT transition. Notice that the two spin-allowed ligand field transitions of $[Cr(NCS)_6]^{3-}$ are at lower energy than those of $[Cr(NH_3)_6]^{3+}$, showing that NCS^- induces a smaller δ_o on Cr^{3+} than does NH_3. Also notice that $[Cr(NH_3)_6]^{3+}$ lacks an intense MLCT band at ~30,000–40,000 cm^{-1}, showing that NH_3 does not have low-lying empty orbitals.

20.1 Determine the configuration (in the form $t_{2g}{}^x$ $e_g{}^y$ or $e^xt_2{}^y$, as appropriate), the number of unpaired electrons, and the ligand-field stabilization energy in terms of Δ_O or Δ_T and P for each of the following complexes using the spectrochemical series to decide, where relevant, which are likely to be high-spin and which low-spin. (a) $[Co(NH_3)_6]^{3+}$? Since the NH_3 ligands are neutral, the cobalt ion in this octahedral complex is Co^{3+}, which is a d^6 metal ion. Ammonia is in the middle of the spectrochemical series but, since the cobalt ion has a 3+ charge, this is a strong field complex and hence is low spin, with S = 0 and no unpaired electrons (the configuration is t^2g^6). The LFSE is 6(0.4Δ_O) = 2.4Δ_O. Note that this is the largest possible value of LFSE for an octahedral complex.

(b) $[Fe(OH_2)_6]^{2+}$? The iron ion in this octahedral complex, which contains only neutral water molecules as ligands, is Fe^{2+}, which is a d^6-metal ion. Since water is lower in the spectrochemical series than NH_3 (i.e., it is a weaker field ligand than NH_3) *and* since the charge on the metal ion is only 2+, this is a weak field complex and hence is high spin, with S = 2 and four unpaired electrons (the configuration is $t_{2g}{}^4e_g{}^2$). The LFSE is 4(0.4Δ_O) – 2(0.6Δ_O) = 0.4Δ_O. Compare this small value to the large value for the low spin d^6 complex in part (a) above.

(c) $[Fe(CN)_6]^{3-}$? The iron ion in this octahedral complex, which contains six negatively charged CN^- ion ligands, is Fe^{3+}, which is a d^5-metal ion. Cyanide ion is a very strong-field ligand, so this is a strong-field complex and hence is low spin, with S = 1/2 and one unpaired electron. The configuration is $t_{2g}{}^5$ and the LFSE is 2.0Δ_O.

(d) $[Cr(NH_3)_6]^{3+}$? The complex contains six neutral NH_3 ligands, so chromium is Cr^{3+}, a d^3 metal ion. The configuration is $t_{2g}{}^3$, and so there are three unpaired electrons and S = 3/2. Note that, for octahedral complexes, only d^4–d^7 metal ions have the possibility of being either high spin or low spin. For $[Cr(NH_3)_6]^{3+}$, the LFSE = $3(0.4\Delta_O) = 1.2\Delta_O$. (For d^1–d^3, d^8, and d^9 metal ions in octahedral complexes, only one spin state is possible.)

(e) $[W(CO)_6]$? Carbon monoxide (i.e., the carbonyl ligand) is neutral, so this is a complex of W(0). The W atom in this octahedral complex is d^6. Since CO is such a strong field ligand (it is even higher in the spectrochemical series than CN^-), $W(CO)_6$ is a strong-field complex and hence is low spin, with no unpaired electrons (the configuration is t_{2g}^3). The LFSE = $6(0.4\Delta_O) = 2.4\Delta_O$.

(f) Tetrahedral $[FeCl_4]^{2-}$? The iron ion in this complex, which contains four negatively charged Cl^- ion ligands, is Fe^{2+}, which is a d^6 metal ion. All tetrahedral complexes are high spin, since Δ_T is much smaller than Δ_O ($\Delta_T = (4/9)\Delta_O$, if the metal ion, the ligands, and the metal-ligand distances are kept constant), so for this complex $S = 2$ and there are four unpaired electrons. The configuration is e^3t^3. The LFSE is $3(0.6\Delta_T) - 3(0.4\Delta_T) = 0.6\ \Delta_T$.

(g) Tetrahedral $[Ni(CO)_4]$? The neutral CO ligands require that the metal center in this complex is Ni^0, which is a d^{10}-metal atom. Regardless of geometry, complexes of d^{10}-metal atoms or ions will never have any unpaired electrons and will always have LFSE = 0, and this complex is no exception.

20.2 **Both H^- and $P(C_6H_5)_3$ are ligands of similar field strength, high in the spectrochemical series. Recalling that phosphines act as π acceptors, is π-acceptor character required for strong-field behaviour? What orbital factors account for the strength of each ligand?**

It is clear that π-acidity cannot be a requirement for a position high in the spectrochemical series, since H^- is a very strong field ligand but is not a π-acid (it has no *low energy* acceptor orbitals of local π-symmetry). However, ligands that are very strong σ-bases will increase the energy of the e_g orbitals in an octahedral complex relative to the t_{2g} orbitals. Thus there are two ways for a complex to develop a large value of Δ_O, by possessing ligands that are π-acids *or* by possessing ligands that are strong σ-bases (of course some ligands, like CN^-, exhibit both π-acidity and moderately strong σ-basicity). A class of ligands that are also very high in the spectrochemical series are alkyl anions, R^- (e.g., CH_3^-). These are not π-acids but, like H^-, are very strong bases.

20.3 **Estimate the spin-only contribution to the magnetic moment for each complex in Exercise 20.1.** The formula for the spin-only moment is $\mu_{SO} = [(N)(N+2)]^{1/2}$. Therefore, the spin-only contributions are:

complex	N	$\mu_{SO} = [(N)(N+2)]^{1/2}$
$[Co(NH_3)_6]^{3+}$	0	0
$[Fe(OH_2)_6]^{2+}$	4	4.9
$[Fe(CN)_6]^{3-}$	1	1.7
$[Cr(NH_3)_6]^{3+}$	3	3.9
$[W(CO)_6]$	0	0
$[FeCl_4]^{2-}$	4	4.9
$[Ni(CO)_4]$	0	0

20.4 **Solutions of the complexes $[Co(NH_3)_6]^{2+}$, $[Co(OH_2)_6]^{2+}$ (both O_h), and $[CoCl_4]^{2-}$ are colored. One is pink, another is yellow, and the third is blue. Considering the spectrochemical series and the relative magnitudes of Δ_T and Δ_O, assign each color to one of the complexes.** The colors of metal complexes are frequently caused by ligand-field transitions involving electron promotion from one subset of *d* orbitals to another (e.g., from t_{2g} to e_g for octahedral complexes or from e to t_2 for tetrahedral complexes). Of the three complexes given, the lowest energy transition probably occurs for $[CoCl_4]^{2-}$, because it is tetrahedral ($\Delta_T = (4/9)(\Delta_O)$) and because Cl^- is a weak-field ligand. This complex is blue, since a solution of it will absorb low energy red light and reflect the complement of red, which is blue. Of the two complexes that are left, $[Co(NH_3)_6]^{2+}$ probably has a higher energy transition than $[Co(OH_2)_6]^{2+}$, since NH_3 is a stronger-field ligand than H_2O (see Table 20.1). The complex $[Co(NH_3)_6]^{2+}$ is yellow because only a small amount of visible light, at the blue end of the spectrum, is absorbed by a solution of this complex. By default, you should conclude that $[Co(OH_2)_6]^{2+}$ is pink.

20.5 **For each of the following pairs of complexes, identify the one that has the larger LFSE:**

(a) $[Cr(OH_2)_6]^{2+}$ or $[Mn(OH_2)_6]^{2+}$

(b) $[Mn(OH2)6]^{2+}$ or $[Fe(OH_2)_6]^{3+}$

(c) $[Fe(OH_2)_6]^{3+}$ or $[Fe(CN)_6]^{3-}$

(d) $[Fe(CN)_6]^{3-}$ or $[Ru(CN)_6]$

(e) tetrahedral $[FeCl_4]^{2-}$ or tetrahedral $[CoCl_4]^{2-}$

(a) The chromium is expected to have a larger crystal field stabilization because of the $t_{2g}3\ e_g1$ configuration [(0.4 × 3) – (0.6 × 1) = 0.6 Δ_O] compared to the Manganese with an electronic configuration $t_{2g}3e_g2$ [(0.4 × 3) – (0.6 × 2) = 0].
(b) Although Fe^{3+} and Mn^{2+} are iso-electronic, the higher charge on the Fe ion leads to higher stabilization energy.
(c) Since water is a weak-field ligand compared to CN^-, the electronic configurations of these two complexes differ. For the $[Fe(CN)_6]^{3-}$ the configuration will be $t_{2g}5\ e_g0$ [(0.4 × 5) = 2 Δ_O] while the corresponding configuration for the aquo complex will be $t_{2g}3\ e_g2$ [(0.4 × 3) – (0.6 × 2) = 0] . Hence $[Fe(CN)_6]^{3-}$ will have higher stabilization.
(d) The LFSE increases down the group and hence the ruthenium complex will have higher stabilization.
(e) In general, tetrahedral complexes form high-spin complexes. Fe^{2+} with an electronic configuration of $e_g2\ t_{2g}3$ will have smaller stabilization [(2 × 0.6) – (2 × 0.4) = 0.2 Δ_O] compared to Co^{2+} with a configuration of $e_g4\ t_{2g}3$ [(4 × 0.6) – (3 × 0.4) = 1.2 Δ_O].

20.6 Interpret the variation, including the overall trend across the 3d series, of the following values of oxide lattice enthalpies (in kJ mol^{-1}). All the compounds have the rock-salt structure: CaO (3460), TiO (3878), VO (3913), MnO (3810), FeO (3921), CoO (3988), NiO (4071)? As in the answer to Exercise S20.3, there are two factors that lead to the values given in this question and plotted below: decreasing ionic radius from left to right across the *d* block, leading to a general increase in ΔH_L from CaO to NiO, and LFSE, which varies in a more complicated way for high-spin metal ions in an octahedral environment, increasing from d^0 to d^3, then decreasing from d^3 to d^5, then increasing from d^5 to d^8, then decreasing again from d^8 to d^{10}. The straight line through the black squares is the trend expected for the first factor, the decrease in ionic radius (the last black square is not a data point, but simply the extrapolation of the line between ΔH_L values for CaO and MnO, both of which have LFSE = 0). The deviations of ΔH_L values for TiO, VO, FeO, CoO, and NiO from the straight line are a manifestation of the second factor, the nonzero values of LFSE for Ti^{2+}, V^{2+}, Fe^{2+}, Co^{2+}, and Ni^{2+}. You will find in Chapter 22 that TiO and VO have considerable metal–metal bonding, and this factor also contributes to their stability.

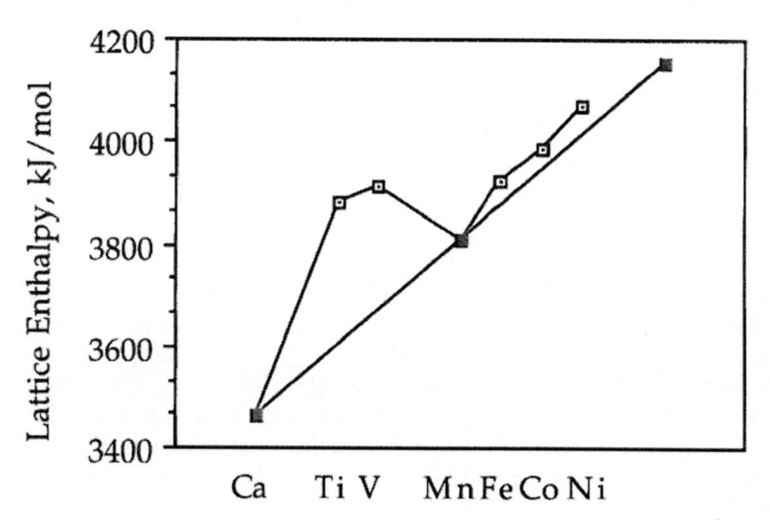

20.7 A neutral macrocyclic ligand with four donor atoms produces a red diamagnetic low-spin d^8 complex of Ni(II) if the anion is the weakly coordinating perchlorate ion. When perchlorate is replaced by two thiocyanate ions, SCN^-, the complex turns violet and is high-spin with two unpaired electrons. Interpret the change in terms of structure? Perchlorate, ClO_4^-, is a very weakly basic anion (consider that $HClO_4$ is a very strong Brønsted acid). Therefore, in the compound containing Ni(II), the neutral macrocyclic ligand, and two ClO_4^- anions, there is probably a four-coordinate square-planar Ni(II) (d^8) complex and two noncoordinated perchlorate anions. Square-planar d^8 complexes are diamagnetic (see Figure 20.10), since they have a configuration (xz, $yz^4(xy)^2(z^2)^2$). When SCN ligands are added, they coordinate to the nickel ion, producing a tetragonal (D_{4h}) complex that has two unpaired electrons (configuration $t_{2g}{}^6e_g{}^2$).

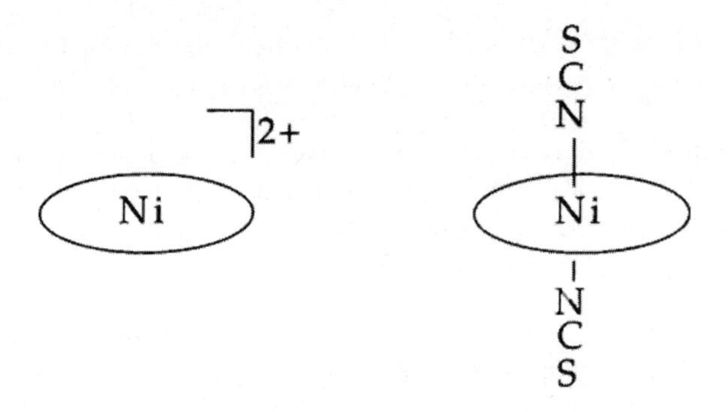

20.8 Bearing in mind the Jahn-Teller theorem, Predict the structure of $[Cr(OH_2)_6]^{2+}$? The main consequence of the Jahn-Teller theorem is that a nonlinear molecule or ion with an orbitally degenerate ground state is not as stable as a distorted version of the molecule or ion if the distortion removes the degeneracy. The high-spin d^4 complex $[Cr(OH_2)_6]^{2+}$ has the configuration $t_{2g}^3e_g^1$, which is orbitally degenerate since the single e_g electron can be in either the d_{z^2} or the $d_{x^2-y^2}$ orbital. Therefore, by the Jahn-Teller theorem, the complex should not have O_h symmetry. A tetragonal distortion, whereby two *trans* metal–ligand bonds are elongated and the other four are shortened, removes the degeneracy. This is the most common distortion observed for octahedral complexes of high-spin d^4, low-spin d^7, and d^9 metal ions, all of which possess e_g degeneracies and exhibit measurable Jahn-Teller distortions. The predicted structure of the $[Cr(OH_2)_6]^{2+}$ ion, with the elongation of the two *trans* Cr–O bonds shown greatly exaggerated, is shown at the right.

20.9 The spectrum of d^1 $Ti^{3+}(aq)$ is attributed to a single electronic transition $e_g \leftarrow t_{2g}$. The band shown in Fig. 20.3 is not symmetrical and suggests that more than one state is involved. Suggest how to explain this observation using the Jahn-Teller theorem? It is suggested that you explain this observation using the Jahn-Teller theorem. The ground state of $Ti(OH_2)_6^{3+}$, which is a d^1 complex, is not one of the configurations that usually leads to an observable Jahn-Teller distortion (the three main cases are listed in the answer to Exercise 20.8, above). However, the electronic excited state of $Ti(OH_2)_6^{3+}$ has the configuration $t_{2g}^0e_g^1$, and so the excited state of this complex possesses an e_g degeneracy. Therefore, the "single" electronic transition is really the superposition of two transitions, one from an O_h ground-state ion to an O_h excited-state ion, and a lower energy transition from an O_h ground-state ion to a lower energy distorted excited-state ion (probably D_{4h}). Since these two transitions have slightly different energies, the unresolved superimposed bands result in an asymmetric absorption peak.

20.10 Write the Russell–Saunders term symbols for states with the angular momentum quantum numbers (L,S): (a) $L = 0$, $S = 5/2$? You should remember that a term, denoted by a capital letter, is related to L in the same way that an orbital, denoted by a lowercase letter, is related to l:

If $l = 0$	1	2	3	4	5	6
orbital = s	p	d	f	g	h	i
If $L = 0$	1	2	3	4	5	6
term = S	P	D	F	G	H	I

The multiplicity of the term, which is always given as a left superscript, can always be determined by using the formula multiplicity = $2S + 1$:

If $S = 0$	1/2	1	3/2	2	5/2
multiplicity $2S + 1 = 1$	2	3	4	5	6

The terms and multiplicities listed above are not a complete list to cover all possibilities for all atoms and ions, but they will cover all possible d^n configurations. As far as the situation $L = 0$, $S = 5/2$ is concerned, the term symbol is ^{6}S. In addition to answering this question, you should try to decide which d^n configurations can give rise to the term. In this case, the only d^n configuration that can give rise to an ^{6}S term is d^5 (e.g., a gas-phase Mn^{2+} or Fe^{3+}ion).

(b) $L = 3$, $S = 3/2$? According to the relations shown above, this set of angular momentum quantum numbers is described by the term symbol ^{4}F. The diagram below, which is another way to depict a microstate, shows how the situation $L = 3$, $S = 3/2$ can arise from a d^3 configuration (e.g., a gas-phase Cr^{3+} ion). It can also arise from d^6 and d^7 configurations.

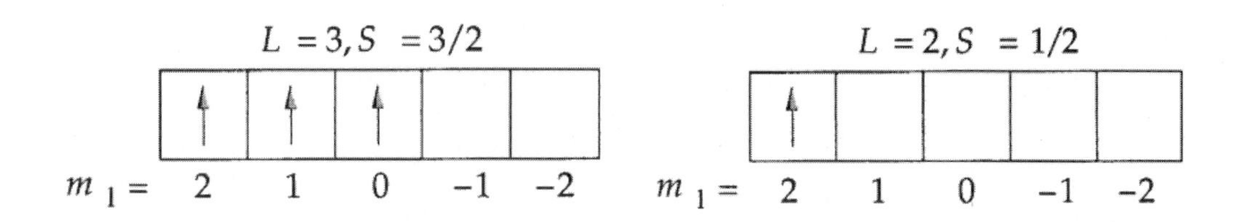

(c) $L = 2$, $S = 1/2$? This set of quantum numbers is described by the term symbol ^{2}D. The microstate diagram above shows how the situation $L = 2$, $S = 1/2$ can arise from a d^1 configuration (e.g., a gas-phase Ti^{3+} ion). It can also arise from d^3, d^5, d^7, and d^9 configurations.

(d) $L = 1$, $S = 1$? This set of quantum numbers is described by the term symbol ^{3}P. It can arise from d^2, d^4, d^6, and d^8 configurations.

20.11 Identify the ground term from each set of terms: (a) ^{3}F, ^{3}P, ^{1}P, ^{1}G? By definition, the ground term has the lowest energy of all of the terms. Recall Hund's two rules: (1) the term with the greatest multiplicity lies lowest in energy; (2) for a given multiplicity, the greater the value of L of a term, the lower the energy. Therefore, the ground term in this case will be a triplet, not a singlet (rule 1). Of the two triplet terms, ^{3}F lies lower in energy than ^{3}P: $L = 3$ for ^{3}F, $L = 1$ for ^{3}P (rule 2). Therefore, the ground term is ^{3}F.

(b) ^{5}D, ^{3}H, ^{3}P, ^{1}G, ^{1}I? The ground term is ^{5}D because this term has a higher multiplicity than the other terms.

(c) ^{6}S, ^{4}G, ^{4}P, ^{2}I? The ground term is ^{6}S because this term has a higher multiplicity than the other terms.

20.12 Give the Russell-Saunders terms of the configurations and identify the ground term? (a) 4 s^1? You can approach this exercise in the way described in Section 20.2 for the d^2 configuration (see Table 20.5). You first write down all possible microstates for the s^1 configuration, then write down the M_L and M_S values for each microstate, then infer the values of L and S to which the microstates belong. In this case, the procedure is not lengthy, since there are only two possible microstates, (0^+) and (0^-). The only possible values of M_L and M_S are 0 and 1/2, respectively. If M_L can only be 0, then L must be 0, which gives an S term (remember that L can take on all values M_L, (M_L–1), …, 0, … –M_L). Similarly, if M_S can only be 1/2 or –1/2, then S must be 1/2, which gives a multiplicity $2S + 1 = 2$. Therefore, the one and only term that arises from a $4s^1$ configuration is ^{2}S.

(b) $3p^2$? This case is more complicated because there are 15 possible microstates:

	$M_S = -1$	$M_S = 0$	$M_S = 1$
$M_L = 2$		$(1^+, 1^-)$	
$M_L = 1$	$(1^-, 0^-)$	$(1^+, 0^-), (1^-, 0^+)$	$(1^+, 0^+)$
$M_L = 0$	$(1^-, -1^-)$	$(1^+, -1^-), (0^+, 0^-), (-1^+, 1^-)$	$(1^+, -1^+)$
$M_L = -1$	$(-1^-, 0^-)$	$(-1^+, 0^-), (-1^-, 0^+)$	$(-1^+, 0^+)$
$M_L = -2$		$(-1^+, -1^-)$	

A term that contains a microstate with $M_L = 2$ must be a D term ($L = 2$). This term can only be a singlet, since if both electrons have $m_l = 1$ they must be spin-paired. Therefore, one of the terms of the $3p^2$ configuration is ^{1}D, which contains five microstates. To account for these, you can cross out $(1^+, 1^-)$, $(-1^+, -1^-)$, and one microstate from each of the other three rows under $M_S = 0$. That leaves ten microstates to be accounted for. The maximum value of M_L of the remaining microstates is 1, so you next consider a P term ($L = 1$). Since M_S can be –1, 0, or 1, this term will be ^{3}P, which contains nine of the ten remaining microstates. The last remaining microstate is one of the original three for which $M_L = 0$ and $M_S = 0$, which is the one and only microstate that belongs to a ^{1}S term ($L = 0$, $S = 0$). Of the three terms ^{1}D, ^{3}P, and ^{1}S that arise from the $3p^2$ configuration, the ground term is ^{3}P, since it has a higher multiplicity than the other two terms (see Hund's rule 1).

20.13 **The gas-phase ion V^{3+} has a ^{3}F ground term. The ^{1}D and ^{3}P terms lie, respectively, 10 642 and 12 920 cm^{-1} above it. The energies of the terms are given in terms of Racah parameters as $E(^3F) = A - 8B$, $E(^3P) = A + 7B$, $E(^1D) = A - 3B + 2C$. Calculate the values of B and C for V^{3+}** The diagram below shows the relative energies of the ^{3}F, ^{1}D, and ^{3}P terms. It can be seen that the 10,642 cm^{-1} energy gap between the ^{3}F and ^{1}D terms is $5B + 2C$, while the 12920 cm^{-1} energy gap between the ^{3}F and ^{3}P terms is $15B$. From the two equations

$$5B + 2C = 10642 \text{ cm}^{-1} \text{ and } 15B = 12920 \text{ cm}^{-1}$$

you can determine that $B = (12920 \text{ cm}^{-1})/(15) = 861.33 \text{ cm}^{-1}$ and $C = 3167.7 \text{ cm}^{-1}$.

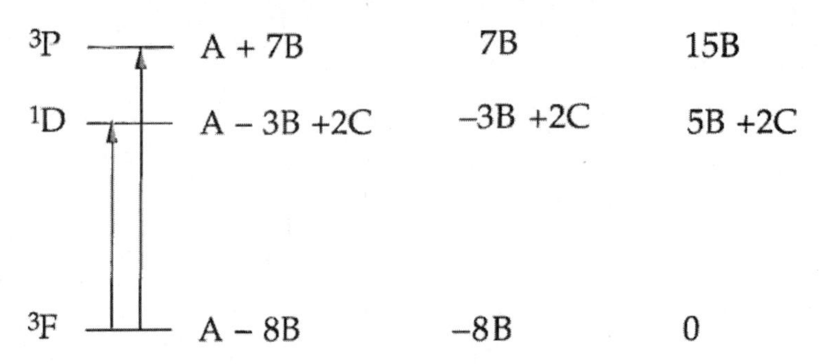

Relative Energies

20.14 **Write the d-orbital configurations and use the Tanabe–Sugano diagrams (*Resource section* 6) to identify the ground term of (a) Low-spin $[Rh(NH_3)_6]^{3+}$?** There are a number of ways to determine the integer n for a given d-block metal. One straightforward procedure is as follows. Count the number of elements from the left side of the periodic table to the metal in question. This will be the number of d electrons *for a metal atom in a complex* (note that an isolated gas-phase metal atom may have a $s^m d^n$ configuration, but the same neutral metal atom *in a complex* will have a d^m configuration). Then subtract the positive charge on the metal ion from this number, leaving the integer n. For example, Rh is the ninth element in period 5, so Rh^0 in a complex has a d^9 configuration, Rh^+ has a d^8 configuration, and so on. Using this procedure, the Rh^{3+} ion in the octahedral complex $[Rh(NH_3)_6]^{3+}$ has a d^6 configuration (9 – 3 = 6). According to the d^6 Tanabe-Sugano diagram (see Resource Section 6), the ground term for a low-spin t_{2g}^6 metal ion is $^1A_{1g}$.
(b) $[Ti(H_2O)_6]^{3+}$? Titanium is the fourth element in period 4, so Ti^0 in a complex has a d^4 configuration. Therefore, the Ti^{3+} ion in the octahedral $[Ti(H_2O)_6]^{3+}$ ion has a d^1 configuration (4 – 3 = 1). A correlation diagram for d^1 metal ions is shown in Figure 20.26 (Resource Section 6 does not include the d^1 Tanabe-Sugano diagram). According to this diagram, the ground term for a t_{2g}^1 metal ion is $^2T_{2g}$.

(c) High-spin $[Fe(H_2O)_6]^{3+}$? Iron is the eighth element in period 4, so Fe^0 in a complex (such as $Fe(CO)_5$) has a d^8 configuration. Therefore, the Fe^{3+} ion in the octahedral $[Fe(H_2O)_6]^{3+}$ ion has a d^5 configuration (8 – 3 = 5). According to the d^5 Tanabe-Sugano diagram, the ground term for a high-spin $t_{2g}{}^3e_g{}^2$ metal ion is $^6A_{1g}$.

20.15 Using the Tanabe-Sugano diagrams in Resource section 6, estimate Δ_O and B for (a) $[Ni(H_2O)_6]^{2+}$ (absorptions at 8500, 15400 and 26000 cm^{-1}) ? According to the d^8 Tanabe-Sugano diagram (Resource Section 6), the absorptions at 8500 cm^{-1}, 13800 cm^{-1}, and 25300 cm^{-1} correspond to the following spin-allowed transitions, respectively: $^3T_{2g} \leftarrow {}^3A_{2g}$, $^3T_{1g} \leftarrow {}^3A_{2g}$, and $^3T_{1g} \leftarrow {}^3A_{2g}$. The ratios 13800/8500 = 1.6 and 25300/8500 = 3.0 can be used to estimate $\Delta_0/B \approx 11$. Using this value of Δ_0/B and the fact that $E/B = \Delta_0/B$ for the lowest-energy transition, Δ_0 = 8500 cm^{-1} and $B \approx 770$ cm^{-1}. Note that B for a gas-phase Ni^{2+} ion is 1080 cm^{-1} (see Table 20.6). The fact that B for the complex is only ~70% of the free ion value is an example of the nephalauxetic effect.
(b) $[Ni(NH_3)_6]^{2+}$ (absorptions at 10750, 17500 and 28200 cm^{-1})? The absorptions for this complex are at 10750 cm^{-1}, 17500 cm^{-1}, and 28200 cm^{-1}. The ratios in this case are 17500/10750 = 1.6 and 28200/10750 = 2.6, and lead to $\Delta_0/B \approx 15$. Thus, Δ_0 = 10,750 cm^{-1} and $B \approx 720$ cm^{-1}. It is sensible that B for $[Ni(NH_3)_6]^{2+}$ is smaller than B for $[Ni(H_2O)_6]^{2+}$, since NH_3 is higher in the nephalauxetic series than is H_2O.

20.16 The spectrum of $[Co(NH_3)_6]^{3+}$ has a very weak band in the red and two moderate intensity bands in the visible to near-UV. How should these transitions be assigned? If this d^6 complex were high spin, the only spin-allowed transition possible would be $^5E_g \leftarrow {}^5T_{2g}$ (refer once again to the d^6 Tanabe-Sugano diagram). On the other hand, if it were low spin, several spin-allowed transitions are possible, including $^1T_{1g} \leftarrow {}^1A_{1g}$, $^1T_{2g} \leftarrow {}^1A_{1g}$, $^1E_g \leftarrow$ A_{1g}, etc. The presence of *two* moderate-intensity bands in the visible/near-UV spectrum of $[Co(NH_3)_6]^{3+}$ suggests that it is low spin. The first two transitions listed above correspond to these two bands. The very weak band in the red corresponds to a spin-forbidden transition such as $^3T_{2g} \leftarrow {}^1A_{1g}$.

20.17 Explain why $[FeF_6]^{3-}$ is colorless whereas $[CoF_6]^{3-}$ is colored but exhibits only a single band in the visible. The d^5 Fe^{3+} ion in the octahedral hexafluoro complex must be high spin. According to the d^5 Tanabe-Sugano diagram (Resource Section 6), a high-spin complex has no higher energy terms of the same multiplicity as the $^6A_{1g}$ ground term. Therefore, since no spin-allowed transitions are possible, the complex is expected to be colorless (i.e., only very weak spin-forbidden transitions are possible). If this Fe(III) complex were low spin, spin-allowed transitions such as $^2T_{1g} \leftarrow {}^2T_{2g}$, $^2A_{2g} \leftarrow {}^2T_{2g}$, etc. would render the complex colored. The d^6 Co^{3+} ion in $[CoF_6]^{3-}$ is also high spin, but in this case a single spin-allowed transition, $^5E_g \leftarrow {}^5T_{2g}$, makes the complex colored and gives it a one-band spectrum.

20.18 The Racah parameter B is 460 cm^{-1} in $[Co(CN)_6]^{3-}$ and 615 cm^{-1} in $[Co(NH_3)_6]^{3+}$. Consider the nature of bonding with the two ligands and explain the difference in nephelauxetic effect? These two ligands are quite different with respect to the types of bonds they form with metal ions. Ammonia and cyanide ion are both σ-bases, but cyanide is also a π-acid. This difference means that NH_3 can form molecular orbitals only with the metal e_g orbitals, while CN^- can form molecular orbitals with the metal e_g and t_{2g} orbitals. The formation of molecular orbitals is the way that ligands "expand the clouds" of the metal *d* orbitals.

20.19 An approximately 'octahedral' complex of Co(III) with ammine and chloro ligands gives two bands with εmax between 60 and 80 dm^3 mol^{-1} cm^{-1}, one weak peak with ε_{max}52 dm^3 mol^{-1} cm^{-1}, and a strong band at higher energy with ε_{max}= 2 x 10^4 dm^3 mol^{-1} cm^{-1}. What do you suggest for the origins of these transitions? Let's start with the intense band at relatively high energy with $\varepsilon_{max} = 2 \times 10^4$ M^{-1} cm^{-1}. This is undoubtedly a spin-allowed charge-transfer transition, since it is too intense to be a ligand field (*d–d*) transition. Furthermore, it is probably an LMCT transition, not an MLCT transition, since the ligands do not have the empty orbitals necessary for an MLCT transition. The two bands with ε_{max} = 60 and 80 M^{-1} cm^{-1} are probably spin-allowed ligand field transitions. Even though the complex is not strictly octahedral, the ligand field bands are still not very intense. The *very* weak peak with ε_{max} = 2 M^{-1} cm^{-1} is most likely a spin-forbidden ligand field transition.

20.20 Ordinary bottle glass appears nearly colorless when viewed through the wall of the bottle but green when viewed from the end so that the light has a long path through the glass. The color is associated with the presence of Fe^{3+} in the silicate matrix. Suggest which transitions are responsible for the color? The Fe^{3+} ions in question are d^5 metal ions. If they were low spin, several spin-allowed ligand field transitions would give the glass a color even when viewed through the wall of the bottle (see the Tanabe-Sugano diagram for d^5 metal ions in

Resource Section 6). Therefore, the Fe^{3+} ions are high spin, and as such have no spin-allowed transitions (the ground state of an octahedral high-spin d^5 metal ion is $^6A_{1g}$, and there are no sextet excited states). The faint green color, which is only observed when looking through a *long* pathlength of bottle glass, is caused by spin-forbidden ligand field transitions.

20.21 Solutions of $[Cr(OH_2)_6]^{3+}$ ions are pale blue–green but the chromate ion, CrO_4^{2-}, is an intense yellow. Characterize the origins of the transitions and explain the relative intensities. The blue-green color of the Cr^{3+} ions in $[Cr(H_2O)_6]^{3+}$ is caused by spin-allowed but Laporte-forbidden ligand field transitions. The relatively low molar absorption coefficient, ε, which is a manifestation of the Laporte-forbidden nature of the transitions, is the reason that the intensity of the color is weak. The oxidation state of chromium in dichromate dianion is Cr(VI), which is d^0. Therefore, no ligand field transitions are possible. The intense yellow color is due to LMCT transitions (i.e., electron transfer from the oxide ion ligands to the Cr(VI) metal center). Charge transfer transitions are intense because they are both spin-allowed and Laporte-allowed.

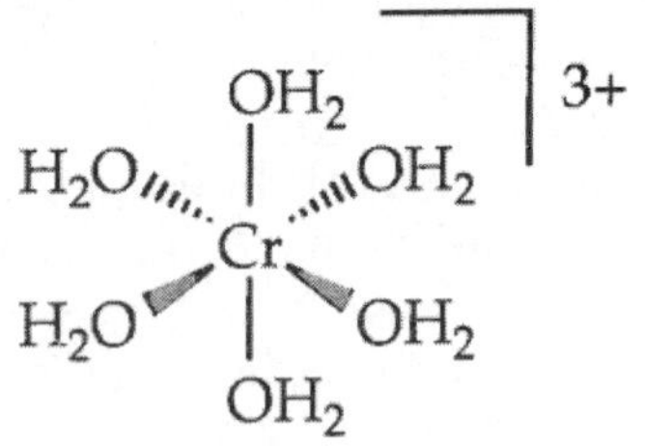

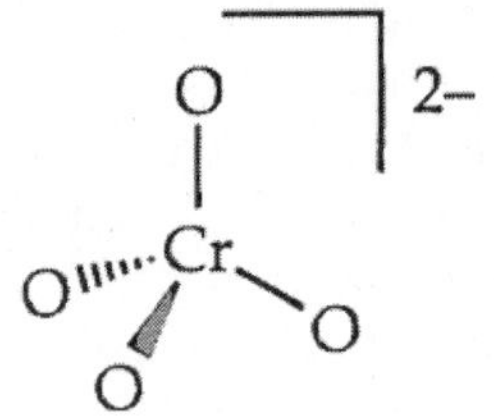

20.22 Classify the symmetry type of the d orbital in a tetragonal C_{4v} symmetry complex, such as $[CoCl(NH_3)_5]^-$, where the Cl lies on the *z*-axis. (a) Which orbitals will be displaced from their position in the octahedral molecular orbital diagram by π interactions with the lone pairs of the Cl^- ligand? (b) Which orbital will move because the Cl^- ligand is not as strong a base as NH_3? (c) Sketch the qualitative molecular orbital diagram for the C_{4v} complex. The d_{z^2} orbital in this complex is left unchanged by each of the symmetry operations of the C_{4v} point group. It therefore has A_1 symmetry. The Cl atom lone pairs of electrons can form π molecular orbitals with d_{xz} and d_{yz}. These metal atomic orbitals are π-antibonding MOs in $[CoCl(NH_3)_5]^{2+}$ (they are nonbonding in $[Co(NH_3)_6]^{3+}$), and so they will be raised in energy relative to their position in $[Co(NH_3)_6]^{3+}$, in which they were degenerate with d_{xy}. Since Cl^- ion is not as strong a σ-base as NH_3 is, the d_{z^2} orbital in $[CoCl(NH_3)_5]^{2+}$ will be at lower energy than in $[Co(NH_3)_6]^{3+}$, in which it was degenerate with $d_{x^2-y^2}$. A qualitative *d*-orbital splitting diagram for both complexes is shown below (L = NH_3).

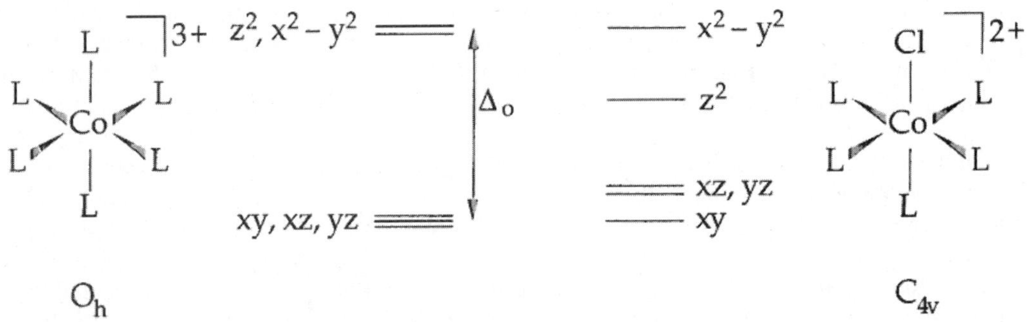

20.23 Consider the molecular orbital diagram for a tetrahedral complex (based on Fig. 20.7) and the relevant d-orbital configuration and show that the purple color of MnO_4^- ions cannot arise from a ligand-field transition. Given that the wavenumbers of the two transitions in MnO_4^- are 18 500 and 32 200 cm^{-1}, explain how to estimate Δ_T from an assignment of the two charge-transfer transitions, even though Δ_T cannot be observed directly. As discussed in Section 20.5, *Charge-transfer bands*, ligand field transitions can occur for *d*-block metal ions with one or more, but fewer than ten, electrons in the metal *e* and t_2 orbitals. However, the oxidation state of manganese in permanganate anion is Mn(VII), which is d^0. Therefore, no ligand field transitionsare possible. The metal *e* orbitals are the LUMOs, and can act as acceptor orbitals for LMCT transitions.

These fully allowed transitions give permanganate its characteristic *intense* purple color. You can see that two possible LMCT transitions are possible, $e \leftarrow a_1$ and $t_2 \leftarrow a_1$. These two transitions give rise to the two absorption bands observed at 18,500 cm^{-1} and 32200 cm^{-1}. The difference in energy between the two transitions, $E(t_2) - E(e) = 13700\ cm^{-1}$, is just equal to Δ_T, as shown in the diagram below:

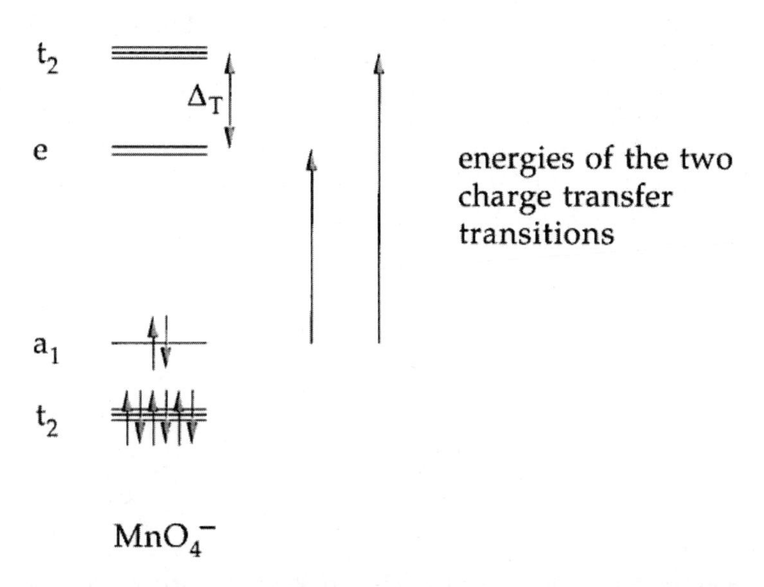

20.24 The lowest energy band in the spectrum of $[Fe(OH_2)_6]^{3+}$ (in 1M $HClO_4$) occurs at lower energy than the equivalent transition in the spectrum of $[Mn(OH_2)_6]^{2+}$. Explain why this is.

Although both Fe^{3+} and Mn^{2+} aqua complexes have high spin d^5 configurations (recall H_2O is a weak field ligand), the extra charge of the iron complexes keeps the e_g and t_{2g} levels close. Accordingly in the lowest energy band in the spectrum of the iron complexes occurs at a lower energy.

Chapter 21 Coordination Chemistry: Reactions of Complexes

Here is the proposed five-coordinate intermediate in substitution reactions of square-planar complexes. The incoming ligand Y adds to the four-coordinate complex MC_2TX (X is the leaving group; T is *trans* to X; the C ligands are *cis* to X). The trigonal bipyramid that forms has T, X, and Y in the equatorial plane. The electronic properties of T dramatically affect the rate of the reaction (the *trans* effect), while the electronic properties of C do not.

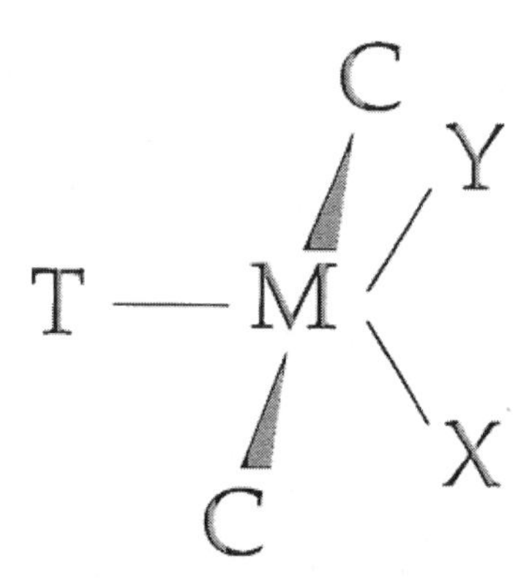

S21.1 Calculate the second-order rate constant for the reaction of *trans*-$[PtCl(CH_3)(PEt_3)_2]$ with NO_2^- in MeOH, for which n_{pt} = 3.22 ? We can use equation 21.6:

$$\log k_2(NO_2^-) = Sn_{Pt}(NO_2^-) + C$$

where S is the nucleophilic discrimination factor for this complex, $n_{Pt}(NO_2^-)$ is the nucleophilicity parameter of nitrite ion, and C is the logarithm of the second order rate constant for the substitution for Cl^- in this complex by MeOH. S was determined to be 0.41 in the example, and $n_{Pt}(NO_2^-)$ is given as 3.22. C is a constant for a given complex and is –0.61 in this case. Therefore, $k_2(NO_2^-)$ can be determined as follows:

$$\log k_2(NO_2^-) = (0.41)(3.22) - 0.61 = 0.71$$

$$k_2(NO_2^-) = 10^{0.71} = 5.1\ M^{-1}s^{-1}$$

S21.2 Given the reactants PPh_3, NH_3, and $[PtCl_4]^{2-}$, Propose efficient routes to both *cis*- and *trans*-$[PtCl_2(NH_3)(PPh_3)]$? For the three ligands in question, Cl^-, NH_3, and PPh_3, the *trans* effect series is $NH_3 < Cl^- < PPh_3$. This means that a ligand *trans* to Cl^- will be substituted at a faster rate than a ligand *trans* to NH_3, and a ligand *trans* to PPh_3 will be substituted at a faster rate than a ligand *trans* to Cl^-. Since our starting material is $[PtCl_4]^{2-}$, two steps involving substitution *of* Cl^- *by* NH_3 or PPh_3 must be used. If you first add NH_3 to $[PtCl_4]^{2-}$, you will produce $[PtCl_3(NH_3)]^-$. Now if you add PPh_3, one of the mutually *trans* Cl^- ligands will be substituted faster than the Cl^- ligand *trans* to NH_3, and the *cis* isomer will be the result.

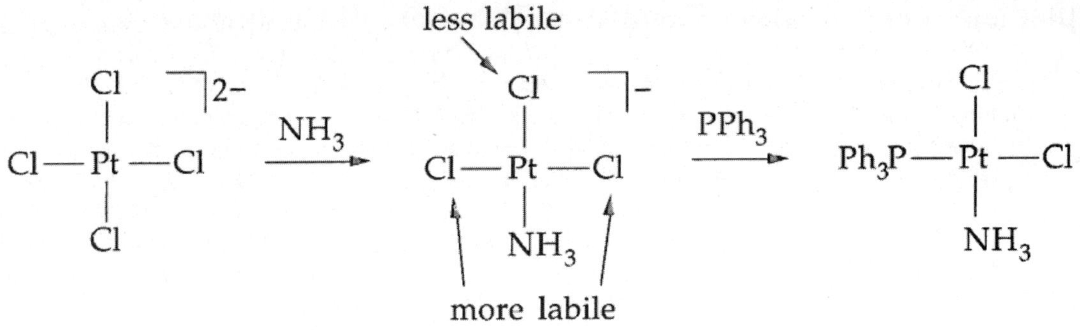

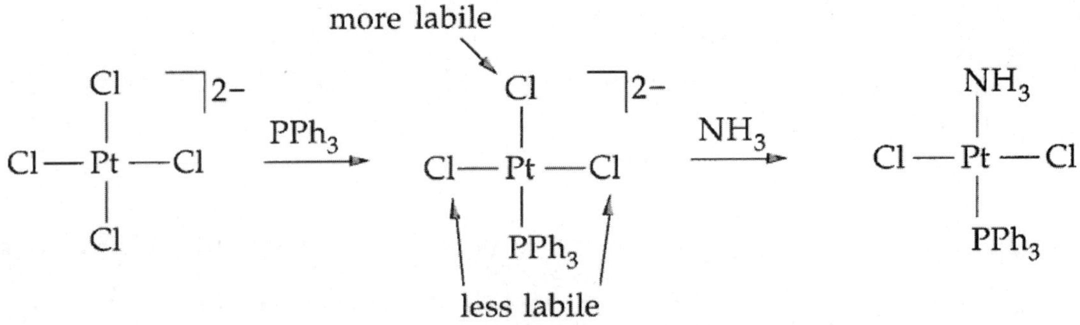

less labile

$[PtCl_4]^{2-}$ $\xrightarrow{NH_3}$ $[PtCl_3(NH_3)]^{-}$ $\xrightarrow{PPh_3}$ $PtCl_2(PPh_3)(NH_3)$

more labile

more labile

$[PtCl_4]^{2-}$ $\xrightarrow{PPh_3}$ $[PtCl_3(PPh_3)]^{2-}$ $\xrightarrow{NH_3}$ $PtCl_2(NH_3)(PPh_3)$

less labile

If you first add PPh_3 to $[PtCl_4]^{2-}$, you will produce $[PtCl_3(PPh_3)]^{-}$. Now if you add NH_3, the Cl^- ligand *trans* to PPh_3 will be substituted faster than one of the mutually *trans* Cl^- ligands, and the *trans* isomer will be the result.

S21.3 Use the data in Table 21.8 to estimate an appropriate value for K_E and calculate k_{r2} for the reactions of V(II) with Cl^- if the observed second-order rate constant is 1.2×10^2 dm^3 mol^{-1} s^{-1}. As discussed in Section 21.6, the Eigen-Wilkins mechanism for substitution in octahedral complexes is:

$$[V(H_2O)_6]^{2+} + Cl^- \rightarrow \{[V(H_2O)_6]^{2+}, Cl^-\}$$

$$\{[V(H_2O)_6]^{2+}, Cl^- \rightarrow [VCl(H_2O)_5]^+ + H_2O$$

The observed rate constant, k_{obs}, is given by:

$$k_{obs} = kK_E$$

and in the case of the substitution of H_2O by Cl^- in $[V(H_2O)_6]^{2+}$ is 1.2×10^2 M^{-1} s^{-1}. You will be able to calculate k if you can estimate a proper value for K_E. Inspection of Table 21.8 shows that $K_E = 1$ M^{-1} for the encounter complex formed by F^- or SCN^- and $[Ni(H_2O)_6]^{2+}$. This value can be used for the reaction in question since (i) the charge and size of Cl^- are similar to those of F^- and SCN^-, and (ii) the charge and size of $[Ni(H_2O)_6]^{2+}$ are similar to those of $[V(H_2O)_6]^{2+}$. Therefore, $k = k_{obs}/K_E = (1.2 \times 10^2\ M^{-1}\ s^{-1})/(1\ M^{-1}) = 1.2 \times 10^2\ s^{-1}$.

21.1 The rate constants for the formation of $[CoX(NH_3)_5]^{2+}$ from $[Co(NH_3)_5OH_2]^{3+}$ for X = Cl_2, Br_2, N_3^-, and SCN^- differ by no more than a factor of two. What is the mechanism of the substitution? If the mechanism of substitution were to be associative, the nature of incoming ligand should affect the rate of the reaction. This is because the rate-limiting step would require the formation of M—X bonds (X = incoming ligand). In the present however, rate of the reaction does not vary much depending on the nature or size of the incoming ligand. Therefore, the more likely mechanism would be *dissociative*.

21.2 If a substitution process is associative, why may it be difficult to characterize an aqua ion as labile or inert? The rate of an associative process depends on the identity of the entering ligand and, therefore, it is not an inherent property of $[M(OH_2)_6]^{n+}$.

21.3 The reactions of $Ni(CO)_4$ in which phosphines or phosphites replace CO to give $Ni(CO)_3L$ all occur at the same rate regardless of which phosphine or phosphite is being used. Is the reaction *d* or *a*? Since the rate of substitution is the same for a variety of different entering ligands L, the activated complex in each case must not include any significant bond making to the entering ligand and the reaction must be *d*. If the rate-determining step included any Ni–L bond making, the rate of substitution would change as the electronic and steric properties of L were changed.

21.4 Write the rate law for formation of $[MnX(OH_2)_5]1$ from the aqua ion and X^-. How would you undertake to determine if the reaction is *d* or *a*? The rate law for this substitution reaction is:

$$\text{rate} = (kK_E[Mn(OH_2)_6^{2+}][X^-])/(1 + K_E[X^-])$$

where K_E is the equilibrium constant for the formation of the encounter complex $\{[Mn(OH_2)_6]^{2+}, X^-\}$, and k is the first-order rate constant for the reaction:

$$\{[Mn(OH_2)_6^{2+}], X^-\} \rightarrow [Mn(OH_2)_5X]^+ + H_2O$$

The rate law will be the same regardless of whether the transformation of the encounter complex into products is dissociatively or associatively activated. However, you can distinguish *d* from *a* by varying the nature of X^-. If k varies as X^- varies, then the reaction is *a*. If k is relatively constant as X^- varies, then the reaction is *d*. Note that k cannot be measured directly. It can be found using the expression $k_{obs} = kK_E$ and an estimate of K_E, as described in Section 21.6.

21.5 Octahedral complexes of metal centers with high oxidation numbers or of d metals of the second and third series are less labile than those of low oxidation number and d metals of the first series of the block. Account for this observation on the basis of a dissociative rate-determining step. If ligand substitution takes place by a *d* mechanism, the strength of the metal-leaving group bond is directly related to the substitution rate. Metal centers with high oxidation numbers will have stronger bonds to ligands than metal centers with low oxidation numbers. Furthermore, period 5 and 6 *d*-block metals have stronger metal ligand bonds (see Section 21.1). Therefore, for reactions that are dissociatively activated, complexes of period 5 and 6 metals are less labile than complexes of period 3 metals, and complexes of metals in high oxidation states are less labile than complexes of metals in low oxidation states (all other things remaining equal).

21.6 A Pt(II) complex of tetramethyldiethylenetriamine is attacked by Cl^- 10^5 times less rapidly than the diethylenetriamine analogue. Explain this observation in terms of an associative rate-determining step. The two complexes are shown below. The ethyl substituted complex presents a greater degree of steric hindrance to an incoming Cl^- ion nucleophile. Since the rate-determining step for associative substitution of X^- by Cl^- is the formation of a Pt–Cl bond, the more hindered complex will react more slowly.

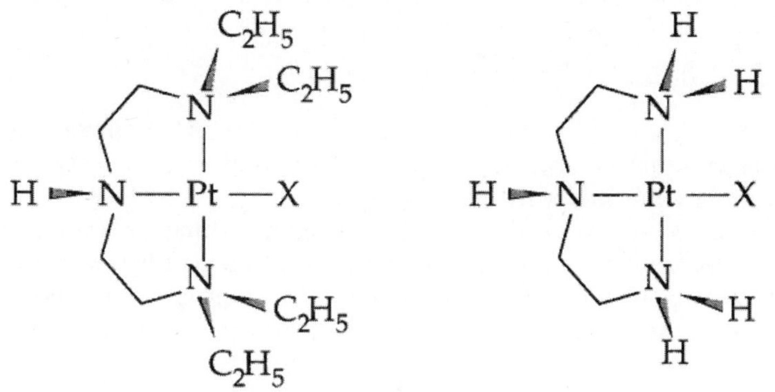

21.7 **The rate of loss of chlorobenzene, PhCl, from $[W(CO)_4L(PhCl)]$ increases with increase in the cone angle of L. What does this observation suggest about the mechanism?** Since the rate of loss of chlorobenzene, PhCl, from the tungsten complex becomes faster as the cone angle of L increases, this is a case of dissociative activation (i.e., steric crowding in the transition state accelerates the rate). This is not at odds with the observation that the rate is proportional to the concentration of the entering phosphane at low phosphane concentrations, since K_E for the equilibrium producing the encounter complex, $C + Y \rightarrow \{CY\}$, is inversely proportional to [Y] (C is the tungsten complex, Y is the entering phosphane; see Section 21.5, *Rate laws and their interpretation*). The overall rate is given by:

$$\text{rate} =+ (kK_E[C][Y])/(1 \quad K_E[Y])$$

and, in the limit of low [Y], this becomes rate ~ $kK_E[C][Y]$.

21.8 **The pressure dependence of the replacement of chlorobenzene (PhCl) by piperidine in the complex $[W(CO)_4(PPh_3)(PhCl)]$ has been studied. The volume of activation is found to be 111.3 $cm^3\ mol^{-1}$. What does this value suggest about the mechanism?** Since the volume of activation is positive (+11.3 $cm^3\ mol^{-1}$), the activated complex takes up more volume in solution than the reactants, as shown in the drawing below. Therefore, the mechanism of substitution must be dissociative. See Section 21.6(c), *steric effects*

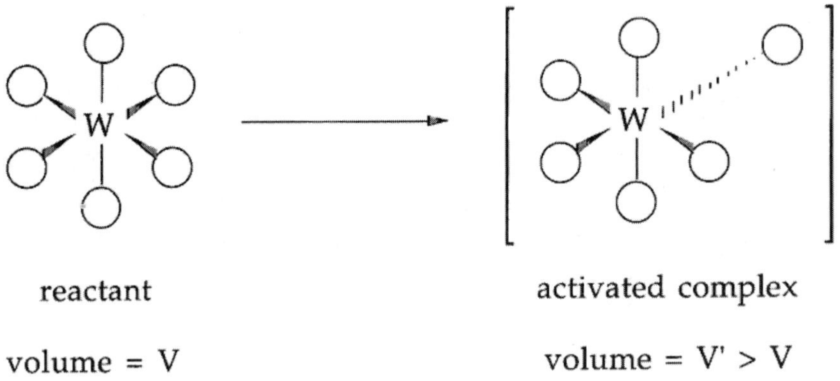

reactant

volume = V

activated complex

volume = V' > V

21.9 **Does the fact that $[Ni(CN)_5]^{3-}$ can be isolated help to explain why substitution reactions of $[Ni(CN)_4]^{2-}$ are very rapid?** The fact that the five-coordinate complex $[Ni(CN)_5]^{3-}$ can be detected does indeed explain why substitution reactions of the four-coordinate complex $[Ni(CN)_4]^{2-}$ are fast. The reason is that, for a detectable amount of $[Ni(CN)_5]^{3-}$ to build up in solution, the forward rate constant k_f must be numerically close to or greater than the reverse rate constant k_r:

$$[Ni(CN)_4]^{2-} + CN^- \rightarrow [Ni(CN)_5]^{3-}$$

If k_f were much smaller than k_r, the equilibrium constant $K = k_f/k_r$ would be small and the concentration of $[Ni(CN)_5]^{3-}$ would be too small to detect. Therefore, since k_f is relatively large, you can infer that rate constants for the association of other nucleophiles are also large, with the result that substitution reactions of $[Ni(CN)_4]^{2-}$ are very fast.

21.10 **Reactions of $[Pt(Ph)_2(SMe_2)_2]$ with the bidentate ligand 1,10-phenanthroline (phen) give $[Pt(Ph)_2phen]$. There is a kinetic pathway with activation parameters $\Delta^{\ddagger}H = 1101\ kJ\ mol^{-1}$ and $\Delta^{\ddagger}S = 142\ J\ K^{-1}\ mol^{-1}$. Propose a mechanism.**

The entropy of activation is positive, ruling out an associative mechanism (associative mechanisms have uniformly negative entropies of activation). A possible mechanism is loss of one dimethyl sulfide ligand, followed by the coordination of 1,10-phenanthroline, giving an unstable, intermediate, a 5-coordinate complex. In the final step, the second dimethyl sulfide ligand leaves and rearranges to form a square-planar complex.

21.11 **A two-step synthesis for *cis*- and *trans*-$[PtCl_2(NO_2)(NH_3)]^-$ starting from $[PtCl_4]^{2-}$?** Starting with $[PtCl_4]^{2-}$, you need to perform two separate ligand substitution reactions. In one, NH_3 will replace Cl^- ion. In the other, NO_2^- ion will replace Cl^- ion. The question is, which substitution to perform first? According to the *trans* effect series shown in Section 21.4, the strength of the *trans* effect on Pt(II) for the three ligands in question is $NH_3 < Cl^- < NO_2^-$. This means that a Cl^- ion *trans* to another Cl^- will be substituted faster than a Cl^- ion *trans* to NH_3, while a Cl^- ion *trans* to NO_2^- will be substituted faster than a Cl^- ion *trans* to another Cl^- ion. If you first add NH_3 to $[PtCl_4]^{2-}$, you will produce $[PtCl_3(NH_3)]^-$. Now if you add NO_2^-, one of the mutually *trans* Cl^- ligands will be substituted faster than the Cl^- ligand *trans* to NH_3, and the *cis* isomer will be the result. If you first add NO_2^- to $[PtCl_4]^{2-}$, you will produce $[PtCl_3(NO_2)]^{2-}$. Now if you add NH_3, the Cl^- ligand *trans* to NO_2^- will be substituted faster than one of the mutually *trans* Cl^- ligands, and the *trans* isomer will be the result. These two-step syntheses are shown below:

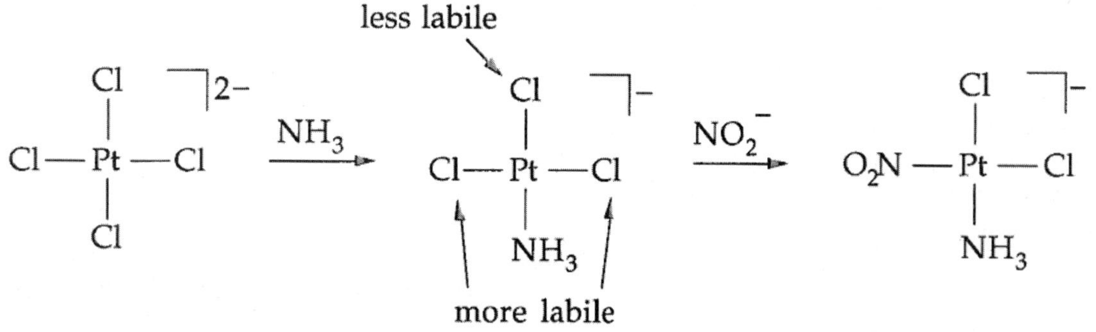

$[PtCl_4]^{2-}$ $\xrightarrow{NH_3}$ $[PtCl_3(NH_3)]^-$ $\xrightarrow{NO_2^-}$ $[PtCl_2(NO_2)(NH_3)]^-$

less labile (Cl *trans* to NH_3); more labile (mutually *trans* Cl)

cis isomer

$[PtCl_4]^{2-}$ $\xrightarrow{NO_2^-}$ $[PtCl_3(NO_2)]^{2-}$ $\xrightarrow{NH_3}$ $[PtCl_2(NO_2)(NH_3)]^-$

more labile (Cl *trans* to NO_2); less labile (mutually *trans* Cl)

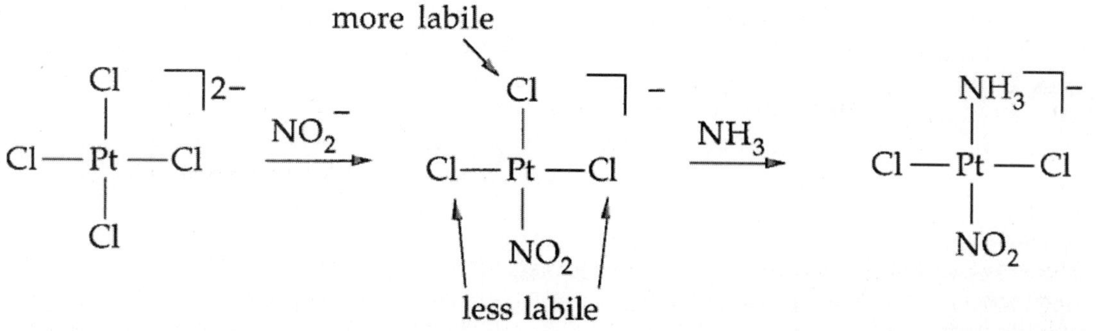

trans isomer

21.12 **How does each of the following affect the rate of square-planar substitution reactions? (a) Changing a *trans* ligand from H to Cl?** Hydride ion lies higher in the *trans* effect series than does chloride ion. Thus if the ligand *trans* to H or Cl is the leaving group, its rate of substitution will be decreased if H is changed to Cl. The change in rate can be as large as a factor of 10^4 (see Table 21.4).

(b) Changing the leaving group from Cl^- to I^-? The rate at which a ligand is substituted in a square-planar complex is related to its position in the *trans* effect series. If a ligand is high in the series, it is a *good* entering group (i.e., a good nucleophile) and a *poor* leaving group. Since iodide ion is higher in the *trans* effect series than chloride ion, it is a poorer leaving group than chloride ion. Therefore, the *iodo* complex will undergo I^- substitution more slowly than the *chloro* complex will undergo Cl^- substitution.

(c) Adding a bulky substituent to a *cis* ligand? This change will hinder the approach of the entering group and will slow the formation of the five-coordinate activated complex. The rate of substitution of a square-planar complex with bulky ligands will be slower than that of a comparable complex with sterically smaller ligands.

(d) Increasing the positive charge on the complex? If all other things are kept equal, increasing the positive charge on a square-planar complex will increase the rate of substitution. This is because the entering ligands are either anions or have the negative ends of their dipoles pointing at the metal ion. As explained in Section 21.3, if the charge density of the complex decreases in the activated complex, as would happen when an anionic ligand adds to a cationic complex, the solvent molecules will be *less* ordered around the complex (the opposite of the process called electrostriction). The increased disorder of the solvent makes ΔS less negative (compare the values of ΔS for the Pt(II) and Au(III) complexes in Table 21.5).

21.13 The rate of attack on Co(III) by an entering group Y is nearly independent of Y with the spectacular exception of the rapid reaction with OH^-. Explain the anomaly. What is the implication of your explanation for the behaviour of a complex lacking Brønsted acidity on the ligands? The general trend is easy to explain: octahedral Co(III) complexes undergo dissociatively activated ligand substitution. The rate of substitution depends on the nature of the bond between the metal and the leaving group, since this bond is partially broken in the activated complex. The rate is independent of the nature of the bond to the entering group, since this bond is formed in a step subsequent to the rate determining step. The anomalously high rate of substitution by OH^- signals an alternate path, that of base hydrolysis, as shown below (see Section 21.7):

$$[CoL_5(EH_n)]^{m+} + OH^- \rightleftharpoons [CoL_5(EH_{n-1})]^{(m-1)+} + H_2O$$

The deprotonated complex $[CoL_5(EH_{n-1})]^{(m-1)+}$ will undergo loss of L faster than the starting complex $[CoL_5(EH_n)]^{m+}$, because the anionic EH_{n-1}^- ligand is a stronger base than the neutral EH_n ligand and can better stabilize the coordinatively unsaturated activated complex. The implication is that a complex without protic ligands will not undergo anomalously fast OH^- ion substitution.

21.14 Predict the products of the following reactions: (a) $[Pt(PR_3)_4]^{2+} + 2Cl^-$? The first Cl^- ion substitution will produce the reaction intermediate $[Pt(PR_3)_3Cl]^+$. This will be attacked by the second equivalent of Cl^- ion. Since phosphanes are higher in the *trans* effect series than chloride ion, a phosphane *trans* to another phosphane will be substituted, giving *cis*-$[PtCl_2(PR_3)_2]$.

(b) $[PtCl_4]^{2-} + 2PR_3$? The first PR_3 substitution will produce the reaction intermediate $[PtCl_3(PR_3)]^-$. This will be attacked by the second equivalent of PR_3. Since phosphanes are higher in the *trans* effect series than chloride ion, the Cl^- ion *trans* to PR_3 will be substituted, giving *trans*-$[PtCl_2(PR_3)_2]$.

(c) *cis*-$[Pt(NH_3)_2(py)_2]^{2+} + 2Cl^-$? You should assume that NH_3 and pyridine are about equal in the *trans* effect series. The first Cl^- ion substitution will produce one of two reaction intermediates, *cis*-$[PtCl(NH_3)(py)_2]^+$ or *cis*-$[PtCl(NH_3)_2(py)]^+$. Either of these will produce the same product when they are attacked by the second equivalent of Cl^- ion. This is because Cl^- is higher in the *trans* effect series than NH_3 or pyridine. Therefore, the ligand *trans* to the first Cl^- ligand will be substituted faster, and the product will be *trans*-$[PtCl_2(NH_3)(py)]$.

21.15 Put in order of increasing rate of substitution by H_2O the complexes: $[Co(NH_3)_6]^{3+}$, $[Rh(NH_3)_6]^{3+}$, $[Ir(NH_3)_6]^{3+}$, $[Mn(OH_2)_6]^{2+}$, $[Ni(OH_2)_6]^{2+}$? The three amine complexes will undergo substitution more slowly than the two aqua complexes. This is because of their higher charge and their low-spin d^6 configurations. You should refer to Table 21.8, which lists ligand field activation energies (LFAE) for various d^n configurations. While low-spin d^6 is not included, note the similarity between d^3 (t_{2g}^3) and low-spin d^6 (t_{2g}^6). A low-spin octahedral d^6 complex has LFAE = $0.4\Delta_O$, and consequently is inert to substitution. Of the three amine complexes listed above, the iridium complex is the most inert, followed by the rhodium complex, which is more inert than the cobalt complex. This is because Δ_O increases on descending a group in the *d* block. Of the two aqua complexes, the Ni complex, with LFAE = $0.2\Delta_O$, undergoes substitution more slowly than the Mn complex, with LFAE = 0. Thus the order of increasing rate is $[Ir(NH_3)_6]^{3+} < [Rh(NH_3)_6]^{3+} < [Co(NH_3)_6]^{3+} < [Ni(OH_2)_6]^{2+} < [Mn(OH_2)_6]^{2+}$.

21.16 State the effect on the rate of dissociatively activated reactions of Rh(III) complexes of each of (a) an increase in the positive charge on the complex? Since the leaving group (X) is invariably negatively charged or the negative end of a dipole, increasing the positive charge on the complex will retard the rate of M–X bond cleavage. For a dissociatively activated reaction, this change will result in a decreased rate.

(b) changing the leaving group from NO_3^- to Cl^-? For the substitution reaction shown, changing the leaving group from nitrate ion to chloride ion results in a decreased rate. The explanation offered in Section 21.6 is that the Co–Cl bond is stronger than the Co–ONO_2 bond. For a dissociatively activated reaction, a stronger bond to the leaving group will result in a decreased rate.

(c) changing the entering group from Cl^- to I^-? This change will have little or no effect on the rate. For a dissociatively activated reaction, the bond between the entering group and the metal is formed subsequent to the rate-determining step.

(d) changing the *cis* ligands from NH_3 to H_2O? These two ligands differ in their σ-basicity. The less basic ligand, H_2O, will decrease the electron density at the metal and will destabilize the coordinatively unsaturated activated complex. Therefore, this change, from NH_3 to the less basic ligand H_2O, will result in a decreased rate.

21.17 Write out the inner- and outer-sphere pathways for reduction of azidopentaamminecobalt(III) ion with V^{2+}(aq). What experimental data might be used to distinguish between the two pathways? The inner-sphere pathway is shown below (solvent = H_2O):

$$[Co(N_3)(NH_3)_5]^{2+} + [V(OH_2)_6]^{2+} \rightarrow [[Co(N_3)(NH_3)_5]^{2+}, [V(OH_2)_6]^{2+}\}$$

$$\{[Co(N_3)(NH_3)_5]^{2+}, [V(OH_2)_6]^{2+}\} \rightarrow \{[Co(N_3)(NH_3)_5]^{2+}, [V(OH_2)_5]^{2+}, H_2O\}$$

$$\{[Co(N_3)(NH_3)_5]^{2+}, [V(OH_2)_5]^{2+}, H_2O\} \rightarrow [(NH_3)_5\underset{Co(III)}{Co}\text{–N=N=N–}\underset{V(II)}{V}(OH_2)_5]^{4+}$$

$$[(NH_3)_5\underset{Co(III)}{Co}\text{–N=N=N–}\underset{V(II)}{V}(OH_2)_5]^{4+} \rightarrow [(NH_3)_5\underset{Co(II)}{Co}\text{–N=N=N–}\underset{V(III)}{V}(OH_2)_5]^{4+}$$

$$[(NH_3)_5\underset{Co(II)}{Co}\text{–N=N=N–}\underset{V(III)}{V}(OH_2)_5]^{4+} \rightarrow [Co(OH_2)_6]^{2+} + [V(N_3)(OH_2)_5]^{2+}$$

The pathway for outer-sphere electron transfer is shown below:

$$[Co(N_3)(NH_3)_5]^{2+} + [V(OH_2)_6]^{2+} \rightarrow \{[Co(N_3)(NH_3)_5]^{2+} [V(OH_2)_6]^{2+}\}$$

$$\{[Co(N_3)(NH_3)_5]^{2+},[V(OH_2)_6]^{2+}\} \rightarrow \{[Co(N_3)(NH_3)_5]^{+}, [V(OH_2)_6]^{3+}\}$$

$$\{[Co(N_3)(NH_3)_5]^{+},[V(OH_2)_6]^{3+}\} \rightarrow \{[Co(OH_2)_6]^{2+}, [V(OH_2)_6]^{3+}$$

In both cases, the cobalt-containing product is the aqua complex, since H_2O is present in abundance and high-spin d^7 complexes of Co(II) are substitution labile. However, something that distinguishes the two pathways is the composition of the vanadium-containing product. If $[V(N_3)(OH_2)_5]^{2+}$ is the product, then the reaction has proceeded via an inner-sphere pathway. If $[V(OH_2)_6]^{3+}$ is the product, then the electron transfer reaction is outer-sphere.

21.18 The compound $[Fe(SCN)(OH_2)_5]^{2+}$ can be detected in the reaction of $[Co(NCS)(NH_3)_5]^{2+}$ with Fe^{2+}(aq) to give Fe^{3+}(aq) and Co^{2+}(aq). What does this observation suggest about the mechanism? The direct transfer of a ligand from the coordination sphere of one redox partner (in this case the oxidizing agent, $[Co(NCS)(NH_3)_5]^{2+}$) to the coordination sphere of the other (in this case the reducing agent, $[Fe(OH_2)_6]^{2+}$) signals an inner-sphere electron transfer reaction. Even if the first formed product $[Fe(NCS)(OH_2)_5]^{2+}$ is short lived and undergoes hydrolysis to $[Fe(OH_2)_6]^{3+}$, its fleeting existence demands that the electron was transferred across a Co–(NCS)–Fe bridge.

21.19 Calculate the rate constants for electron transfer in the oxidation of $[V(OH_2)_6]^{2+}$ (E° (V^{3+}/V^{2+}) = –0.255 V) and the oxidants (a) $[Ru(NH_3)_6]^{3+}$ ($E^\circ(Ru^{3+}/Ru^{2+})$ = + 0.07 V), (b) $[Co(NH_3)_6]^{3+}$ ($E^\circ(Co^{3+}/Co^{2+})$ = +0.10 V). Comment on the relative sizes of the rate constants.

Using the Marcus-Cross relationship (Equation 21.14), we can calculate the rate constants. In this equation $[k_{12} = [k_{11}\cdot k_{22}\cdot K_{12}\cdot f_{12}]^{1/2}$, the values of k_{11} and k_{22} can be obtained from Table 21.12. We can assume f_{12} to be unity. The redox potential data allows us to calculate K_{12}, since ε^o = [RT/nF]lnK. The value of ε^o can be calculated by subtracting the anodic reduction potential (the Cr^{3+}/Cr^{2+} couple serves as the anode) from the cathodic one. For calculation details, see exercise 21.20 below.

(a) $[Ru(NH_3)_6]^{3+}$ $k = 4.53 \times 10^3$ dm^{3}mol^{-1}s^{-1}

(b) $[Co(NH_3)_6]^{3+}$ $k = 1.41 \times 10^{-2}$ dm^{3}mol^{-1}s^{-1}

Relative sizes? The reduction of the Ru complex is more thermodynamically favoured than the reaction of the Co complex, and it is faster as evident from its larger k value.

21.20 **Calculate the rate constants for electron transfer in the oxidation of $[Cr(OH_2)_6]^{2+}$ (E^O–(Cr^{3+}/Cr^{2+}) = –0.41 V) and each of the oxidants $[Ru(NH_3)_6]^{3+}$ (E^O(Ru^{3+}/Ru^{2+}) = +0.07 V), $[Fe(OH_2)_6]^{3+}$ (E^O(Fe^{3+}/Fe^{2+}) = +0.77 V) and $[Ru(bpy)_3]^{3+}$ (E^O(Ru^{3+}/Ru^{2+}) = +1.26 V). Comment on the relative sizes of the rate constants**
Using the Marcus-Cross relationship (Equation 21.14), we can calculate the rate constants. In this equation $[k_{12} = [k_{11} \cdot k_{22} \cdot K_{12} \cdot f_{12}]^{1/2}$, the values of k_{11} and k_{22} can be obtained from Table 21.12. We can assume f_{12} to be unity. The redox potential data allows us to calculate K_{12}, since ε^o = [RT/nF]lnK. The value of ε^o can be calculated by subtracting the anodic reduction potential (the Cr^{3+}/Cr^{2+} couple serves as the anode) from the cathodic one.

(a) k_{11} (Cr^{3+}/Cr^{2+}) = 1×10^{-5} dm^{3}mol^{-1}s^{-1}; k_{22} (Ru^{3+}/Ru^{2+} for the hexamine complex) = 6.6×10^3 dm^{3}mol^{-1}s^{-1}; f_{12} = 1; $K_{12} = e^{[nF\,\varepsilon^o/RT]}$ where ε^o = 0.07 V – (–0.41V) = 0.48 V; n = 1; F = 96485 C; R = 8.31 Jmol^{-1}K^{-1} and T = 298 K. Using these values we get $K_{12} = 1.32 \times 10^8$. Substitution of these values in the Marcus-Cross relationship gives $k_{12} = 2.95 \times 10^3$ dm^{3}mol^{-1}s^{-1}.

(b) k_{11} (Cr^{3+}/Cr^{2+}) = 1×10^{-5} dm^{3}mol^{-1}s^{-1}; k_{22} (Fe^{3+}/Fe^{2+} for the aqua complex) = 1.1 dm^{3}mol^{-1}s^{-1}; f_{12} = 1; $K_{12} = e^{[nF\,\varepsilon^o/RT]}$ where ε^o = 0.77 V – (–0.41V) = 1.18 V; n = 1; F = 96485 C; R = 8.31 Jmol^{-1}K^{-1} and T = 298 K. Using these values we get $K_{12} = 9.26 \times 10^{19}$. Substitution of these values in the Marcus-Cross relationship gives $k_{12} = 3.19 \times 10^7$ dm^{3}mol^{-1}s^{-1}.

(c) k_{11} (Cr^{3+}/Cr^{2+}) = 1×10^{-5} dm^{3}mol^{-1}s^{-1}; k_{22} (Ru^{3+}/Ru^{2+} for the bipy complex) = 4×10^8 dm^{3}mol^{-1}s^{-1}; f_{12} = 1; $K_{12} = e^{[nF\,\varepsilon^o/RT]}$ where ε^o = 1.26 V – (–0.41V) = 1.67 V; n = 1; F = 96485 C; R = 8.31 Jmol^{-1}K^{-1} and T = 298 K. Using these values we get $K_{12} = 1.81 \times 10^{28}$. Substitution of these values in the Marcus-Cross relationship gives $k_{12} = 8.51 \times 10^{15}$ dm^{3}mol^{-1}s^{-1}.

Note: The assumption that f_{12} = 1 is used in all cases; however, a more precise value would be needed for cases (b) and (c) because of higher values of the equilibrium constant.

21.21 **The photochemical substitution of $[W(CO)_5(py)]$ (py = pyridine) with triphenylphosphine gives $W(CO)_5(P(C_6H_5)_3)$. In the presence of excess phosphine, the quantum yield is approximately 0.4. A flash photolysis study reveals a spectrum that can be assigned to the intermediate $W(CO)_5$. What product and quantum yield do you predict for substitution of $[W(CO)_5(py)]$ in the presence of excess triethylamine? Is this reaction expected to be initiated from the ligand field or MLCT excited state of the complex?** Since the intermediate is believed to be $[W(CO)_5]$, the properties of the entering group (triethylamine versus triphenylphosphane) should not affect the quantum yield of the reaction, which is a measure of the rate of formation of $[W(CO)_5]$ from the excited state of $[W(CO)_5(py)]$. The product of the photolysis of $[W(CO)_5(py)]$ in the presence of excess triethylamine will be $[W(CO)_5(NEt_3)]$, and the quantum yield will be 0.4. This photosubstitution is initiated from a ligand field excited state, not an MLCT excited state. A metal-ligand charge transfer increases the oxidation state of the metal, which would strengthen, not weaken, the bond between the metal and a σ-base like pyridine.

21.22 **From the spectrum of $[CrCl(NH_3)_5]^{2+}$ shown in Fig. 20.32, propose a wavelength for photoinitiation of reduction of Cr(III) to Cr(II) accompanied by oxidation of a ligand.** The intense band at ~250 nm is an LMCT transition (specifically a Cl^--to-Cr^{3+} charge transfer). Irradiation at this wavelength should produce a population of $[CrCl(NH_3)_5]^{2+}$ ions that contain Cr^{2+} ions and Cl atoms instead of Cr^{3+} ions and Cl^- ions. Irradiation of the complex at wavelengths between 350 and 600 nm will not lead to photoreduction. The bands that are observed between these two wavelengths are ligand field transitions: the electrons on the metal are rearranged, and the electrons on the Cl^- ion are not involved.

Chapter 22 *d*-Metal Organometallic Chemistry Metals

Here is the structure of tricarbonyl(η^5-cyclopentadienyl)-manganese(I) (the oxygen atoms of the carbonyl ligands are not shaded). This molecule contains two important types of ligands frequently found in organometallic compounds, π-acids (the CO ligands) and unsaturated organic molecules or molecular fragments that donate their π electrons to the metal center (the η^5-C_5H_5 ligand).

S22.1 **Is $Mo(CO)_7$ likely to be stable?** A Mo atom (group 6) has six valence electrons, and each CO ligand is a two-electron donor. Therefore, the total number of valence electrons on the Mo atom in this compound would be 6 + 7(2) = 20. Since organometallic compounds with more than 18 valence electrons on the central metal are never stable, $Mo(CO)_7$ is *not* likely to exist. The compound $Mo(CO)_6$, with exactly 18 valence electrons, is very stable. Note that throughout Chapter 22, the authors do not always write the formulas for organometallic complexes in square brackets. According to the rules of inorganic nomenclature, *all* complexes should be written in square brackets. Nevertheless, the authors are following a common, informal set of rules: neutral organometallic complexes are written without square brackets, while cationic or anionic organometallic complexes are written with them.

S22.2 **What is the electron count for and oxidation number of platinum in the anion of Zeise's salt, $[PtCl_3(C_2H_4)]^-$? Treat CH_2=CH_2 as a neutral two-electron donor.** The Pt atom (Group 10) has ten valence electrons, each Cl atom is a one-electron donor, ethylene is a two-electron donor, and one electron must be added for the –1 charge of the complex. Therefore, the total number of valence electrons on the Pt atom in this complex is 10 + 3(1) + 2 + 1 = 16. This complex deviates from the 18-electron rule, as do many four-coordinate period 4 and 5 d^8 complexes (see Section 22.3). As a consequence of being two electrons short of 18, this complex undergoes ligand substitution by an associative mechanism (i.e., with the formation of a five-coordinate 18-electron intermediate).

The name of this complex, which has two ligand types, is trichloro(ethylene)□alatinate(II) or trichloro(ethylene) alatinate(1–). You should review the nomenclature rules for metal complexes. Complexes are named with their ligands in alphabetical order. The prefixes di-, tri-, tetra-, etc., do not count in the alphabetical order. Thus *c* before *e* in tri*c*hloro-(*e*thylene) alatinate(II). The structure of this complex is square planar, the usual structure for period 4 and 5 d^8 metal ions. The structure of $[PtCl_3(C_2H_4)]^-$ is shown below. The geometry around the Pt atom is square planar, although the complex anion is not planar (the ethylene ligand is perpendicular to the $PtCl_3$ plane). Because of back-donation, the C=C bond distance in the complex, 1.375 Å, is slightly longer than the C=C bond distance in ethylene, 1.337 Å.

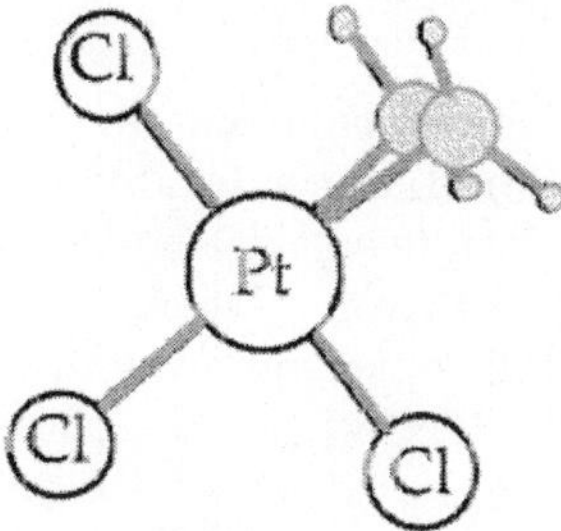

S22.3 **What is the formal name of $[Ir(Br)_2(CH_3)(CO)(PPh_3)_2]$?** Dibromocarbonylmethylbis(triphenylphosphine)iridium(III).

S22.4 Which of the two iron compounds $Fe(CO)_5$ and $[Fe(CO)_4(PEt_3)]$ will have the higher CO stretching frequency? Which will have the longer M–C bond? $Fe(CO)_5$ will have the higher CO stretching frequency and the longer bond. As noted in Example 22.4, PET_3 causes increased backbonding, lower CO stretching frequency and a shorter Fe-C bond.

S22.5 Show that both are 18-electron species. (a) $[(\eta^6\text{-}C_7H_8)Mo(CO)_3]$ (49)? The $\eta^6\text{-}C_7H_8$ provides 6 electrons, as does the formally neutral Mo atom. Each carbonyl provides another 2 electrons, giving a total of 18. **(b) $[(\eta^7\text{-}C_7H_7)Mo(CO)_3]^+$(51)?** The $\eta^7\text{-}C_7H_7^+$, group provides 6 electrons, leaving a formally neutral Mo atom and the 3 carbonyl groups each providing a further 6, giving a total of 18.

S22.6 Propose a synthesis for $Mn(CO)_4(PPh_3)(COCH_3)$ starting with $[Mn_2(CO)_{10}]$, PPh_3, Na and CH_3I. Consider the reactions of carbonyl complexes discussed in Section 22.18. If you use $Mn_2(CO)_{10}$ as the source of manganese, you can reductively cleave the Mn–Mn bond with sodium, forming $Na[Mn(CO)_5]$:

$$Mn_2(CO)_{10} + 2\,Na \rightarrow 2Na[Mn(CO)_5]$$

oxidation number: 0 0 +1 −1

The anionic carbonyl complex is a relatively good nucleophile. When it is treated with CH_3I, it displaces I^- to form $Mn(CH_3)(CO)_5$ (sometimes written as $CH_3Mn(CO)_5$):

$$Na[Mn(CO)_5] + CH_3I \rightarrow Mn(CH_3)(CO)_5 + NaI$$

Many alkyl-substituted metal carbonyls undergo a migratory insertion reaction when treated with basic ligands. The alkyl group (methyl in this case) migrates from the Mn atom to an adjacent C atom of a CO ligand, leaving an open coordination site for the entering group (PPh_3 in this case) to attack:

$$Mn(CH_3)(CO)_5 + PPh_3 \rightarrow Mn(CO)_4(PPh_3)(COCH_3)$$

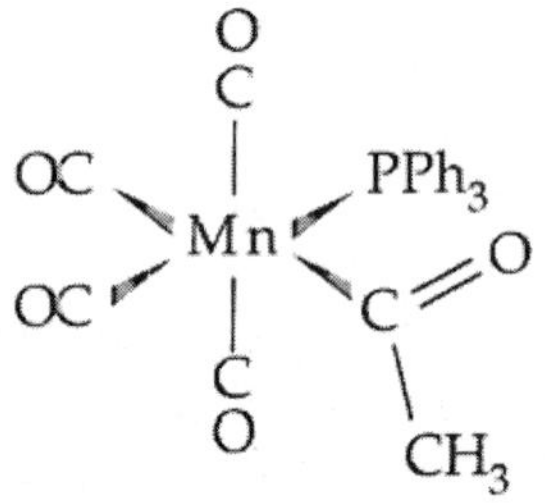

The structure of $Mn(CO)_4(PPh_3)(COCH_3)$

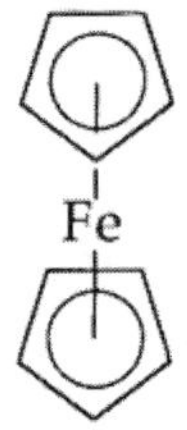

The structure of ferrocene

S22.7 The IR spectrum of $[Ni_2(\eta5\text{-}Cp)_2(CO)_2]$ has a pair of CO stretching bands at 1857 cm^{-1} (strong) and 1897 cm^{-1} (weak). Does this complex contain bridging or terminal CO ligands, or both? (Substitution of $\eta5\text{-}C_5H_5$ ligands for CO ligands leads to small shifts in the CO stretching frequencies for a terminal CO ligand.) The CO stretching bands at 1857 cm^{-1} and 1897 cm^{-1} are both lower in frequency than typical terminal CO ligands (for terminal CO ligands, $\nu(CO) > 1900\ cm^{-1}$ (see Figure 22.4)). Therefore, it seems likely that it only contains bridging CO ligands. The presence of two bands suggests that the bridging CO ligands are probably *not* collinear, since only one band would be observed if they were (see Figure 22.12)

S22.8 By using the same molecular orbital diagram, comment on whether the removal of an electron from $[Fe(\eta5\text{-}Cp)_2]$ to produce $[Fe(\eta5\text{-}Cp)_2]^+$ should produce a substantial change in M–C bond length relative to neutral ferrocene. Neutral ferrocene (shown above) contains 18 valence electrons (8 from Fe and 10 from the two Cp ligands). The MO diagram for ferrocene is shown in Figure 22.13. Eighteen electrons will fill it up to the second a_1' orbital. Since this orbital is the HOMO, oxidation of ferrocene will result in removal of an electron from it, leaving the 17-electron ferricinium cation, $[FeCp_2]^+$. If this orbital were strongly bonding, removal of an electron would result in weaker Fe–C bonds. If this orbital were strongly antibonding, removal of an electron would result in stronger Fe–C bonds. However, this orbital is essentially nonbonding (see Figure 22.13). Therefore, oxidation of $FeCp_2$ to $[FeCp_2]^+$ will not produce a substantial change in the Fe–C bond order or the Fe–C bond length.

S 22.9 The compound [$Fe_4(Cp)_4(CO)_4$] is a dark-green solid. Its IR spectrum shows a single CO stretch at 1640 cm^{-1}. The ^{1}H NMR spectrum is a single line even at low temperatures. From this spectroscopic information and the CVE, propose a structure for [$Fe_4(Cp)_4(CO)_4$]. There are four relevant pieces of information given. First of all, the fact that the compound is highly colored suggests that it contains metal–metal bonds. Second, the composition can be used to determine the cluster valence electron count, which can be used to predict which polyhedral structure is likely:

Fe_4	$4(8e^-)$	= 32 valence e^-
Cp_4	$4(5e^-)$	= 20 valence e^-
$(CO)_4$	$4(2e^-)$	= 8 valence e^-
Total		= 60 valence e^- (a tetrahedral cluster)

Third, the presence of only one line in the ^{1}H NMR spectrum suggests that the Cp ligands are all equivalent and are η^5. Finally, the single CO stretch at 1640 cm^{-1} suggests that the CO ligands are triply bridging (see Figure 22.4) and that they form a relatively high symmetry array (otherwise there would be more bands; see Table 22.12). A likely structure for $Fe_4Cp_4(CO)_4$ is shown below, with only one of the Cp ligands and only one of the triply bridging CO ligands shown for simplicity.
The tetrahedral Fe_4 core of this cluster exhibits six equivalent Fe–Fe bond distances. Only one of the η^5-C_5H_5 ligands and one of the triply bridging CO ligands is shown. Each Fe atom has a Cp ligand, and each of the four triangular faces of the Fe_4 tetrahedron is capped by a triply bridging CO ligand.

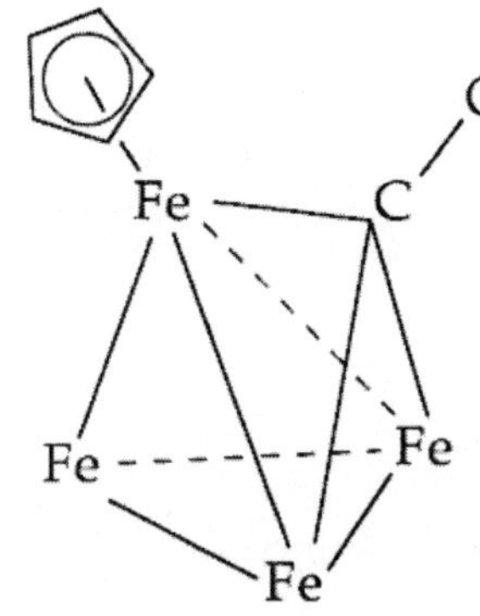

S22.10 If $Mo(CO)_3L_3$ is desired, which of the ligands $P(CH_3)_3$ or $P(t\text{-}Bu)_3$ would be preferred? Give reasons for your choice. Since this is a highly substituted complex, the effects of steric crowding must be considered. This is especially true in this case, since the two ligands in question should be very similar electronically. The cone angle for $P(CH_3)_3$, given in Table 22.11, is 118°. The cone angle for $P(t\text{-}Bu)_3$, also given in the table, is 182°. Therefore, because of its smaller size, PMe_3 would be preferred.

S22.11 Assess the relative substitutional reactivities of indenyl and fluorenyl (86) compounds? Fluorenyl compounds are more reactive than indenyl since they have two aromatic resonance forms when they bond in the η^3 mode.

S22.12 Show that the reaction is an example of reductive elimination?

$$PdCl_2(Ph)(PPh_3)_2 \longrightarrow PdCl_2(PPh_3)_2 + PhCl$$

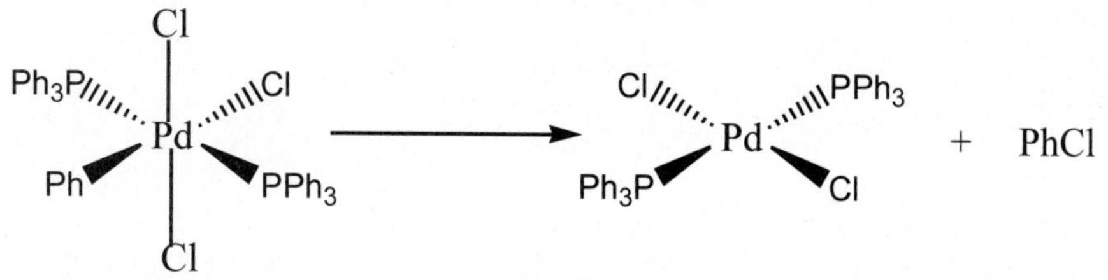

The six-coordinate palladium starting material is an 18-electron Pd(IV) species. The four-coordinate product is a 16-electron Pd(II) species. The decrease in both coordination number and oxidation number by 2 identifies the reaction as reductive elimination.

S22.13 Explain why $[Pt(PEt_3)_2(Et)(Cl)]$ readily decomposes, whereas $[Pt(PEt_3)_2(Me)(Cl)]$ does not? The ethyl group in $[Pt(PEt_3)_2(Et)(Cl)]$ is prone to *β*-hydride elimination, whereas the methyl group in $[Pt(PEt_3)_2(Me)(Cl)]$ is not.

22.1 Name the species, draw the structures of, and give valence electron counts to the metal atoms? Do any of the complexes deviate from the 18-electron rule? If so, how is this reflected in their structure or chemical properties?
(a) $Fe(CO)_5$? Pentacarbonyliron(0), 18e$^-$; **(b) $Mn_2(CO)_{10}$?** decacarbonyldimanganese(0), 18e$^-$; **(c) $V(CO)_6$?** hexcarbonylvanadium(0), 17e$^-$ because it is easily reduced; **(d) $[Fe(CO)_4]^{2-}$?** tetracarbonylferrate(–2), 18e$^-$; **(e) $La(\eta^5\text{-}Cp^*)_3$?** tris(pentamethylcyclopentadienyl)lanthanum(III), 18e$^-$; **(f) $Fe(\eta^3\text{-allyl})(CO)_3Cl$?** allyltricarbonylchloroiron(II), 18e$^-$; **(g) $Fe(CO)_4(PEt_3)$?** tetracarbonyltriethylphosphineiron(0); **(h) $Rh(CO)_2(Me)(PPh_3)$?** dicarbonylmethyltriphenylphosphinerhodium(I),16e$^-$ b/c of the square planar geometry; **(i) $Pd(Cl)(Me)(PPh_3)_2$?** chloromethylbis(triphenylphosphine)palladium(II), 16e$^-$ because it is square planar; **(j) $Co(\eta^5\text{-}C_5H_5)(\eta^4\text{-}C_4Ph_4)$?** cyclopentadienyltetraphenylcylcobutadinecobalt(I), 18e$^-$; **(k) $[Fe(\eta^5\text{-}C5H_5)(CO_2)]^-$?** dicarbonylcyclopentadienylferrate(0), 18e$^-$; **(l) $Cr(\eta^6\text{-}C_6H_6)(\eta^6\text{-}C_7H_8)$?** benzenecycloheptatrienechromium(0), 18e$^-$; **(m) $Ta(\eta^5\text{-}C_5H_5)_2Cl_3$?** trichlorobiscyclopentadineyltanatalum(V), 18e$^-$; **(n) $Ni(\eta^5\text{-}C_5H_5)NO$?** cyclopentadineylnitrosylnickel(0),18e$^-$.

22.2 Sketch an η^2 interaction of 1,4-butadiene with a metal atom and (b) do the same for an η^4 interaction.

dihapto

tetrahapto

22.3 What hapticities are possible for the interaction of each of the following ligands with a single *d*-block metal atom such as cobalt? (a) C_2H_4? Ethylene coordinates to *d*-block metals in only one way, using its π-electrons to form a metal–ethylene σ-bond (there may also be a significant amount of back donation, if ethylene is substituted with electron-withdrawing groups; see Section 22.9). Therefore, C_2H_4 is always η^2, as shown below.

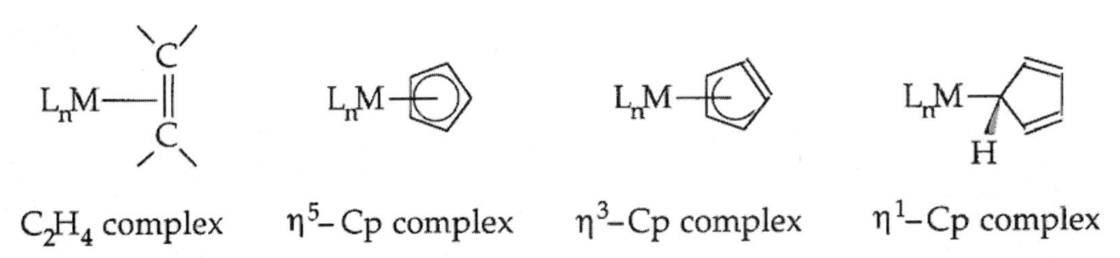

(b) Cyclopentadienyl? This is a very versatile ligand that can be η^5 (a five-electron donor), η^3 (a three-electron donor similar to simple allyl ligands), or η^1 (a one-electron donor similar to simple alkyl and aryl ligands). These three bonding modes are shown above.
(c) C_6H_6? This is also a versatile ligand, which can form η^6, η^4, and η^2 complexes. In such complexes, the ligands are, respectively, six-, four-, and two-electron donors. These three bonding modes are shown below.

η^6–C_6H_6 complex η^4–C_6H_6 complex η^2–C_6H_6 complex

(d) Cyclooctadiene? This ligand can form η^2 (two-electron donor) and η^4 (four-electron donor) complexes.
(e) Cyclooctatetraene? This ligand contains four C=C double bonds, any combination of which can coordinate to a *d*-block (or *f*-block) metal. Thus *cyclo*-C_8H_8 can be η^8 (an eight-electron donor), η^6 (a six-electron donor), η^4 (a four-electron donor), and η^2 (a two-electron donor, in which it would resemble an η^2-ethylene complex, except that three of the four C=C double bonds would remain uncoordinated).

22.4 **Draw plausible structures and give the electron count of (a) $Ni(\eta^3\text{-}C_3H_5)_2$ (b) $Co(\eta^4\text{-}C_4H_4)(\eta^5\text{-}C_5H_5)$ (c) $Co(\eta^3\text{-}C_3H_5)(CO)_3$. If the electron count deviates from 18, is the deviation explicable in terms of periodic terms.** Bis(allyl)nickel(0) has the structure shown below. Each allyl ligand, which is planar, is a three-electron donor, so the number of valance electrons around the Ni atom (group 10) is 10 + 2(3) = 16. Sixteen-electron complexes are very common for group 9 and group 10 elements, especially for Rh^+, Ir^+, Ni^{2+}, Pd^{2+}, and Pt^{2+} (all of which are d^8).
(b) $Co(\eta^4\text{-}C_4H_4)(\eta^5\text{-}C_5H_5)$? This complex has the structure shown below. Since the η^4-C_4H_4 ligand is a four-electron donor and the η^5–C_5H_5 ligand is a five-electron donor, the number of valence electrons around the Co atom (group 9) is 9 + 4 + 5 = 18.
(c) $Co(\eta^3\text{-}C_3H_5)(CO)_3$? This complex has the structure shown below. The electron count for the Co atom is:

Co	9 valence e^-
$\eta^3-C_3H_5$	3
3CO	6
Total	18 e^-

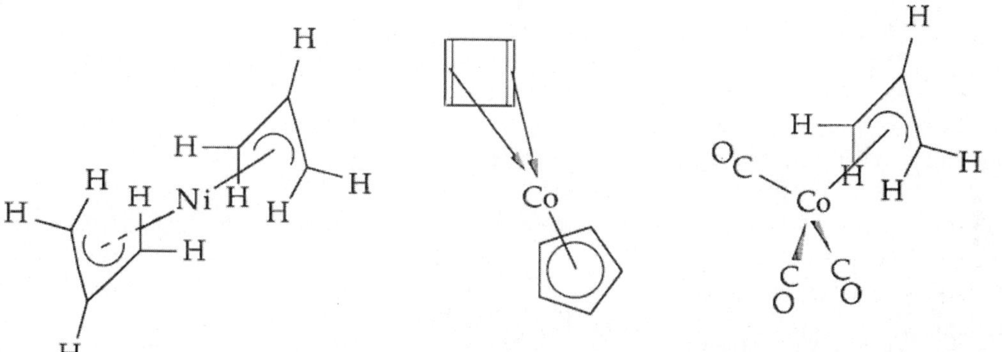

bis(allyl)nickel(0) (cyclobutadiene)CpCo (allyl)tricarbonylcobalt(C

22.5 **State the two common methods for the preparation of simple metal carbonyls and illustrate your answer with chemical equations. Is the selection of method based on thermodynamic or kinetic considerations?**
As discussed in Section 22.18, the two principal methods are (1) direct combination of CO with a finely divided metal and (2) reduction of a metal salt in the presence of CO under pressure. Two examples are shown below, the preparation of hexacarbonylmolybdenum(0) and octacarbonyldicobalt(0). Other examples are given in the text.

$$(1)\ Mo(s) + 6\,CO(g) \rightarrow Mo(CO)_6(s) \quad \text{(high temperature and pressure required)}$$

$$(2)\ 2\,CoCO_3(s) + 2\,H_2(g) + 8\,CO(g) \rightarrow Co_2(CO)_8(s) + 2\,CO_2 + 2\,H_2O$$

The reason that the second method is preferred is kinetic, not thermodynamic. The atomization energy (i.e., sublimation energy) of most metals is simply too high for the first method to proceed at a practical rate.

22.6 Suggest a sequence of reactions for the preparation of $Fe(CO)_3(dppe)$, given iron metal, CO, dppe ($Ph_2PCH_2CH_2PPh_2$), and other reagents of your choice.
The most general way to prepare ligand-substituted metal carbonyl complexes is to treat the parent binary metal carbonyl, in this case $Fe(CO)_5$, with the ligand of choice, in this case diphos (1,2-bis(diphenylphosphinoethane)). The two-step reaction sequence is:

$$Fe(s) + 5\,CO(g) \rightarrow Fe(CO)_5(l) \quad \text{(high temperature and pressure required)}$$

$$Fe(CO)_5(l) + diphos(s) \rightarrow Fe(CO)_3(diphos)(s) + 2\,CO(g)$$

The second step would require a slightly elevated temperature. A convenient way to achieve this would be to perform the ligand substitution in a refluxing solvent such as THF.

22.7 Suppose that you are given a series of metal tricarbonyl compounds having the respective symmetries C_{2v}, D_{3h}, and C_s. Without consulting reference material, which of these should display the greatest number of CO stretching bands in the IR spectrum? Check your answer and give the number of expected bands for each by consulting Table 22.7.
In general, the lower the symmetry of an $M(CO)_n$ fragment, the greater the number of CO stretching bands in the IR spectrum. Therefore, given complexes with $M(CO)_3$ fragments that have C_{3v}, D_{3h}, or C_s symmetry, the complex that has C_s symmetry will have the greatest number of bands. When you consult Table 22.7, you will see that an $M(CO)_3$ fragment with D_{3h} symmetry will give rise to one band, one with C_{3v} symmetry will give rise to two bands, and one with C_s symmetry will give rise to three bands.

22.8 Provide plausible reasons for the differences in IR wavenumbers between each of the following pairs:
(a) $Mo(PF_3)_3(CO)_3$ 2040, 1991 cm^{-1} versus $Mo(PMe_3)_3(CO)_3$ 1945, 1851 cm^{-1}? The two CO bands of the trimethylphosphane complex are 100 cm^{-1} or more lower in frequency than the corresponding bands of the trifluorophosphane complex. This is because PMe_3 is primarily a σ-donor ligand, while PF_3 is primarily a π-acid ligand (PF_3 is the ligand that most resembles CO electronically). The CO ligands in $Mo(PF_3)_3(CO)_3$ have to compete with the PF_3 ligands for electron density from the Mo atom for backdonation. Therefore, less electron density is transferred from the Mo atom to the CO ligands in $Mo(PF_3)_3(CO)_3$ than in $Mo(PMe_3)_3(CO)_3$. This makes the Mo–C bonds in $Mo(PF_3)_3(CO)_3$ weaker than those in $Mo(PMe_3)_3(CO)_3$, but it also makes the C–O bonds in $Mo(PF_3)_3(CO)_3$ stronger than those in $Mo(PMe_3)_3(CO)_3$, and stronger C–O bonds will have higher CO stretching frequencies.
(b) $MnCp(CO)_3$ 2023, 1939 cm^{-1} vs. $MnCp^*(CO)_3$ 2017, 1928 cm^{-1}? The two CO bands of the Cp* complex are slightly lower in frequency than the corresponding bands of the Cp complex. Recall that the Cp* ligand is η^5-C_5Me_5. It is a stronger donor ligand than Cp, because of the inductive effect of the five methyl groups. Therefore, the Mn atom in the Cp* complex has a greater electron density than in the Cp complex, and hence there is a greater degree of backdonation to the CO ligands in the Cp* complex than in the Cp complex. As explained in part **(a)**, above, more back-donation leads to lower stretching frequencies.

22.9 The compound $Ni_3(C_5H_5)_3(CO)_2$ has a single CO stretching absorption at 1761 cm^{-1}. The IR data indicate that all C_5H_5 ligands are pentahapto and probably in identical environments. (a) On the basis of these data, propose a structure. The structure below is consistent with the IR spectroscopic data. This D_{3h} complex has all three η^5-C_5H_5 ligands in identical environments. Furthermore, it has two collinear bridging CO ligands, which fits the single CO stretching band at a relatively low frequency. Although it was not given, you should expect this complex to be highly colored, because of the presence of metal–metal bonds.

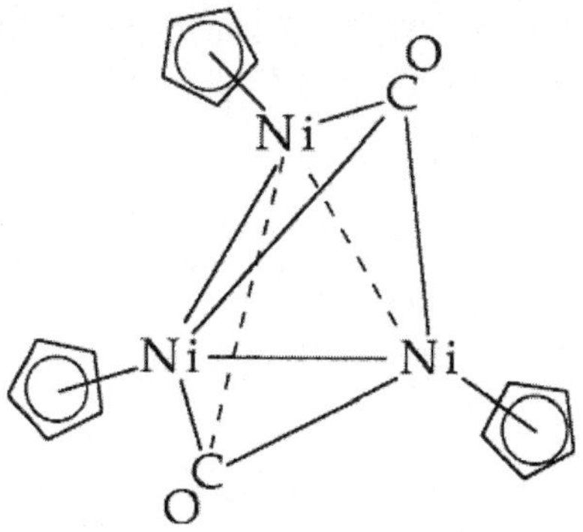

(b) Does the electron count for each metal in your structure agree with the 18-electron rule? If not, is nickel in a region of the periodic table where deviations from the 18-electron rule are common? The three Ni atoms are in identical environments, so you only have to determine the number of valence electrons for one of them:

Ni	10 valence e^-
$\eta^5 - C_5H_5$	5
2 Ni – Ni	2
1/3 (2 CO)	4/3
Total	18 1/3

Therefore, the Ni atoms in this trinuclear complex do not obey the 18-electron rule. Deviations from the rule are common for cyclopentadienyl complexes to the right of the *d* block. For example, the stable complex $(\eta^5\text{-}C_5H_5)_2Co$ is a 19-electron compound.

22.10 Decide which of the two complexes $W(CO)_6$ or $IrCl(CO)(PPh_3)_2$ should undergo the faster exchange with ^{13}CO. Justify your answer. The 18-electron tungsten complex undergoes ligand substitution by a dissociative mechanism. The rate-determining step involves cleavage of a relatively strong W–CO bond. In contrast, the 16-electron iridium complex undergoes ligand substitution by an associative mechanism, which does not involve Ir–CO bond cleavage in the activated complex. Accordingly, $IrCl(CO)(PPh_3)_2$ undergoes faster exchange with ^{13}CO than does $W(CO)_6$.

22.11 Which metal carbonyl in each of (a) $[Fe(CO)_4]^{2-}$ or $[Co(CO)_4]^-$ (b) $[Mn(CO)_5]^-$ or $[Re(CO)_5]^-$ should be the most basic toward a proton? What are the trends on which your answer is based? (a) The dianionic complex $[Fe(CO)_4]^{2-}$ should be the more basic. The trend involved is the greater affinity for a cation that a species with a higher negative charge will have, all other things being equal. In this case the "other things" are (1) same set of ligands, (2) same structure (tetrahedral), and (3) same electron configuration (d^{10}). For a more detailed explanation, see *Metal basicity* in Section 22.18(e).

(b) The rhenium complex is the more basic. The trend involved is the greater M–H bond enthalpy for a period 6 metal ion relative to a period 4 metal ion in the same group, all other things being equal. In this case the "other things" are (1) same set of ligands, (2) same structure (trigonal bipyramidal), and (3) same metal oxidation number (–1). Remember that in the *d* block, bond enthalpies such as M–M, M–H, and M–R *increase* down a group. This behavior is opposite to that exhibited by *p*-block elements.

22.12 Using the 18-electron rule as a guide, indicate the probable number of carbonyl ligands in (a) $W(\eta^6\text{-}C_6H_6)(CO)_n$, (b) $Rh(\eta^5\text{-}C_5H_5)(CO)_n$, and (c) $Ru_3(CO)_n$.

(a) 3 (W contributes 6 electrons and the benzene ring gives 6 electrons; 3 carbonyls at 2 electrons each gives us 18 electrons for the complex); **(b)** 2 (Rh contributes 9 electrons and the Cp ring gives 5 electrons; 2 COs at 2 electrons each yields an 18-electron complex; **(c)** 12 (This is a cluster compounds with 3 Ru in the center. Table 22.9 predicts a CVE count of 48 electrons. Each Ru contributes 8 electrons, giving 24 electrons; to reach 48 we need 12 COs that each contribute 2 electrons.

22.13 Propose two syntheses for $MnMe(CO)_5$, both starting with $Mn_2(CO)_{10}$, with one using Na and one using Br_2? You may use other reagents of your choice. (i) Reduce $Mn_2(CO)_{10}$ with Na to give $Mn(CO)_5^-$; react with MeI to give $MnMe(CO)_5$. (ii), Oxidize with Br_2. to give $MnBr(CO)_5$; displace the bromide with MeLi to give $MnMe(CO)_5$.

22.14 Give the probable structure of the product obtained when $Mo(CO)_6$ is allowed to react first with LiPh and then with the strong carbocation reagent, $CH_3OSO_2CF_3$.
The product of the first reaction contains a –C(=O)Ph ligand formed by the nucleophilic attack of Ph^- on one of the carbonyl C atoms:

$$Mo(CO)_6 + PhLi \rightarrow Li[Mo(CO)_5(COPh)]$$

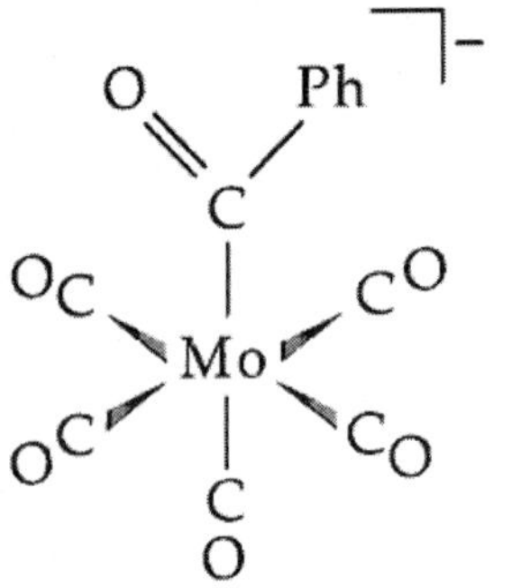

$[Mo(CO)_5(COPh)]^-$ 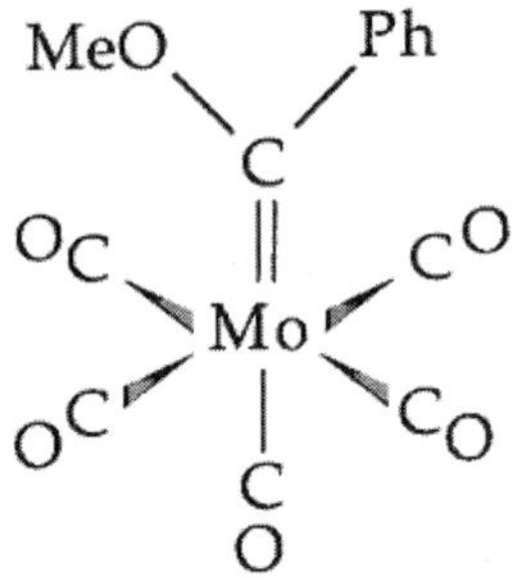 $Mo(CO)_5(C(OMe)Ph)$

The most basic site on this anion is the acyl oxygen atom, and it is the site of attack by the methylating agent $CH_3OSO_2CF_3$:

$$Li[Mo(CO)_5(COPh)] + CH_3OSO_2CF_3 \rightarrow Mo(CO)_5(C(OCH_3)Ph) + LiSO_3CF_3$$

The final product is a carbene (alkylidene) complex. Since the carbene C atom has an oxygen-containing substituent, it is an example of a Fischer carbene (see Section 22.19). The two reaction products are shown above.

22.15 $Na[W(\eta^5\text{-}C_5H_5)(CO)_3]$ reacts with 3-chloroprop-1-ene to give a solid, A, which has the molecular formula $W(C_3H_5)(C_5H_5)(CO)_3$. Compound A loses carbon monoxide on exposure to light and forms compound B, which has the formula $W(C_3H_5)(C_5H_5)(CO)_2$. Treating compound A with hydrogen chloride and then potassium hexafluorophosphate, K^+PF^{6-}, results in the formation of a salt, C. Compound C has the molecular formula $[W(C_3H_6)(C_5H_5)(CO)_3]PF_6$. Use this information and the 18-electron rule to identify the compounds A, B, and C. Sketch a structure for each, paying particular attention to the hapticity of the hydrocarbon.

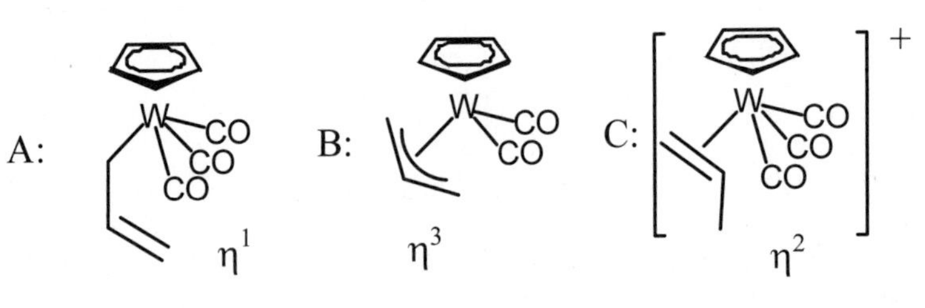

A: η^1 B: η^3 C: η^2

22.16 Treatment of $TiCl_4$ at low temperature with EtMgBr gives a compound that is unstable above 270°C. However, treatment of $TiCl_4$ at low temperature with MeLi or $LiCH_2SiMe_3$ gives compounds that are stable at room temperature. Rationalize these observations.
The compounds formed are TiR_4; notice that neither the methyl or trimethylsilyl groups have β hydrogens, so unlike the ethyl compound, the low energy β-hydride elimination decomposition reaction is not available to them.

22.17 Suggest syntheses of (a) $[Mo(\eta^7\text{-}C_7H_7)(CO)_3]BF_4$ from $Mo(CO)_6$ and (b) $[IrCl_2(COMe)(CO)(PPh_3)_2]$ from $[IrCl(CO)(PPh_3)_2]$? (a) Reflux $Mo(CO)_6$ with cycloheptatriene to give $[Mo(\eta^6\text{-}C_7H_8)(CO)_3]$; treat with the trityl tetrafluoroborate to abstract a hydride and give $[Mo(\eta^7\text{-}C_7H_7)(CO)_3]BF_4$. (b) React $[IrCl(CO)(PPh_3)_2]$ with MeCl to give (oxidative addtion) and then expose to CO atmosphere to induce the migration to give $[IrCl_2(COMe)(CO)(PPh_3)_2]$.

22.18 When $Fe(CO)_5$ is refluxed with cyclopentadiene compound A is formed which has the empirical formula $C_8H_6O_3Fe$ and a complicated ^1H NMR spectrum. Compound A readily loses CO to give compound B with two ^1H-NMR resonances, one at negative chemical shift (relative intensity one) and one at around 5ppm (relative intensity 5). Subsequent heating of B results in the loss of H_2 and the formation of compound C. Compound C has a single ^1H-NMR resonance and the empirical formula $C_7H_5O_2Fe$. Compounds A, B, and C all have 18 valence electrons: identify them and explain the observed spectroscopic data.

+ $Fe(CO)_5$ → (−2CO) A → (−CO) B → (−H)

$[\pi\text{-}C_5H_5Fe(CO)_2]_2$
C

A: Fe, OC, CO, CO

B: Fe, OC, CO, H

The compound B shows two ^1H NMR resonance, one at negative chemical shift due to Fe-H proton and the other at around 5ppm due to aromatic Cp ring protons. The compound C shows a single ^1H NMR resonance because of aromatic Cp ring protons.

22.19 When $Mo(CO)_6$ is refluxed with cyclopentadiene compound D is formed which has the empirical formula $C_8H_5O_3Mo$ and an absorption in the IR spectrum at 1960 cm^{-1}. Compound D can be treated with bromine to yield E or with Na/Hg to give compound F. There are absorptions in the IR spectra of E and F at 2090 and 1860 cm^{-1}, respectively. Compounds D, E, and F all have 18 valence electrons: identify them and explain the observed spectroscopic data.
Loss of CO results in reduction of C_5H_6 to the cyclopentadienyl ion, at the expense of Mo, which is oxidized to Mo(I), and gives D, with one predicted CO stretch, which must be the "piano-stool" complex $C_5H_5Mo(CO)_3$. D is oxidized by Br2 to $C_5H_5Mo(CO)_3Br$, and reduced by Na/Hg to the sodium salt, $C_5H_5Mo(CO)_3Na$. Relative to the IR band for $C_5H_5Mo(CO)_3$, the oxidized species has a band at higher energy reflecting a shorter C=O bond due to the higher bond-order expected when the transition metal is oxidized to a higher charge, and demands more electron density from C, which demand is passed along to O, and raises the bond order, as well as the partial plus charge on oxygen (due to a higher participation of the canonical resonance structure –C(triple bond)O^+). Reduction has the opposite effect, weakening the C=O bond and giving a relatively reduced C=O with some negative charge build-up on the O and a lower bond order, and hence a lower frequency stretch.

22.20 Which compound would you expect to be more stable, $RhCp_2$ or $RuCp_2$? Give a plausible explanation for the difference in terms of simple bonding concepts?
Refer to Figure 22.13, the MO diagram for metallocenes. The number of valence electrons for the two complexes differs by one; $RhCp_2$ has 19 electrons, while $RuCp_2$ has 18 electrons. The MOs of ruthenocene are filled up to the very weakly bonding a_1' orbital. Its electron configuration, starting with the e_2' orbitals, is $(e_2')^4(a_1')^2$. Rhodocene, with its extra electron, has an $(e_2')^4(a_1')^2(e_1'')^1$ configuration. Since the e_1'' orbitals are antibonding, the Rh–C bonds in rhodocene will be longer than the Ru–C bonds in ruthenocene (refer to Table 22.8 and compare the Fe–C and Co–C bond distances in $FeCp_2$ and $CoCp_2$, which are 2.06 Å and 2.12 Å, respectively, and which are isoelectronic with $RuCp_2$ and $Rh(Cp_2)$). Therefore, ruthenocene is the more stable of these two complexes.

22.21 Give the equation for a workable reaction for the conversion of $Fe(\eta^5\text{-}C_5H_5)_2$ to $Fe(\eta^5\text{-}C_5H_5)(\eta^5\text{-}C_5H_4COCH_3)$ and (b) $Fe(\eta^5\text{-}C_5H_5)(\eta^5\text{-}C_5H_4CO_2H)$ (a) The Cp rings in cyclopentadienyl complexes behave like simple aromatic compounds such as benzene, and so are subject to typical reactions of aromatic compounds such as Freidel-Crafts alkylation and acylation. If you treat ferrocene with acetyl chloride and some aluminum(III) chloride as a catalyst, you will obtain the desired compound:

$$Fe(\eta^5 - C_5H_5)_2 + CH_3COCl \rightarrow Fe(\eta^5 - C_5H_5)(\eta^5 - C_5H_4COCH_3) + HCl$$

(b)

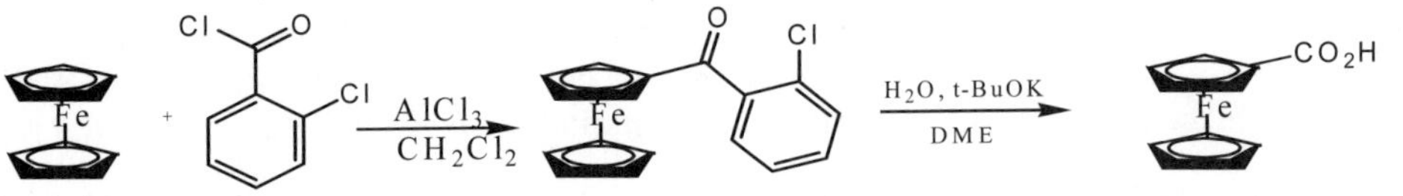

Fe + Cl, O, Cl $\xrightarrow[CH_2Cl_2]{AlCl_3}$ Fe, O, Cl $\xrightarrow[DME]{H_2O,\ t\text{-BuOK}}$ Fe, CO_2H

22.22 Sketch the a_1' symmetry-adapted orbitals for the two eclipsed C_5H_5 ligands stacked together with D_{5h} symmetry. Identify the s, p, and d orbitals of a metal atom lying between the rings that may have nonzero overlap, and state how many a_1' molecular orbitals may be formed.
The symmetry-adapted orbitals of the two eclipsed C_5H_5 rings in a metallocene are shown in Resource Section 5, the d_{z2} orbital on the metal has a_1' symmetry and so can form MOs with this symmetry-adapted orbital. Resource Section 5 shows that the s orbital on the metal also has a_1' symmetry, so it can form MOs with the d_{z2} orbital as well as with the symmetry-adapted orbital. Therefore, three a_1' MOs will be formed, since three orbitals of a_1' symmetry are available on the metal and the ligands. Figure 22.13 shows the energies of the three a_1' MOs: one is the most stable orbital of a metallocene involving metal-ligand bonding; one is relatively nonbonding (it is the HOMO in $FeCp_2$); and one is a relatively high-lying antibonding orbital.

22.23 The compound $Ni(\eta^5\text{-}C_5H_5)_2$ readily adds one molecule of HF to yield $[Ni(\eta^5\text{-}C_5H_5)(\eta^4\text{-}C_5H_6)]^+$ whereas $Fe(\eta^5\text{-}C_5H_5)_2$ reacts with strong acid to yield $[Fe(\eta^5\text{-}C_5H_5)_2H]^+$. In the latter compound the H atom is attached to the Fe atom. Provide a reasonable explanation for this difference.
Protonation of $FeCp_2$ at iron does not change its number of valence electrons: both $FeCp_2$ and $[FeCp_2H]^+$ are 18-electron complexes:

$FeCp_2$		$[FeCp_2H]^+$	
Fe	8 valence e^-	Fe	8 valence e^-
2 Cp	10 e^-	2 Cp	10 e^-
		H	1 e^-
		+1 charge	$-1\,e^-$
Total	18 e^-		18 e^-

By the same token, since $NiCp_2$ is a 20-electron complex, the hypothetical metal-protonated species $[NiCp_2H]^+$ would also be a 20-electron complex. On the other hand, protonation of $NiCp_2$ at a Cp carbon atom produces the 18-electron complex $[NiCp(\eta^4\text{-}C_5H_6)]^+$. Therefore, the reason that the Ni complex is protonated at a carbon atom is that a more stable (i.e., 18-electron) product is formed:

$[NiCp(\eta^4 - C_5H_6)]^+$	
Ni	10 valence e^-
Cp	5 e^-
$\eta^4 - C_5H_6$	4 e^-
+1 charge	$-1\,e^-$
Total	18 e–

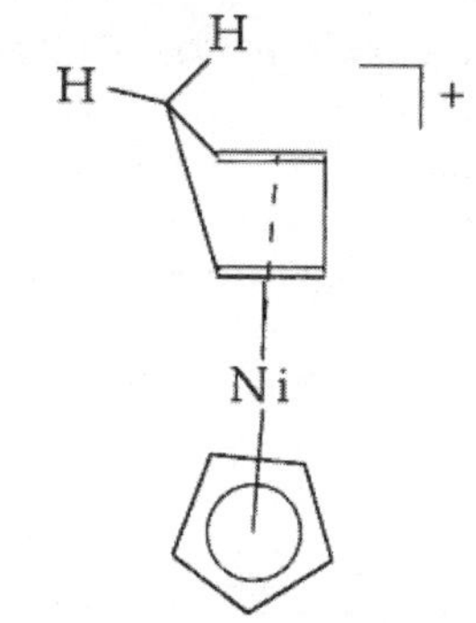

22.24 Write a plausible mechanism, giving your reasoning, for the reactions: (a) $[Mn(CO)_5(CF_2)]^+ + H_2O \rightarrow [Mn(CO)_6]^+ + 2HF$? The $=CF_2$ carbene ligand is similar electronically to a Fischer carbene (see Section 22.19). The electronegative F atoms render the C atom subject to nucleophilic attack, in this case by a water molecule. The $[(CO)_5Mn(CF_2(OH_2))]^+$ complex can then eliminate two equivalents of HF, as shown in the mechanism below:

$$[(CO)_5Mn{=}CF_2]^+ + H_2O \rightarrow [(CO)_5Mn{-}CF_2{-}OH_2]^+ \xrightarrow{-HF}$$

$$[(CO)_5Mn{-}CF{=}OH]^+ \xrightarrow{-HF} [(CO)_5Mn{-}C{\equiv}O]^+ = [Mn(CO)_6]^+$$

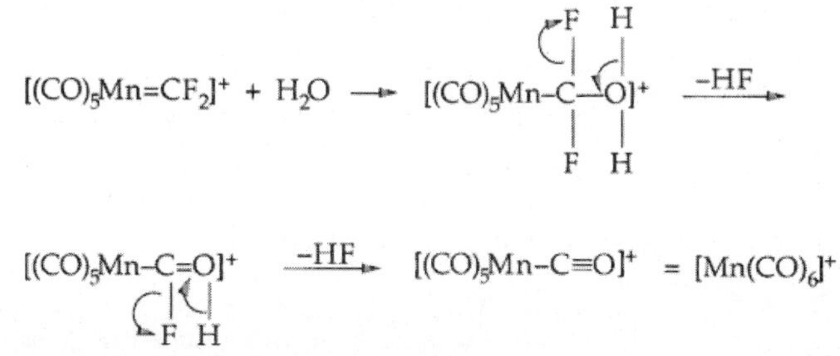

(b) $Rh(C_2H_5)(CO)(PR_3)_2 \rightarrow RhH(CO)(PR_3)_2 + C_2H_4$? This is an example of a β-hydrogen elimination reaction, discussed at the beginning of Section 22.8. This reaction is believed to proceed through a cyclic intermediate involving a (3c,2e) M–H–C interaction, as shown in the mechanism below:

$$(PR_3)_2(CO)Rh{-}C_2H_5 \rightarrow (PR_3)_2(CO)Rh{-}CH_2{-}CH_3 \text{ (with Rh}\cdots\text{H–C interaction)} \rightarrow$$

$$(PR_3)_2(CO)RhH(\eta^2\text{-}CH_2{=}CH_2) \rightarrow RhH(PR_3)_2(CO) + C_2H_4$$

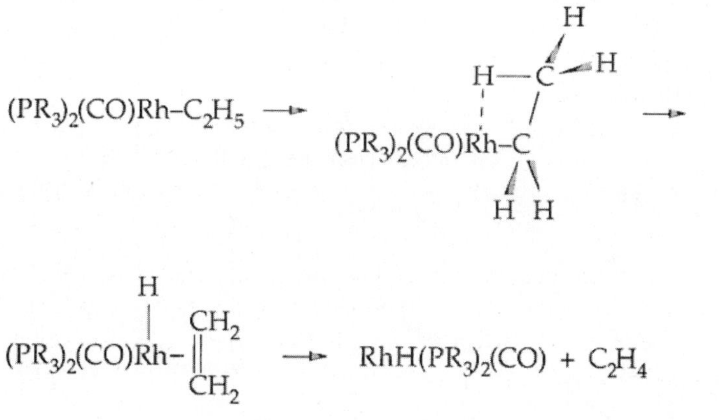

22.25 Given mechanism of CO insertion, what rate constant can be extracted from rate data?

$$RMn(CO)_5 \underset{k_{a'}}{\overset{k_a}{\rightleftharpoons}} (RCO)Mn(CO)_4$$

$$(RCO)Mn(CO)_4 + L \xrightarrow{k_b} (RCO)MnL(CO)_4$$

with the rate given by:

$$\text{rate} = (k_a k_b[RMn(CO)_5][L])/(k_{a'} + k_b[L])$$

When [L] is very high, the denominator becomes approximately equal to k_b[L] and the rate expression simplifies to:

$$\text{rate} = k_a[\text{RMn(CO)}_5]$$

Therefore, when [L] is very high, the first-order rate constant k_a can be determined from rate vs. [RMn(CO)$_5$] data.

22.26 (a) What cluster valence electron (CVE) count is characteristic of octahedral and trigonal prismatic complexes?
According to Table 22.9, an octahedral M_6 will have 86 cluster valence electrons and a trigonal prismatic M_6 cluster will have 90.
(b) Can these CVE values be derived from the 18-electron rule? No. As discussed in Section 22.2(a), the bonding in smaller clusters can be explained in terms of local M–M and M–L electron pair bonding and the 18-electron rule, but the bonding in octahedral M_6 and larger clusters cannot.
(c) Determine the probable geometry of $[Fe_6(C)(CO)_{16}]^{2-}$ and $[Co_6(C)(CO)_{15}]^{2-}$? The iron complex has 86 CVEs, while the cobalt complex has 90 (see the calculations below). Therefore, the iron complex probably contains an octahedral Fe_6 array, while the cobalt complex probably contains a trigonal-prismatic Co_6 array.

6 Fe	$6\times8\ e^- =$	$48\ e^-$	6 Co	$6\times9\ e^- =$	$54\ e^-$
C		$4\ e^-$	C		$4\ e^-$
16 CO	$16\times2\ e^- =$	$32\ e^-$	15 CO	$15\times2\ e^- =$	$30\ e^-$
2 – charge		$2\ e^-$	2 – charge		$2\ e^-$
Totals		$86\ e^-$			$90\ e^-$

22.27 Based on isolobal analogies, choose the groups that might replace the group in boldface in
(a) $Co_3(CO)_9$CH→ OCH_3, $N(CH_3)_2$, or $SiCH_3$? Isolobal groups have the same number and shape valence orbitals *and* the same number of electrons in those orbitals. The CH group has three sp^3 hybrid orbitals that each contain a single electron. The $SiCH_3$ group has three similar orbitals similarly occupied, so it is isolobal with CH and would probably replace it in $Co_3(CO)_9CH$ to form $Co_3(CO)_9SiCH_3$. In contrast, the OCH_3 and $N(CH_3)_2$ groups are not isolobal with CH. Instead, they have, respectively, three sp^3 orbitals that each contain a pair of electrons and two sp^3 orbitals that each contain a pair of electrons.
(b) $(OC)_5MnMn(CO)_5$ → I, CH_2, or CCH_3? The $Mn(CO)_5$ group has a single σ-type orbital that contains a single electron. An iodine atom is isolobal with it, since it also has a singly occupied σ orbital. Therefore, you can expect the compound $MnI(CO)_5$ to be reasonably stable. In contrast, the CH_2 and CCH_3 are not isolobal with $Mn(CO)_5$. The CH_2 group has either a doubly occupied σ orbital and an empty *p* orbital or a singly occupied σ orbital and a singly occupied *p* orbital. The CCH_3 group has three singly occupied σ orbitals (note that it is isolobal with $SiCH_3$).

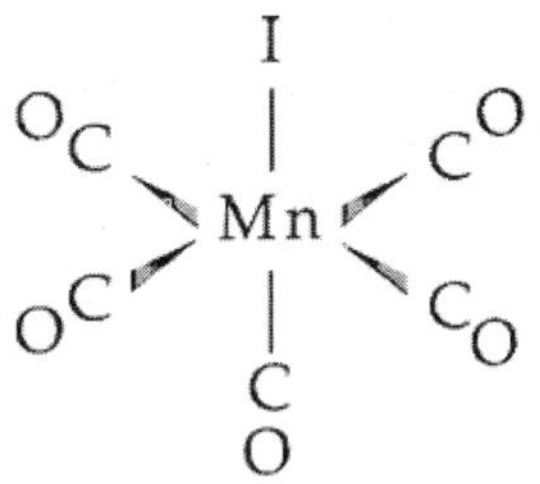

$MnI(CO)_5$ is a stable complex

22.28 Ligand substitution reactions on metal clusters are often found to occur by associative mechanisms, and it is postulated that these occur by initial breaking of an M-M bond, thereby providing an open coordination site for the incoming ligand. If the proposed mechanism is applicable, which would you expect to undergo the fastest exchange with added ^{13}CO, $Co_4(CO)_{12}$ or $Ir_4(CO)_{12}$? Suggest an explanation. If the rate-determining step in the substitution is cleavage of one of the metal–metal bonds in the cluster, the cobalt complex will exhibit faster exchange. This is because metal–metal bond strengths *increase* down a group in the *d* block. Therefore, Co–Co bonds are weaker than Ir–Ir bonds, all other things like geometry and ligand types kept the same.

Chapter 23 The *f*-Block Metals

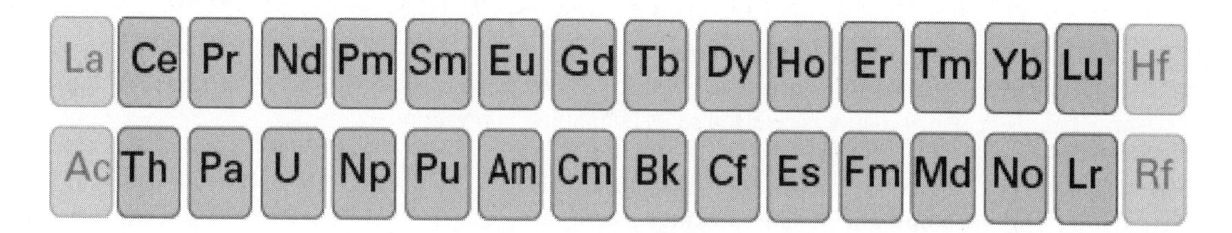

There are two series, each of 14 elements, in the ***f* block.** The 4f elements are normally referred to as the **lanthanoids** (formerly and still commonly "the lanthanides") and the 5f elements as the **actinoids** (the "actinides"). There is a striking uniformity in the properties of the 4f elements and greater diversity in the chemistry of the 5f elements. The lanthanoids are sometimes referred to as the "rare earth elements"; however, that name is inappropriate because they are not particularly rare, except for promethium, which has no stable isotopes.

S23.1 Derive the ground state of the Tm^{3+} ion.
Tm^{3+} is a $4f^{12}$ system. Therefore, $M_L = +5$ and S = 1. The term will be ^3H. According to the clebsch-cordan series, the total angular momentum of a term with L = 5 and S = 1 will be J = 6, 5, 4. According to Hunds rules for a more than half full shell the level with the highest value of J lies lowest (in this case J = 6), so we can expect its term symbol to be $^3H_6$.

S23.2 The product of the reaction above is in fact a hydride bridged dimer (9). Suggest a strategy to ensure that the hydride is monomeric. A simple approach would be to use the Cp rings substituted by bulky groups. In such a case, the formation of dihydrogen complex will be hindered. On the other hand, if the proposed mechanism does not involve bridging hydrogen complex, the rate of formation and nature of the products will not change.

S23.3 Use the Frost diagrams and data in *Resource section* 2 to determine the most stable uranium ion in acid aqueous solution in the presence of air and give its formula. From the Frost diagram in Figure 23.10, it can be seen that the most stable oxidation state of uranium in aqueous acid is U^{4+} (that is, it has the most negative free energy of formation, the quantity plotted on the *y* axis). However, the reduction potentials for the UO_2^{2+}/U^{4+} and UO_2^{2+}/UO_2^{2+} couples are quite small, 0.380 and 0.170 V, respectively (the UO_2^{+}/U^{4+} potential, therefore, is 0.275 V). Therefore, since the O_2/H_2O reduction potential is 1.229 V, the most stable uranium ion is UO_2^{2+} if sufficient oxygen is present:

$$2U^{4+} + O_2 + 2\,H_2O \rightarrow 2\,UO_2^{2+} + 4\,H^+ \quad E^\circ = 1.229\text{ V} - 0.275\text{ V} = 0.0954\text{ V}$$

23.1 (a) Give a balanced equation for the reaction of any of the lanthanoids with aqueous acid. (b) Justify your answer with redox potentials and with a generalization on the most stable positive oxidation states for the lanthanoids. (c) Name the two lanthanoids that have the greatest tendency to deviate from the usual positive oxidation state and correlate this deviation with electronic structure? (a) Balanced equation All of the lanthanide metals are very electropositive and reduce water (or protons) to H_2 while being oxidized to the trivalent state. If Ln is used as the symbol for a generic lanthanide element, the balanced equation is:

$$2\,Ln(s) + 6\,H_3O^+(aq) \rightarrow 2\,Ln^{3+}(aq) + 3\,H_2(g) + 6\,H_2O(l)$$

(b) Redox potentials The potentials for the Ln^0/Ln^{3+} oxidations in acid solution range from a low of 1.99 V for europium to 2.38 V for lanthanum, a remarkably self-consistent set of values spanning 15 elements. In fact, europium is the only lanthanide with a potential lower than 2.22 V. Since the potential for the H_3O^+/H_2 reduction is 0 V, the E° values for the equation shown in part **(a)** for all lanthanides range from 1.99 to 2.38 V, a very large driving force.

(c) Two unusual lanthanides The usual oxidation state for the lanthanide elements in aqueous acid is 3+. There are two lanthanides that deviate slightly from this trend. The first is Ce^{4+}, which, while being a strong oxidizing agent, is kinetically stable in aqueous acid. Since Ce^{3+} is $4f^1$ and Ce^{4+} is $4f^0$, you can see that the special stability of tetravalent cerium is because of its stable $4f^0$ electron configuration. The second deviation is Eu^{2+}, which is a strong reducing agent but exists in a growing number of compounds. Since Eu^{2+} is $4f^7$ and Eu^{3+} is $4f^6$, you can see that the special stability of divalent europium is because of its stable $4f^7$ electron configuration.

23.2 Explain the variation in the ionic radii between La^{3+} and Lu^{3+}.
The ionic radius of Lu^{3+} is significantly smaller than La^{3+} because of the incomplete shielding of the 4f electrons of the former. This is also sometimes referred to as lanthanide contraction.

23.3 From a knowledge of their chemical properties, speculate on why Ce and Eu were the easiest lanthanoids to isolate before the development of ion-exchange chromatography. In exercise 23.1 we explained why Ce^{4+} and Eu^{2+} are relatively stable with respect to all other tetravalent and divalent lanthanide ions. These unusual oxidation states were used in separation procedures, since their charge/radius ratios are very different than those of the typical Ln^{3+} ions. Tetravalent cerium can be precipitated as $Ce(IO_3)_4$, leaving all other Ln^{3+} ions in solution because their iodates are soluble. Divalent europium, which resembles Ca^{2+}, can be precipitated as $EuSO_4$, leaving all other Ln^{3+} ions in solution because their sulfates are soluble.

23.4 How would you expect the first and second ionization energies of the lanthanoids to vary across the series? Sketch the graph that you would get if you plotted the third ionization energy of the lanthanoids versus atomic number? Identify elements at any peaks or troughs and suggest a reason for their occurrence? First and second IEs would show general increase across the lanthanoids (higher atomic number means higher ionization energy). With the third, anomalies arise; these are a peak at Eu (gives a f^6 configuration, therefore requires removal of electron from a stable, half-filled f^7) and a trough at Gd (gives the stable f^7 configuration).

23.5 Derive the ground state of the Tb^{3+} ion
Tm^{3+} is a $4f^8$ system. Therefore $M_L = +3$ and $S = 3$. The term will be ^7F. The expected term symbol is $^7F_6$

23.6 Predict the magnetic moment of a compound containing the Tb^{3+} ion.
From the previous problem, we know the spin (S) and angular quantum numbers (L) for the ion are 3. Because of the more than half-filled state of the f-subshell the J = L+S = 6. Accordingly the g_j value can be calculated as shown below:

$$g_j = 1 + \frac{3(3+1)-3(3+1)+ 6(6+1)}{2 \text{ x } 6(6+1)} = 1.5$$

Therefore,

$$\mu = g_j\{J(J+1)\}^{1/2}\mu_B = 1.5\ \{6(6+1)\}^{1/2}\mu_B = 9.72\ \mu_B$$

23.7 Explain why stable and readily isolable carbonyl complexes are unknown for the lanthanoids? Stable carbonyl compounds need back-bonding from metal orbitals of the appropriate symmetry. With the lanthanoids, the 5*d* orbitals are empty, and the 4*f* orbitals do not extend beyond the [Xe] core and cannot participate in bonding.

23.8 Suggest a synthesis of neptunocene from $NpCl_4$?
$Np(COT)_2$ (COT = $C_8H_8^{2-}$) were prepared by the reaction in THF solution of K_2COT with $NpCl_4$ under inert atmosphere.

$$NpCl_4 + 2\ K_2COT \xrightarrow[\text{THF, — 40°C}]{} Np(COT)_2 + 4KCl$$

23.9 Account for the similar electronic spectra of Eu^{3+} complexes with various ligands and the variation of the electronic spectra of Am^{3+} complexes as the ligand is varied. There is a considerable difference in the interactions of the 4*f* orbitals of the lanthanide ions and the 5*f* orbitals of the actinide ions with ligand orbitals. In the case of the 4*f* orbitals, interactions with ligand orbitals are negligible. Therefore, splitting of the 4*f* subshell by the ligands is also negligible and does not vary as the ligands vary. Since the colors of lanthanide ions are because of 4*f*–4*f* electronic transitions, the colors of Eu^{3+} complexes are invariant as a function of ligand. In contrast, the 5*f* orbitals of the actinide ions interact strongly with ligand orbitals, and the splitting of the 5*f* subshell, as well as the color of the complex, varies as a function of ligand.

23.10 Predict a structure type for BkN based on the ionic radii $r(Bk^{3+})$ = 96 pm and $r(N^{3-})$ = 146 pm. The radius ratio is 96/146 = 0.657. Using Table 3.6, the rock-salt structure is predicted as this ratio falls between 0.414 and 0.732.

23.11 Describe the general nature of the distribution of the elements formed in the thermal neutron fission of 235U, and decide which of the following highly radioactive nuclides are likely to present the greatest radiation hazard in the spent fuel from nuclear power reactors: (a) ^{39}Ar, (b) ^{228}Th, (c) ^{90}Sr, (d) ^{144}Ce. The thermal neutron fission of ^{235}U yields two fragment nuclei of unequal mass, i.e., the fission is asymmetrical. Several elements in the neighborhood of mass 95 are formed in ~5% each, and similarly, several elements in the neighborhood of mass 135 are formed in ~5% each (see Box 22.3). Therefore, in terms of abundance, the isotopes ^{39}Ar and ^{228}Th will not present a large radiation hazard because their masses are far from the two peaks around masses 95 and 135. On the other hand, the isotopes ^{90}Sr and ^{144}Ce, with masses close to 95 and 135, respectively, will be present in relatively large abundance and therefore will pose a significant radiation hazard.

Chapter 24 Solid-state and Materials Chemistry

The top figure shows the ideal structure of AgCl with no defects (the Cl^- ions are the larger open spheres). All of the Ag^+ ions, including the highlighted black one, are in octahedral holes of Cl^- ions. In a real sample of AgCl, some of the Ag^+ ions are displaced from octahedral sites to tetrahedral sites, as shown for the black Ag^+ ion in the bottom figure. As discussed in Chapter 3, this is an example of a Frenkel defect, which is a type of intrinsic point defect. In this chapter we explore extended defects in materials that influence properties such as mechanical strength and conductivity.

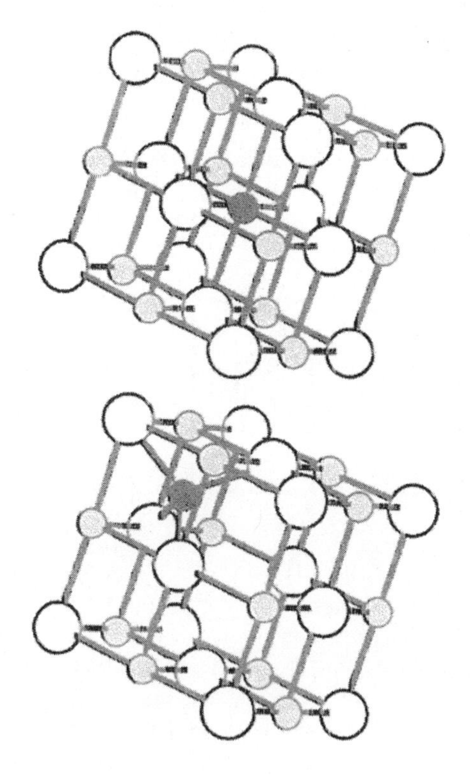

S24.1 **Synthesis for Sr_2MoO_4?** Place the following in a sealed tube at high temperature, 6 SrO(s) + Mo(s) + 2 MoO_3(s) yielding 3 Sr_2MoO_4.

S24.2 **Why does increased pressure reduce the conductivity of K^+ more than that of Na^+ in β-alumina?** Increasing the pressure on a crystal compresses it, reducing the spacings between ions (for example, subjecting NaCl to a pressure of 24,000 atm reduces the Na–Cl distance from 2.82 Å to 2.75 Å). In a rigid lattice such as β-alumina, larger ions migrate more slowly than smaller ions with the same charge (as discussed in the example). At higher pressures, with smaller conduction plane spacings, *all* ions will migrate more slowly than they do at atmospheric pressure. However, larger ions will be impeded to a greater extent by smaller spacings than will smaller ions. This is because the ratio (radius of migrating ion)/(conduction plane spacing) changes more, per unit change in conduction plane spacing, for a large ion than for a small ion. Therefore, increased pressure reduces the conductivity of K^+ more than that of Na^+ because K^+ is larger than Na^+.

S24.3 **Rationalize the observation that $FeCr_2O_4$ is a normal spinel?** In the normal AB_2O_4 spinel structure, the A^{2+} ions (Fe^{2+} in this example) occupy tetrahedral sites and the B^{3+} ions (Cr^{3+}) occupy octahedral sites. The fact that $FeCr_2O_4$ exhibits the normal spinel structure can be understood by comparing the ligand field stabilization energy of high-spin octahedral Fe^{2+} and octahedral Cr^{3+}. With six *d* electrons, LFSE = 0.4 Δ_0 for high-spin Fe^{2+}. With only three *d* electrons, LFSE = 1.2 Δ_0 for Cr^{3+}. Since the Cr^{3+} ions experience more stabilization in octahedral sites than do the Fe^{2+} ions, the normal spinel structure is more stable than the inverse spinel structure, which would exchange the positions of the Fe^{2+} ions with half of the Cr^{3+} ions (see Section 24.7c).

24.1 **NiO doped with Li_2O?** The electronic conductivity of the solid increases owing to formation of $Ni_{1-x}Li_x)O$ containing Ni(III) and a free electron that can hop through the structure from Ni(II) to Ni(III), thus promoting increased conductivity in the structure.

24.2 **What is a crystallographic shear plane?** A crystallographic shear plane may be considered a defect or a way of describing a new structure (see the end of Section 24.3). When crystallographic shear planes are distributed randomly throughout the solid, they are called Wadsley defects. A continuous range of composition is possible, because a new, discrete phase has not been formed. For example, tungsten oxide can have a composition ranging from WO_3 to $WO_{2.93}$. However, at W/O ratios higher than 1/(2.93), the shear planes are distributed in a nonrandom, periodic manner (i.e., a new stoichiometric phase has been formed, for example $W_{20}O_{58}$ (W/O = 1/(2.90))). Thus crystallographic shear planes are defects when disordered and give a new phase with a new structure when ordered.

24.3 **How might you distinguish between a solid solution and a series of discrete crystallographic shear plane structures?** A solid solution would contain a random collection of crystallographic shear planes, whereas a series of discrete structures would contain ordered arrays of crystallographic shear planes. These two possibilities are represented in the figures below.

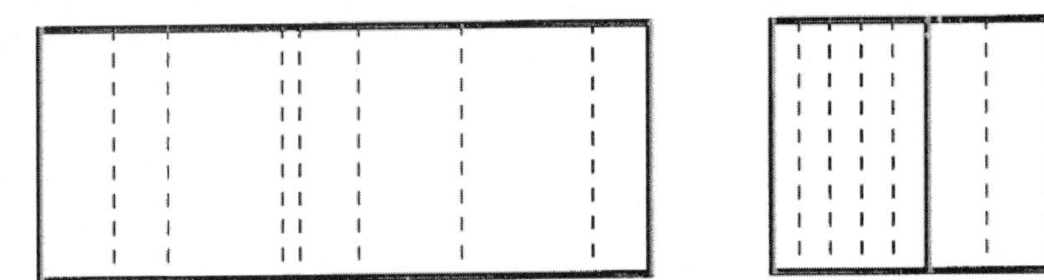

solid solution (a random array of shear planes)

discrete phases (a series of ordered arrays of shear planes)

Because of the lack of long-range order, the solid solution would give rise to an electron micrograph showing a random distribution of shear planes. In addition, the solid solution would not give rise to new X-ray diffraction peaks. In contrast, the ordered phases of the solid on the right would be detectable by electron microscopy and by the presence of a series of new peaks in the X-ray diffraction pattern, arising from the evenly spaced shear planes.

24.4 **Wurtzite crystal structure and bottleneck?** Sketch the wurtzite crystal structure and locate on the sketch the interstitial sites to which a cation might migrate. What is the nature of the bottleneck between the normal and interstitial sites? The wurtzite structure is shown in Figure 3.35. The normal sites for cations in this structure are the tetrahedral holes (shown for Zn^{2+}). Cations in this structure are found in 1/2 of the tetrahedral holes but unlike the zinc blende structure that anions are packed on an expanded hcp array. The bottleneck for transport out of the tetrahedral hole involves the space formed by three close packed anions (see Figure 3.17). In order for the cation to migrate, it must pass through this opening within adjacent hexagonal close packed layers.

24.5 **Synthesis of?** (a) **$MgCr_2O_4$** [heat $(NH_4)_2Mg(CrO_4)_2$ gradually to 1100-1200°C], (b) **$SrFeO_3Cl$** [heat SrO + $SrCl_2$ + Fe_2O_3 in correct stoichiometric proportions in a sealed tube], (c) **Ta_3N_5** [heat Ta_2O_5 under NH_3 at 700°C].

24.6 **Products of reactions? (a)** $LiNiO_2$, **(b)** Sr_2WMnO_6 (= $Sr(W_{0.5}Mn_{0.5})O_3$, a perovskite). See Section 24.1 for more details.

24.7 **Where might intercalated Na^+ ions reside in the ReO_3 structure?** The unit cell for ReO_3, which is shown in Figure 24.16(a), is reproduced in the figure on the left below (the O^{2-} ions are the open spheres). As you can see, the structure is very open, with a very large hole in the centre of the cell (if the Re–O distance is a, the distance from an O^{2-} ion to the center of the cell is 1.414a). An ReO_3 unit cell containing an Na^+ ion (large, heavily shaded sphere) at its centre is shown on the right below.

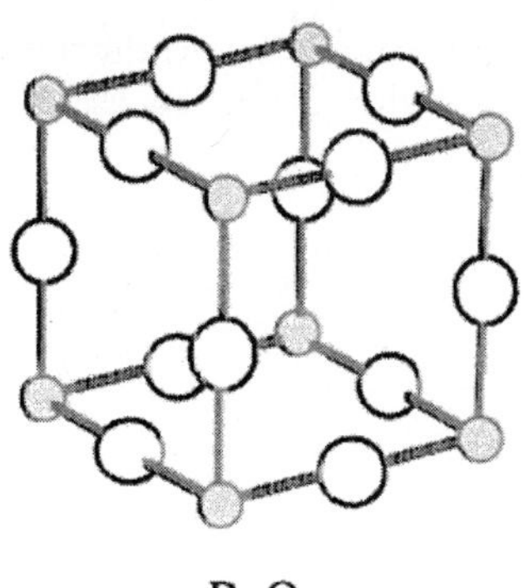

ReO_3

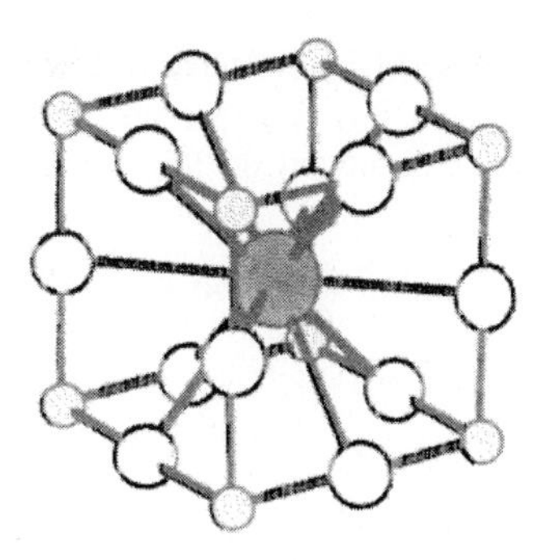

Na_xReO_3

24.8 **Antiferromagnet ordering?** As discussed in Section 20.8, in a ferromagnetic substance the spins on different metal centres (atoms or ions) are coupled into a *parallel* alignment. In the diagram on the left below, each arrow represents the spin on one metal centre. Ordered regions of a sample, called domains, are shown in the bottom figure. In an antiferromagnetic substance, the spins on different metal centres are coupled into an *antiparallel* alignment. As the temperature approaches absolute zero, the net magnetic moment of a ferromagnetic domain becomes very large, while that of an antiferromagnetic domain goes to zero.

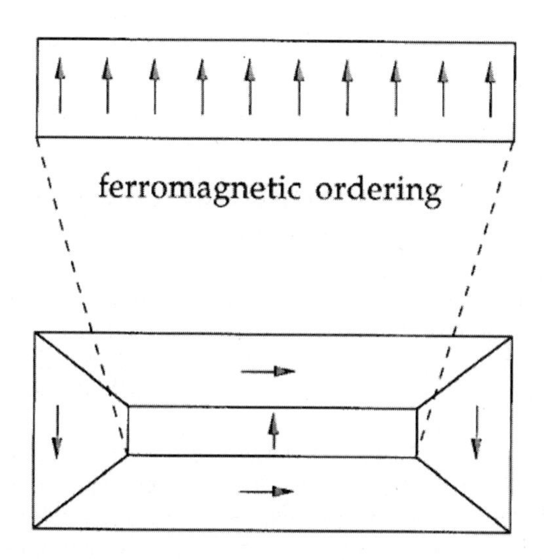

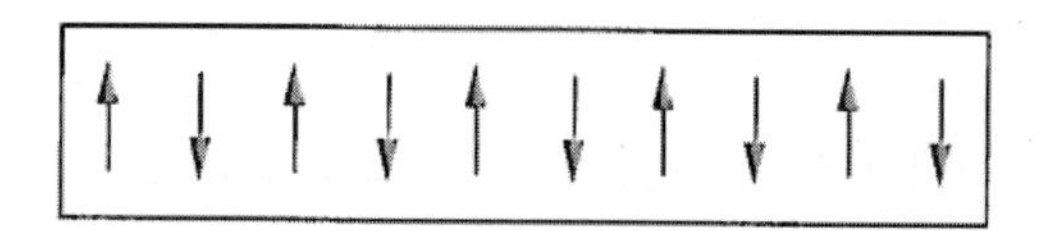

antiferromagnetic ordering

One possible domain structure for a ferromagnetic crystal. The arrows represent the net magnetic moment for each domain. The crystal as a whole has a net magnetic moment.

The direction of the net moment = →

24.9 **Magnetic measurements on ferrite?** In the inverse spinel structure, $Fe[CoFe]O_4$ with Co^{2+} and Fe^{3+}, three unpaired electrons are expected from the d^7 octahedral Co^{2+}. The structure appears to be the inverse spinel with 3.4 spins per formula unit observed.

24.10 **Using LFSEs determine site preference for A= Ni(II) and B = Fe(III)?** Using Table 24.3 for these A and B cations, we see better ligand field stabilization and a strong preference for inverse spine, as found in practice.

24.11 **High-temperature superconductors?** The main condition to meet the criteria involves the presence of variable oxidation state coppers. Thus all except $Gd_2Ba_2Ti_2Cu_2O_{11}$, which contains only Cu^{2+}, are superconductors.

24.12 **Classify oxides as glass-forming or nonglass-forming?** BeO, B_2O_3, and to some extent GeO_2, since they involve metalloid and nonmetal oxides. Transition metal and rare earth oxides are typically nonglass-forming oxides, many of which have crystalline phases.

24.13 **Which metal sulfides might be glass forming?** Transition metal and rare earth sulfides are typically nonglass-forming oxides, many of which have crystalline and layered phases. Any metal sulfide glasses would involve metalloid and nonmetal sulfides.

24.14 **Examples of spinels containing? (a) sulfide?** $Zn(II)Cr(III)_2S_4$ **(b) fluoride?** Li_2NiF_4

24.15 **Synthesis of $LiTaS_2$?** The product can be synthesized through direct reaction of TaS_2 with BuLi, or through electrochemical insertion of Li ions. See Section 24.10 for more details.

24.16 **Oxotetrahedral species in framework structures?** Be, Ga, Zn, and P all form stable tetrahedral units with oxygen, as noted in Section 24.12.

24.17 **Formulas for structures isomorphous with SiO_2 containing Al, P, B, and Zn replacing Si?** The following zeotypes are possible, as described in Section 24.12c: $(AlP)O_4$, $(BP)O_4$, $(ZnP_2)O_6$.

24.18 **Mass percent of hydrogen in $NaBH_4$ and hydrogen storage?** The mass percent of hydrogen in $NaBH_4$ is approximately 10.7 % and comparable to other complex hydrides discussed in Section 24.14b. Given this mass percentage it is a good candidate to consider but a problem for its use is that elevated temperatures for decomposition are required to liberate the hydrogen gas (typically above 500°C).

24.19 **Formula for this lithium aluminium magnesium dihydride and structures?** As noted in Section 24.14a, magnesium hydride can be doped with various metals including Al and Li to obtain solid solutions with the formula $Mg_{1-x}(M)_xH_{2y}$ (M=Al, Li) which have slightly lower decomposition temperatures. The Li and Al will be incorporated as metal hydride solid solutions and not directly doped on Mg sites in the structure shown in Figure 24.59.

24.20 **Color intensity differences in Egyptian blue pale versue blue-green spinel?** The Cu site in Egyptian blue is square planar and thus has a center of symmetry. The *d–d* transitions that give rise to the color are thus symmetry-forbidden and less intense. In copper aluminate spinel blue, the site is tetrahedral with no center of symmetry and the transitions are not symmetry-forbidden. This leads to the increased intensity of blue color in the spinel. For more on spin-allowed versus spin-forbidden transitions in electronic spectra see Section 20.6.4.

24.21 **Order of band gaps?** BN > C(diamond) > AlP > InSb. See Sections 3.19, 24.18, and 24.19 for more details.

24.22 **Fulleride structures?** In Na_2C_{60}, all of the tetrahedral holes are filled with sodium cations within the close-packed array of fulleride anions. Consider the analogy to fluorite shown in Figure 3.38. In Na_3C_{60}, all of the tetrahedral holes and all of the octahedral holes are filled with sodium cations within the fcc lattice of fulleride anions, as shown in Figure 24.69.

Chapter 25 Nanomaterials, Nanoscience, and Nanotechnology

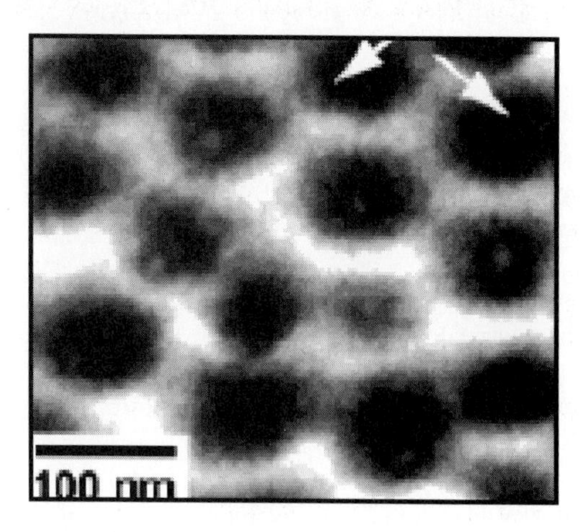

This chapter is concerned with a new class of inorganic materials called **nanomaterials** that contain at least one critical dimension between 1 and 100 nm. Nanomaterials have been generating enormous excitement across many disciplines including chemistry, physics, biology, engineering, and materials science owing to their unique, often improved, materials properties and tunable structures. The SEM figure adjacent shows CdS nanocrystals self-assembled on a laser-embossed, templated surface within wells that have zeptolitre volumes. This work by Odom and coworkers illustrates an important combination of both top-down and bottom-up methods to achieve novel optical nanomaterials with potential for sensing, solar cells and LEDs.

S25.1 **Synthesis of core-shell nanoparticles?** The formation of the core particles is carried out in the same manner as above. The shell is then formed by controlling the heterogeneous nucleation of shell material on the core particles (which represents a type of growth of the particles). The thermodynamic driving force is adjusted to a level that allows for heterogeneous nucleation of the shell material on the core but prevents homogeneous nucleation of the shell material. This heterogeneous nucleation occurs slowly (the driving force is small) and the best results are obtained when growth occurs uniformly on all core particles.

S25.2 **Hemispherical imprint for 2 nm nanoparticle?** Working the problem in reverse, we start with $d = 2 \times 10^{-9}$ nm.

The volume of NaCl is therefore

$$V = \frac{4\pi}{3}\left(\frac{d}{2}\right)^3 = \frac{4\pi}{3}\left(\frac{2\times 10^{-9}\ \text{m}}{2}\right)^3 = 4.19\times 10^{-27}\ \text{m}^3$$

The mass equivalent is

$$m = 4.19\times 10^{-27}\ \text{m}^3 \times 2170\ \text{kg m}^{-3} = 9.09\times 10^{-24}\ \text{kg}$$

The number of moles is

$$n = \frac{9.09\times 10^{-21}\ \text{g}}{58.442\ \text{g mol}^{-3}} = 1.56\times 10^{-22}\ \text{mol}$$

The volume of the solution is

$$V_{solution} = \frac{1.56\times 10^{-22}\ \text{mol}}{1.00\times 10^{3}\ \text{mol m}^{-3}} = 1.56\times 10^{-25}\ \text{m}^{-3}$$

Finally, if the volume of each hemispherical imprint is half that of a sphere of radius r, then the radius of the hemispherical imprint should be

$$r = \left(\frac{3}{4\pi}(2\times 1.56\times 10^{-25}\ \text{m}^{-3})\right)^{1/3} = 4.21\times 10^{-9}\ \text{m}$$

Therefore, the radius is 4.21 nm and the diameter is 8.42 nm.

S25.3 **Host for QDs?** MCM-41 is a better choice as it has a range of pore sizes from 2 to 10 nm that can better accommodate QDs with diameters ranging from 2 to 8 nm that exceed typical ZSM-5 pores (less than 1 nm).

25.1 **(a) Surface areas?** 3.14×10^2 nm^2 versus 3.14×10^6 nm^2 (a factor of 10^4). **(b) Nanoparticles based on size?** The 10 nm particle can be considered a nanoparticle, while the 1000 nm particle cannot. **(c) Nanoparticles based on properties?** Considering a surface plasmon, a nanoparticle should exhibit properties different from those of a molecule or an extended solid. The particle would have to exhibit a standing wave behavior of this surface property.

25.2 **Electron length and quantum confinement?** A characteristic length of an electron-hole pair is the exciton Bohr radius, or the average physical separation between an electron-hole pair. If the exciton Bohr radius is on the order of the physical size of the particle, which can occur in $\approx$2 10- nm particles, then quantum confinement will occur since the two particles are forced to be closer than they prefer.

25.3 **Why are QDs better for bioimaging?** The quantum dots exhibit broadband absorption and single wavelength emission, which can be tailored by engineering the size or surface termination of the quantum dot. Organic fluorophores typically exhibit narrow-band absorption and single wavelength emission. In general, one light source can be used to excite different quantum dots, but each organic fluorophore requires it own excitation source.

25.4 **Band energies for QD versus bulk semiconductor?** The energies of the band edges for a QD nanocrystal are more widely separated (i.e., there is a larger bandgap) than those of the similar bulk semiconductor. Also, the states are localized in quantum dots and are characterized by no linear momentum.

25.5 **(a) Top-down versus bottom-up?** The "top-down" approach requires one to "carve out" nanoscale features from a larger object. The "bottom-up" approach requires one to "build up" nanoscale features from smaller entities. Lithography is a standard approach to "top-down" and thin film deposition of quantum wells or superlattices is a standard approach to "bottom-up."
(b) Advangtages and disadvantages? Top-down methods allow for precise control over the spatial relationships between nanoscale entities, but they are limited to the design of rather large nanoscale items. Bottom-up methods allow for the precise spatial control over atoms and molecules relative to each other, but the long-range spatial arrangement is often difficult to realize.

25.6 **(a) What are SPMs?** Scanning probe microscopy is a method to image the microscale (and below) features of a material by scanning a very small probe over the surface and measuring some physical interaction between the tip and the material. **(b) SPM and a specific nanomaterial?** The local magnetic domains of magnetic nanomaterials, such as nanorods of iron oxide, can be imaged using magnetic force microscopy, where the probe is sensitive to localized magnetic fields.

25.7 **SEM versus TEM?** In scanning electron microscopy, an electron beam is scanned over a material, and an image is typically generated by recording the intensity of secondary or back-scattered electrons. In transmission electron microscopy, an electron beam is transmitted through the materials, and the image is that is collected is simply the spatial variation in the number of transmitted electrons. The sample preparation for SEM is simply to ensure that the material is conductive (perhaps requiring a coating). In TEM, the sample needs to be made transparent to the electron beam, which occurs for very thin samples.

25.8 **(a) Steps in solutions synthesis of nanoparticles?** The species must first be solvated; then stable nuclei of nanometer dimensions must be formed and finally; growth of particles to the final desired size will occur.
(b) Why should the last two steps occur independently? The last two steps should be independent so that nucleation fixes the total number of particles and growth leads to a controlled size and a narrow size distribution. **(c) Stabilizers?** These molecules prevent surface oxidation and aggregation, and they limit traps for the holes and electrons, thus promoting improved quantum yields and luminescence.

25.9 **Vapor-phase versus solution-based techniques?** (a) Vapor-phase techniques typically lead to large sizes because it is more difficult to control the nucleation/growth regimes and because temperatures are typically higher. Because it is difficult to add stabilizers/surfactants to the vapor phase, agglomeration and excessive growth during nucleation often occur. (b) Vapor-phase synthesis tends to lead to more agglomeration. Without stabilizers, the particles tend to agglomerate into soft agglomerates. If the temperature is high, these agglomerates can sinter into hard agglomerates.

25.10 **(a) What is a core-shell nanoparticle?** A schematic of the core and shell of the nanoparticle is shown to the right. **(b) How are they made?** In both cases, one could cause nucleation in one solution and then grow the particle in another solution. **(c) Purpose?** In biosensing, the dielectric property of the shell can control the surface plasmon of the core. The shell can be affected by the environment. In drug delivery, the shell could react with a specific location and the core could be used as a treatment (drug).

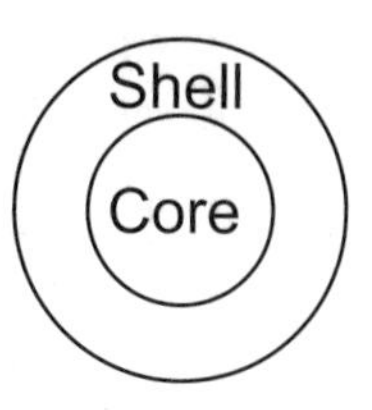

25.11 **(a) Homogeneous versus heterogeneous?** Homogeneous nucleation leads to solid formation throughout the vapour phase. In heterogeneous nucleation, solid formation occurs only at specific locations such as a preexisting solid surface. **(b) thin film?** Heterogeneous nucleation is preferred in thin-film growth such that the solid forms a film on the solid substrate. **(c) nanoparticles?** Homogeneous nucleation is preferred for nanoparticle synthesis (unless it is shell formation, where heterogeneous nucleation on the core is preferred).

25.12 **PVD versus CVD?** In chemical vapour deposition, the atomic species of interest are bound chemically to other species; they are therefore molecular and are stable chemical species. Also, their energies are typically rather low (thermal energy only). In physical vapour deposition, the atomic species of interest are typically atomic (or polyatomic), (poly)ionic, radicals, and clusters; they are therefore somewhat unstable—the atomic species are most stable. Also, their energies can range from thermal to plasma energies.

25.13 **(a) Purpose of QD layers?** Multiple layers of quantum dots can, if they do not interact, simply increase the intensity of any optical absorption or emission. They can also, if they do interact, be used to form quantum cascade lasers. **(b) Limitations?** The limitations come from the requirements on how coherent the interface between the two materials must be and on how easy it is to grow flat layers of one and QDs of the other. In general, it is somewhat strict.

25.14 **(a) Applications of quatum wells?** Quantum wells are used in lasers and optical sensors. **(b) Why are they used over other materials?** Quantum wells are used because they exhibit properties that are not observed in molecular or traditional solid state materials. **(c) How are they made?** Quantum wells are typically made using molecular beam epitaxy as described in Section 24.7.

25.15 **Superlattices and improved properties?** By developing superlattices of two materials, such as AlN and TiN, that have different elastic constants and that have a large number of interfaces spaced on the nanoscale, much improved mechanical properties (hardness values) can be realized—better than those of either of the two parent materials.

25.16 **(a) Self-assembly? (a)** Self-assembly offers methods to bridge bottom-up and top-down approaches to synthesis of nanomaterials. **(b) In nanotechnology?** Self-assembly offers a route to assemble nanosized particles into macroscopic structures.

25.17 **Common features of self-assembly?** Looking at the definitions given in Section 25.6, the common features are: molecular or nanoscale subunits; spontaneous assembly of the subunits; noncovalent interactions between the assembled subunits; and longer-range structures arising from the assembly process.

25.18 **Static versus dynamic self-assembly?** Static self-assembly is when a system self-assembles to a stable state, either a local or global equilibrium. Dynamic self-assembly is when the system is oscillating between states and is dissipating energy in the process. Liquid crystal self-assembly is a static process. An oscillating chemical reaction is an example of a dynamic process.

25.19 **Compare SAMs and cell membranes?** Surfactants are molecules that contain both hydrophilic and hydrophobic moieties that influence materials architectures. Our cell membranes are composed of phospholipid bilayers that contain anionic, hydrophilic phosphate polar head groups, and twin hydrophobic, hydrocarbon tails. Like the Au-organothiol films, these self-assemble owing to specific hydrophobic and hydrophilic interactions. In the SAM, Au–S interactions and dispersion forces between hydrocarbon tails mediate self-assembly and film stability.

25.20 **What is morphosynthesis?** Morposynthesis is the control of nanoarchitectures in inorganic materials through changes in synthesis parameters. Ethylenediamine (en) has been used to control dimensionality and achieve ZnS nanowires.

25.21 **(a) What are the two classes of inorganic-organic nanocomposites?** Class I corresponds to hybrid materials where no covalent or ionic bonds are present. In class II, at least some of the components are linked through strong chemical bonds. **(b) Examples?** An example of class I is block copolymers. An example of class II is polymer/clay nanocomposites.

25.22 **(a) Why is dispersion important in nanocomposites?** Highly dispersed inorganic nanoparticles lead to increased exposed surface areas that afford control over the interfaces between the inorganic and organic components, thus allowing for the tuning of materials properties including stress and strain. **(b) Why is dispersion difficult?** The often nonpolar organic polymers do not have strong interactions with the polar or ionic inorganic components, which leads to problems with inhomogeneity and materials failure.

25.23 **A bionanomaterial and its application?** PPF/PPF-DA is an injectable bionanomaterial used for bone-tissue engineering. See Section 24.16 for other examples.

25.24. **(a) Biomimetics?** Biomimetics for nanotechnology involves designing nanomaterials that mimic biological systems. **(b) example of biomimetics?** Cellulose fibers in paper have been used to template the growth of titanium oxide nanotubes.

25.25 **Bionanocomposites and improved mechanical strength?** To achieve biomimetics you need trabecular and cortical bone tissues, which have different mechanical properties that combine compressibility and tensile strength. An understanding of the chemistry that limits composite interactions is crucial in the development of artificial bone. More specifically, hybrid alumoxane nanoparticles dispersed within PPF/PPF-DA have shown enhanced strength.

Chapter 26 Catalysis

There are many chemical parallels between homogeneous and heterogeneous catalysis, despite the different terminology used to describe them and the different physical methods used to study them. For example, soluble rhodium complexes and rhodium metal both react with hydrogen. In the case of the soluble complex, we call the reaction an oxidative addition; in the case of rhodium metal, we call the reaction a chemisorption. In both cases, however, reactive Rh–H bonds are formed.

H
Ph_3P Rh H
Cl PPh_3
Sol

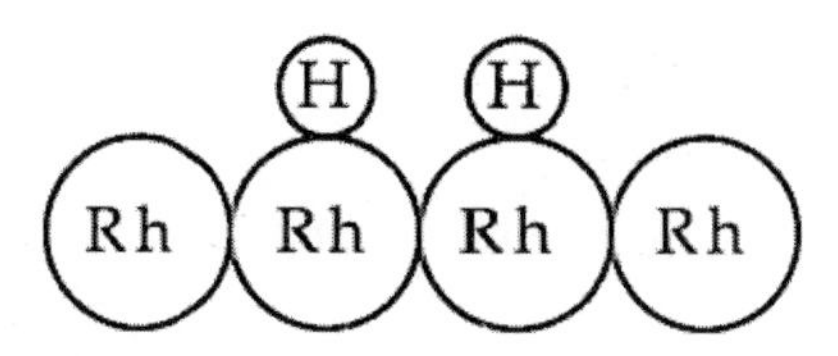

S26.1 **The effect of added phosphine on the catalytic activity of $RhH(CO)(PPh_3)_3$?** The compound $RhH(CO)(PPh_3)_3$ is an 18-electron complex that must lose a phosphine ligand before it can enter the catalytic cycle:

$$\underset{(18\ e^-)}{RhH(CO)(PPh_3)_3} \rightleftharpoons \underset{(16\ e^-)}{RhH(CO)(PPh_3)_2} + PPh_3$$

The coordinatively unsaturated complex $RhH(CO)(PPh_3)_2$ can add the alkene that is to be hydroformylated. Added phosphine will shift the above equilibrium to the left, resulting in a lower concentration of the catalytically active 16-electron complex. Thus, you can predict that the rate of hydroformylation will be decreased by added phosphine.

S26.2 **γ-Alumina heated to 900°C, cooled, and exposed to pyridine vapor?** Heating alumina to 900°C results in complete dehydroxylation. Therefore, only Lewis acid sites are present in γ-alumina heated to 900°C. The IR spectrum of a sample of dehydroxylated γ-alumina exposed to pyridine will exhibit bands near 1465 cm^{-1}. It will not exhibit bands near 1540 cm^{-1}, since these are due to pyridine that is hydrogen bonded to the surface, and no surface OH groups are present in dehydroxylated alumina.

S26.3 **Would a pure silica analog of ZSM-5 be an active catalyst for benzene alkylation (see Figure 26.24)?** A pure silica analog of ZSM-5 would contain Si–OH groups, which are only moderate Brønsted acids, and not the strongly acidic Al–OH_2 groups found in aluminosilicates. Only a strong Brønsted acid such as an Al–OH_2 group in an aluminosilicate can protonate an alkene to form the carbocations that are necessary intermediates in benzene alkylation. Therefore, a pure silica analog of ZSM-5 would not be an active catalyst for benzene alkylation.

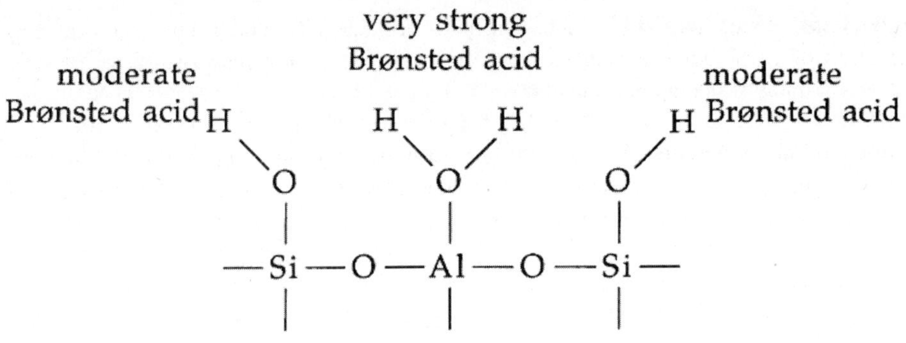

very strong Brønsted acid; moderate Brønsted acid; moderate Brønsted acid

(H–O, H–O–H, O–H groups attached to –Si–O–Al–O–Si– framework)

S26.4 Demonstate that the polymerization of propene with a simple Cp_2ZrCl_2 catalyst would give rise to atactic polypropene? Polymerization of mono-substituted alkenes introduces sterogenic centres along the carbon chain at every other position. Without R groups attached to the Zr centre there is no preference for specific binding of new alkenes during polymerization and thus the repeat is random or atactic (shown below).

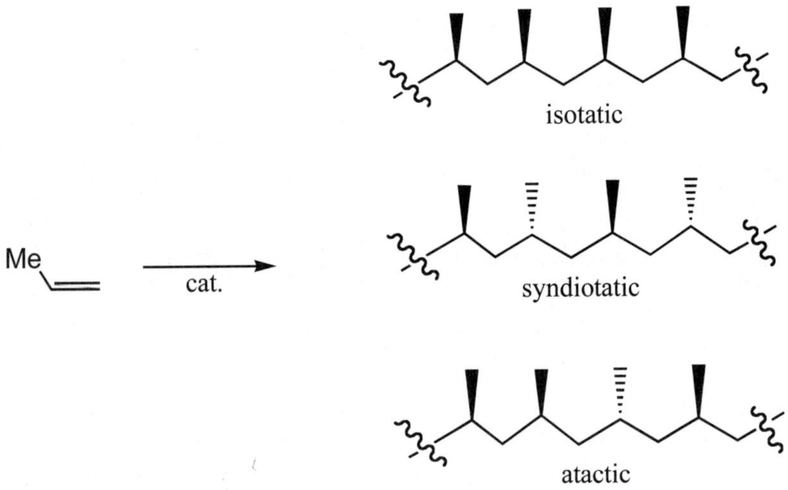

Me (propene) —cat.→ isotatic; syndiotatic; atactic

As you can see, the atactic is the most random, meaning that nothing is driving the stereochemistry of the polymer. Cp_2ZrCl_2 has simple cyclopentadienes as ancillary ligands, which generally do not have enough steric bulk to drive a specific isomer; therefore primarily atactic polypropene is produced. Placing bulking alkyl groups on the Cp rings offers one way to obtain a specific geometry and thus mediate the materials properties of polypropene.

26.1 Which of the following constitute catalysis? (a) H_2 and C_2H_4 in contact with Pt? This is an example of genuine catalysis. The formation of ethane ($H_2 + C_2H_4 \rightarrow C_2H_6$) has a very high activation barrier. The presence of platinum causes the reaction to proceed at a useful rate. Furthermore, the platinum can be recovered unchanged after many turnovers, so it fits the two criteria of a catalyst, a substance that increases the rate of a reaction but is not itself consumed.

(b) H_2 plus O_2 plus an electrical arc? This gas mixture will be completely converted into H_2O once the electrical arc is struck. Nevertheless, it does not constitute an example of catalysis. The arc provides activation energy to initiate the reaction. Once the reaction is started, the heat liberated by the reaction provides enough energy to sustain the reaction. The activation energy of the reaction has *not* been lowered by an added substance, so catalysis has not occurred.

(c) The production of Li_3N and its reaction with H_2O? This is not an example of catalysis. Lithium and nitrogen are both consumed in the formation of Li_3N, which occurs at an appreciable rate even at room temperature. Water and Li_3N react rapidly to produce NH_3 and LiOH at room temperature. As in the formation of Li_3N, both substances are consumed. Since both reactions have intrinsically low activation energies, a catalyst is not necessary.

26.2 **Define the following terms? (a) Turnover frequency?** This is defined differently for homogeneous and heterogeneous catalysis. In both cases, however, it is really the same thing, the *amount* of product formed per unit time per unit *amount* of catalyst. In homogeneous catalysis, the turnover frequency is the rate of formation of product, given in mol L^{-1} s^{-1}, divided by the concentration of catalyst used, in mol L^{-1}. This gives the turnover frequency in units of s^{-1}. In heterogeneous catalysis, the turnover frequency is typically the amount of product formed per unit time, given in mol s^{-1}, divided by the number of moles of catalyst present. In this case, the turnover frequency also has units of s^{-1}. Since one mole of a finely divided heterogeneous catalyst is more active than one mole of the same catalyst with a small surface area, the turnover frequency is sometimes expressed as the amount of product formed per unit time divided by the surface area of the catalyst. This gives the turnover frequency in units of mol s^{-1} cm^{-2}. Often one finds the turnover frequency for commercial heterogeneous catalysts expressed in rate per gram of catalyst.

(b) Selectivity? This is a measure of how much of the desired product is formed relative to undesired by-products. Unlike enzymes (see Chapter 27), man-made catalysts rarely are 100% selective. The catalytic chemist usually has to deal with the often difficult problem of separating the various products. The separations, by distillation, fractional crystallization, or chromatography, are generally expensive and are always time-consuming. Furthermore, the by-products represent a waste of raw materials. Recall that an expensive rhodium hydroformylation catalyst is sometimes used industrially instead of a relatively inexpensive cobalt catalyst because the rhodium catalyst is more selective.

(c) Catalyst? The first line of Chapter 26 reads, "A catalyst is a substance that increases the rate of a reaction but is not itself consumed." This does not mean that the "catalyst" that is added to the reaction mixture is left unchanged during the course of the reaction. Frequently, the substance that is added is a *catalyst precursor* that is transformed under the reaction conditions into the active catalytic species.

(d) Catalytic cycle? This is a sequence of chemical reactions, each involving the catalyst, that transform the reactants into products. It is called a cycle because the actual catalytic species involved in the first step is regenerated during the last step. Note that the concept of a "first" and "last" step may lose its meaning once the cycle is started.

(e) Catalyst support? In cases where a heterogeneous catalyst does not remain a finely divided pure substance with a large surface area under the reaction conditions, it must be dispersed on a support material, which is generally a ceramic like γ-alumina or silica gel. In some cases, the support is relatively inert and only serves to maintain the integrity of the small catalyst particles. In other cases, the support interacts strongly with the catalyst and may affect the rate and selectivity of the reaction.

26.3 **Classify the following as homogeneous or heterogeneous catalysis? (a) The increased rate of SO_2 oxidation in the presence of NO?** The balanced equation for SO_2 oxidation is:

$$2SO_2(g) + O_2(g) \longrightarrow 2SO_3(g)$$

All of these substances, as well as the catalyst NO, are gases. Since they are all present in the same phase, this is an example of homogeneous catalysis.

(b) The hydrogenation of oil using a finely divided Ni catalyst? In this case, the balanced equation is:

$$\text{RHC=CHR}' + H_2 \longrightarrow \text{RH}_2\text{C-CH}_2\text{R}'$$

The reactants and the products are all present in the liquid phase (the hydrogen is dissolved in the liquid oil), but the catalyst is a solid. Therefore, this is an example of heterogeneous catalysis.

(c) The conversion of D-glucose to a D,L mixture by HCl? The catalyst for the racemization of D-glucose is HCl (really H_3O^+), which is present in the same aqueous phase as the D-glucose. Therefore, since the substrate and the catalyst are both in the same phase, this is homogeneous catalysis.

$$\begin{array}{c}CHO\\ |\\ H—C—OH\\ |\\ HO—C—H\\ |\\ H—C—OH\\ |\\ H—C—OH\\ |\\ CH_2OH\end{array} \quad \underset{}{\overset{H_3O^+}{\rightleftharpoons}} \quad \begin{array}{c}CHO\\ |\\ H—C—OH\\ |\\ HO—C—H\\ |\\ H—C—OH\\ |\\ HO—C—H\\ |\\ CH_2OH\end{array}$$

D–Glucose L–Glucose

26.4 **Which of the following processes would be worth investigating? (a) The splitting of H_2O into H_2 and O_2?** A catalyst does not affect the free energy (ΔG) of a reaction, only the activation free energy ($\Delta G^{\ddagger}$). A thermodynamically unfavorable reaction (one with a positive ΔG) will not result in a useful amount of products unless energy in the form of light or electric current is added to the reaction mixture. For example, if $\Delta G \approx +17$ kJ mol^{-1}, $K_{eq} \approx 10^{-3}$. Therefore, it would not be a worthwhile endeavour to try to develop a catalyst to split water into hydrogen and oxygen because water is a thermodynamically stable compound (i.e., it is stable with respect to its elements; $\Delta_f G°(H_2O) = -237$ kJ mol^{-1}).

(b) The decomposition of CO_2 into C and O_2? As in part (a), you would be trying to catalyze the decomposition of a very stable compound into its constituent elements. It would be a waste of time.

$$CO_2(g) \longrightarrow C(s) + O_2(g) \qquad \Delta G° = +394 \text{ kJ/mol}$$

(c) The combination of N_2 with H_2 to produce NH_3? This would be a very worthwhile reaction to try to catalyze efficiently at 80°C. Ammonia is a stable compound with respect to nitrogen and hydrogen. Furthermore, it is a compound that is important in commerce, since it is used in many types of fertilizers. $\Delta G°$ is only –16.5 kJ mol^{-1} for the reaction $N_2 + 3H_2 \rightarrow 2NH_3$. The high temperatures usually required for ammonia synthesis (~400°C) make ΔG less negative and result in a smaller yield. An efficient *low temperature* synthesis of ammonia from nitrogen and hydrogen that could be carried out on a large scale would probably make you and your industrialist rich.

(d) The hydrogenation of double bonds in vegetable oil? The reaction $RHC{=}CHR + H_2 \rightarrow RH_2C{-}CH_2R$ has a negative $\Delta G°$. Therefore, the hydrogenation of vegetable oil would be a candidate for catalyst development. However, there are many homogeneous and heterogeneous catalysts for olefin hydrogenation, including those that operate at low temperatures (i.e., ~80°C). For this reason, the process can be readily set up with existing technology.

26.5 **Why does the addition of PPh_3 to $RhCl(PPh_3)_3$ reduce the hydrogenation turnover frequency?** The catalytic cycle for homogeneous hydrogenation of alkenes by Wilkinson's catalyst, $RhCl(PPh_3)_3$, is shown in Figure 26.5. Let us focus on the dominant path. The catalytic species that enters the cycle is not $RhCl(PPh_3)_3$ but $RhCl(PPh_3)_2(Sol)$ (Sol = a solvent molecule), formed by the following equilibrium:

$$RhCl(PPh_3)_3 + \text{Sol.} \rightleftharpoons RhCl(PPh_3)_2(\text{Sol}) + PPh_3$$

Therefore, added PPh_3 will shift this equilibrium to the left, resulting in a lower concentration of the active catalytic species. Note that this is the only reaction that added PPh_3 could affect, since no other step involves free PPh_3.

26.6 Explain the trend in rates of H_2 absorption by various olefins catalyzed by $RhCl(PPh_3)_3$? The data show that hydrogenation is faster for hexene than for *cis*-4-methyl-2-pentene (a factor of 2910/990 ≈ 3). It is also faster for cyclohexene than for 1-methylcyclohexene (a factor of 3160/60 ≈ 53). In both cases, the alkene that is hydrogenated more slowly has a greater degree of substitution and so is sterically more demanding: hexene is a monosubstituted alkene while *cis*-4-methyl-2-pentene is a disubstituted alkene; cyclohexene is a disubstituted alkene, while 1-methylcyclohexene is a trisubstituted alkene. According to the catalytic cycle for hydrogenation shown in Figure 26.5, different alkenes could affect the equilibrium (b) → (c) or the reaction (c) → (d). A sterically more demanding alkene could result in (i) a smaller equilibrium concentration of (c) ($RhClH_2L_2$(alkene)), or (ii) a slower rate of conversion of (c) to (d) ($RhClHL_2$(alkyl)(Sol)). Since the reaction (c) → (d) is the rate determining step for hydrogenation by $RhCl(PPh_3)_3$, either effect would lower the rate of hydrogenation.

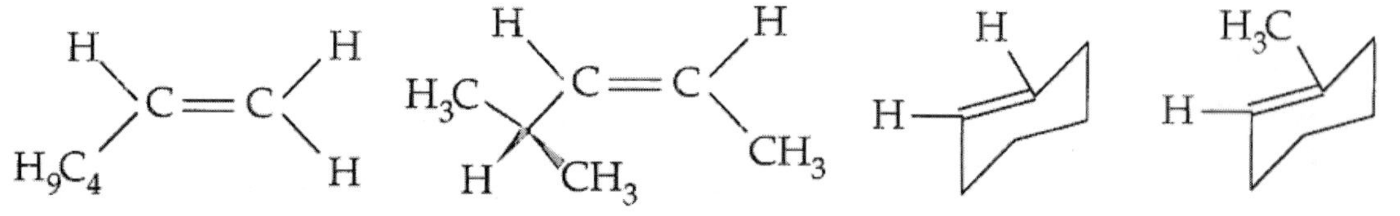

2910 L mol^{-1} s^{-1} 990 L mol^{-1} s^{-1} 3160 L mol^{-1} s^{-1} 60 L mol^{-1} s^{-1}

26.7 Hydroformylation catalysis with and without added $P(n\text{-}Bu)_3$? Since compound (E) in Figure 26.8 is observed under catalytic conditions, its formation must be faster than its transformation into products. If this were not true, a spectroscopically observable amount of it would not build up. Therefore, the transformation of (E) into $CoH(CO)_4$ must be the rate-determining step in the absence of added $P(n\text{-}Bu)_3$. In the presence of added $P(n\text{-}Bu)_3$, neither compound (E) nor its phosphine-substituted equivalent $Co(C(=O)C_4H_9)\text{–}(CO)_3(P(n\text{–}Bu)_3)$ is observed, requiring that their formation must be slower than their transformation into products. Thus, in the presence of added $P(n\text{-}Bu)_3$, the formation of either (A) or (E), or their phosphine-substituted equivalents is the rate-determining step.

26.8 How does starting with MeCOOMe instead of MeOH lead to ethanoic anhydride instead of ethanoic acid using the Monsanto acetic acid process? The key difference here is that iodide ions produce MeI and $MeCO_2$. The reaction of the ethanoate ion with the acetyl iodide (produced at the end of the catalytic cycle) leads to ethanoic anhydride. See the catalytic cycle in Figure 26.13.

26.9 Suggest a reason why? (a) Ring opening alkene metathesis polymerization (ROMP) proceeds? ROMP can result in reduced steric strain, thereby providing a thermodynamic driving force for the reaction.

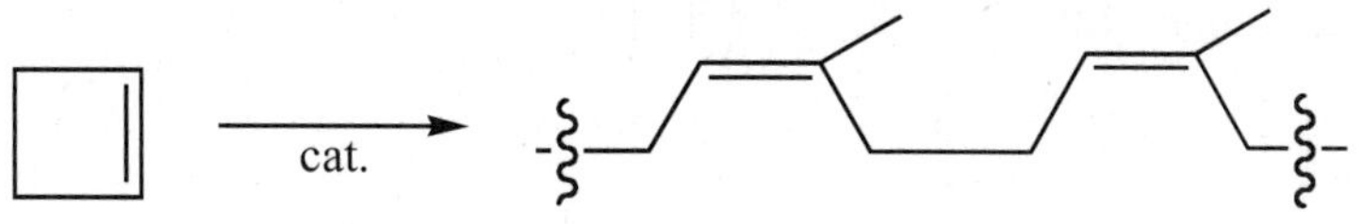

cat.

(b) Ring-closing metathesis (RCM) reaction proceeds? RCM results in the loss of ethane, and by removing this gas, the position of equilibrium can easily be driven to favour the ring product.

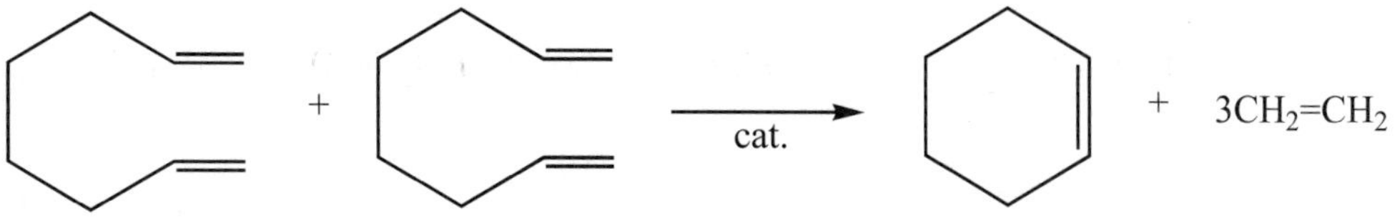

+ cat. + $3CH_2{=}CH_2$

26.10 **(a) Attack by dissolved hydroxide?** The stereochemistry of this compound would be the same as structure C in Figure 26.11 **(b) Attack by coordinated hydroxide?** The stereochemistry of this compound would match that of structure E given in Figure 26.11 **(c) Can one differentiate the stereochemistry?** Yes. Detailed stereochemical studies indicate that the hydration of the alkene-Pd(II) complexes occurs by attack of water from the solution on the coordinated ethene rather than the insertion of coordinated OH.

26.11 **(a) Enhanced acidity?** When Al^{3+} replaces Si^{4+} on lattice site charge is balanced by H_3O^+ increasing the acidity of the solid catalyst.
(b) Three other ions? Other 3+ ions including Ga^{3+}, Co^{3+}, and Fe^{3+} will have the same effect as Al^{3+}.

26.12 **Why is the platinum-rhodium in automobile catalytic converters dispersed on the surface of a ceramic rather than used in the form of thin foil?** Special measures are required in heterogeneous catalysis to ensure that the reactants achieve contact with catalytic sites. For a given amount of catalyst, as the surface area increases, there are a greater number of catalytic sites. A thin foil of platinum-rhodium will not have as much surface area as an equal amount of small particles finely dispersed on the surface of a ceramic support. The diagram below shows this for a "foil" of catalyst that is 1000 times larger in two dimensions than in the third dimension (i.e., the thickness is 1; note that the diagram is not to scale). If the same amount of catalyst is broken into cubes that are 1 unit on a side, the surface area of the catalyst is increased by nearly a factor of three.

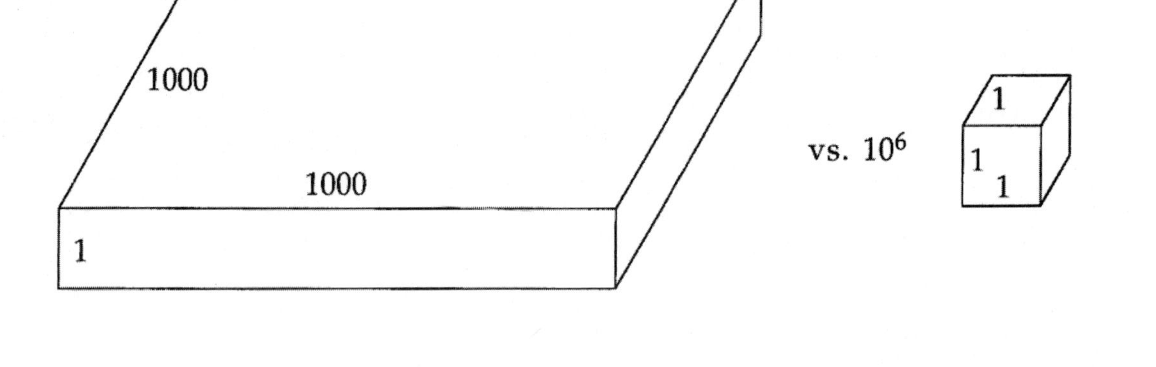

total area = 2,004,000 total area = 6,000,000

26.13 **Devise a plausible mechanism to explain the deuteration of 3,3-dimethylpentane?** There are *two* observations that must be explained. The first is that ethyl groups are deuterated before methyl groups. The second observation is that a given ethyl group is completely deuterated before another one incorporates *any* deuterium. Let's consider the first observation first. The mechanism of deuterium exchange is probably related to the reverse of the last two reactions in Figure 26.20, which shows a schematic mechanism for the hydrogenation of an olefin by D_2. The steps necessary for deuterium substitution into an alkane are shown below, and include the dissociative chemisorption of an R–H bond, the dissociative chemisorption of D_2, and the dissociation of R–D. This can occur many times with the same alkane molecule to effect complete deuteration.

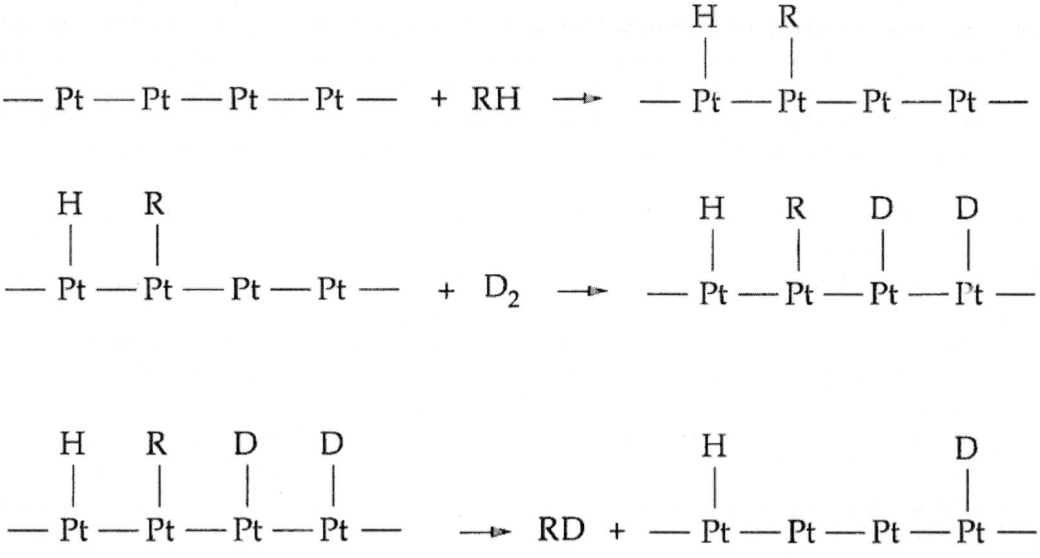

Since the $-CH_2CH_3$ groups are deuterated before the $-CH_3$ groups, one possibility is that dissociative chemisorption of a C–H bond from a $-CH_2-$ group is faster than dissociative chemisorption of a C–H bond from a $-CH_3$ group. The second observation can be explained by invoking a mechanism for rapid deuterium exchange of the methyl group in the chemisorbed $-CHR(CH_3)$ group ($R = C(CH_3)_2(C_2H_5)$). The scheme below shows such a mechanism. It involves the successive application of the equilibrium shown in Figure 26.20. If this equilibrium is maintained more rapidly than the dissociation of the alkane from the metal surface, the methyl group in question will be completely deuterated before dissociation takes place.

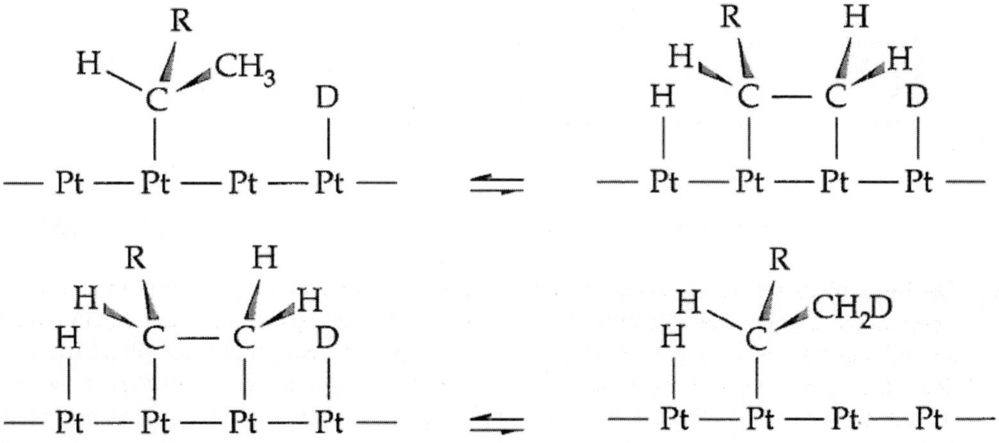

Another possibility, not shown in the scheme above, is that the terminal $-CH_3$ groups undergo more rapid dissociative chemisorption than the sterically more hindered internal $-CH_3$ groups.

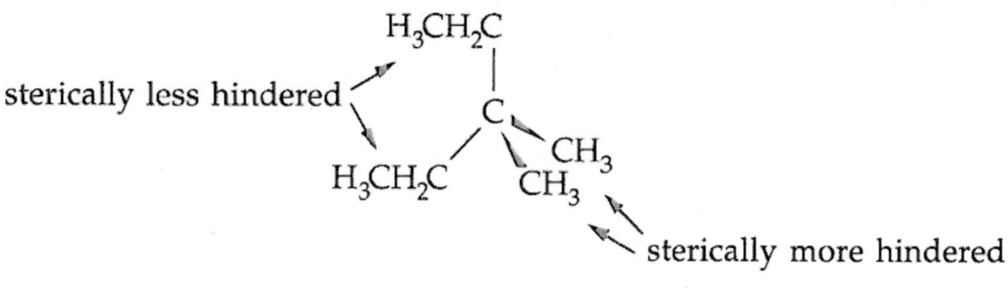

sterically less hindered

sterically more hindered

26.14 Why does CO decrease the effectiveness of Pt in catalyzing the reaction $2H^+(aq) + 2e^- \rightarrow H_2(g)$? The reduction of hydrogen ions to H_2 probably involves the formation of surface hydride species similar to the ones shown in the figure on the opening page of this chapter of the *Solutions Manual*. Dissociation of H_2 by reductive elimination would complete the catalytic cycle. According to Figure 26.28, platinum not only has a strong tendency to chemisorb H_2, but it also has a strong tendency to chemisorb CO. If the surface of platinum is covered with CO, the number of catalytic sites available for H^+ reduction will be greatly diminished and the rate of H_2 production will decrease.

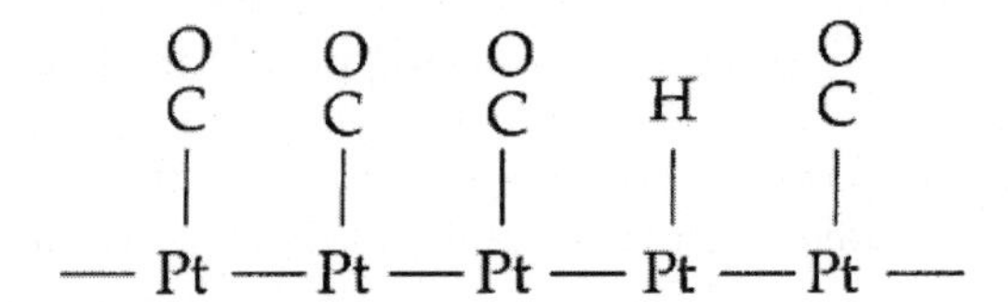

26.15 Describe the role of an electrocatalyst? While platinum is the most efficient electrocatalyst for accelerating oxygen reduction at the fuel cell cathode, it is expensive (read section 26.17 Electrocatalysis). Current research is focused on the efficiency of a platinum monolayer by placing it on a stable metal or alloy clusters, your book mentions the use of the alloy Pt_3N. An example would be a platinum monolayer fuel-cell anode electrocatalyst, which consists of ruthenium nanoparticles with a sub-monolayer of platinum. Other areas of research include using tethered metalloporphyrin complexes for oxygen activation and subsequent reduction.

Chapter 27 Biological Inorganic Chemistry

Bioinorganic chemistry is concerned with the roles of metal ions in biological systems. Many of the spectroscopic tools that inorganic chemists use to study the kinds of complexes discussed in earlier chapters can be used to study metalloproteins and other metal-containing biomolecules. However, the structure of the active sites in proteins cannot be determined very accurately using X-ray diffraction. Therefore, bioinorganic chemists often study the structures of low molecular weight model complexes and make the following inference: if there is a spectral congruence between the model and the protein, it is assumed that there is a structural congruence as well. Tris(pyrazolyl)borate, commonly called Tp, is a monoanionic ligand (boron bears a negative charge) that donates six electrons to the metal center. Tp complexes are quite versatile and a wide variety of Tp derivatives with differing steric and electronic properties have been synthesized, many of which, when coordinated to zinc, model zinc metalloproteins such as carbonic anhydrase both structurally and chemically.

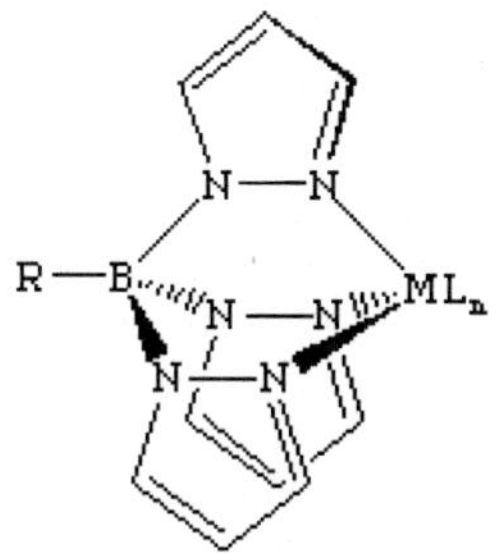

S27.1 Is Iron (II) expected to be present in the cell as uncomplexed ions? Uncomplexed Fe^{2+} is present at low concentrations ($\geq 10^{-7}$ M). Fe^{2+} is low in the Irving-Williams series. By contrast, Fe^{3+} is strongly complexed (relative to Fe^{2+}) by highly specific ligands such as ferritin and by smaller polyanionic ligands, particularly citrate. Free Fe^{2+} is therefore easily oxidized.

S27.2 Unusual coordination of Mg? The protein's 3D structure can place any particular atom in a suitable position for axial coordination and prevent water molecules (which would be the natural choice for Mg^{2+}, a hard cation) from occupying that site. For a refresher on hard/soft acid/base theory see Section 4.12.

S27.3 Why does saline contain NaCl? The need for osmotic balance requires that the saline fluid being administered have the same concentration of solutes as blood plasma. Blood plasma (the external medium with respect to red blood cells) contains high levels of Na^+ and Cl^-, but low levels of K^+ and other anions: use of NaCl rather than another salt is therefore essential in order to maintain the electrical potential.

S27.4 Explain the significance of the Calcium ion pumps activation by calmodulin? Calmodulin does not bind to the pump unless Ca^{2+} is coordinated. The binding of calmodulin is thus a signal informing the pump that the cytoplasmic Ca^{2+} level has risen above a certain level.

S27.5 Why are iron-porphyrin complexes unable to bind O_2 reversibly? Dioxygen is a highly oxidizing ligand and without steric protection; the Fe(II) gets oxidized to Fe(III), yielding an oxo-Iron(III) porphyrin complex. Generally this oxidation yields a dimeric Fe(III) porphyrin complex with a bridging oxygen atom, shown below.

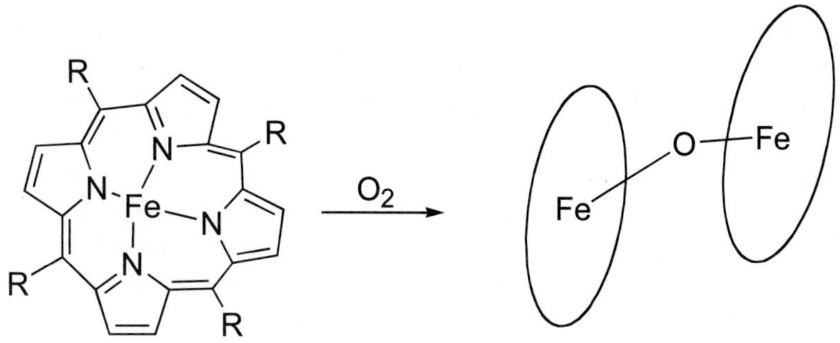

μ-oxo porphyrin dimer

S27.6 **What is the nature of binding at Cu blue centers as indicated by the EPR spectrum?** There is greater covalence in blue Cu centres than in simple Cu(II) compounds. The unpaired electron is delocalized onto the cysteine sulphur and spends more time away from the Cu nucleus. In general, late, low-valent transition metals have more covalent character. In most of the Cu blue centres, the reduced form is Cu(I), which is a soft acid, so it makes sense that it bonds well with sulfur-containing ligands such as a cysteine residue.

S27.7 **What is the nature of an active site with Copper (III)?** Cu(III) is d^8. It is likely to be highly oxidizing, and probably diamagnetic with square-planar geometry. Recall from Section 7.4 that square planar complexes are favoured for d^8 transition metals.

S27.8 **Why mercury is so toxic because of the action of enzymes containing cobalamin?** Species such as CH_3Hg^+ and $(CH_3)_2Hg$ are hydrophobic and can penetrate cell membranes. Cobalamins are very active methyl transfer reactions, which can methylate anything in the cell, which is bad news for the human body.

S27.9 **Suggest experiments that could establish the structure of the MoFe cofactor?** Spectroscopic measurements that are metal specific, such as EPR, which detects the number of unpaired electrons, could be used on both enzyme and cofactor. Also, single-crystal X-ray diffraction and EXAFS would reveal bond distances and angles between Fe or Mo and the sulphur ligands; both these techniques can be carried out on the enzyme and cofactor dissolved in DMF. Of all of these, single-crystal X-ray diffraction is our most powerful technique for determining structure.

S27.10 **Why might Cu sensors be designed to bind Cu(I) rather than Cu(II)?** Cu(I) has an almost unique ability to undergo linear coordination by sulphur-containing ligands. The only other metals with this property are Ag(I), Au(I), and Hg(II), but these are not common in biology. Binding as Cu(II) would be less specific because it adopts more common geometries. Also, the cell redox potential is quite low, which favours the presence of Cu(I) rather than Cu(II).

27.1 **Lanthanides versus calcium?** Calcium-binding proteins can be studied using lanthanide ions (Ln^{3+}) because, like calcium ions, they are hard Lewis acids and prefer coordination by hard bases such as anionic oxygen-containing ligands like carboxylates. For more on hard/soft acid/base theory, see Section 4.12. The larger size of lanthanides is compensated for by their higher charge, although they are likely to have higher coordination numbers. Many lanthanide ions have useful electronic and magnetic properties. For example, Gd^{3+} has excellent fluorescence, which is quenched when it binds close to certain amino acids such as trptophan.

27.2 **Substituting Co^{2+} for Zn^{2+}?** Co(II) commonly adopts distorted tetrahedral and five-coordinate geometries typical of Zn(II) in enzymes, and when substituted for Zn(II) in the parent enzyme, it generally retains catalytic activity. Zn(II) is d^{10} and therefore colorless, however, Co(II) is d^7, and its peaks in the UV-Vis spectrum are quite intense and report on the structure and ligand-binding properties of the native Zn sites. Furthermore, it is almost impossible to use NMR to study the active site of zinc enzymes because of all of the interferences and overlaps one gets with the amino acids themselves. Fortunately, Co(II) is paramagnetic, enabling Zn enzymes to be studied by EPR.

27.3 **Compare the acid/base catalytic activities of Zn(II), Fe(III), and Mg(II)?** Acid strengths (coordinated water molecules) lie in the order Fe(III) > Zn(II) > Mg(II); see Section 4.4. Ligand binding rates are Mg(II) > Zn(II) > Fe(III). Mg(II) is usually six-coordinate, so it can accommodate more complex reactions (such as rubisco).

27.4 **Propose a physical method for the determination of Fe(V)?** Since Iron 5+ theoretically has the electron configuration of $[Ar]3d^3$; this means the ion would have 3 unpaired electrons. The logical chose would be either EPR (electron paramagnetic resonance) or Mossbauer spectroscopy. See Chapter 8, Physical techniques in inorganic chemistry, for more discussion on both of these instruments.

27.5 **Interpret the Mossbauer spectra of ferredoxin?** Ferredoxins contain FeS clusters, see Figure 27.26. In the spectrum, the oxidized spectrum is consistent with the iron atoms having essentially all the same valence, intermediate between the 3+ and the 2+ states, with one pair spin-up and the other pair spin-down. Thus two peaks occur. The reduction spectrum is indicative of an electron going to one pair of iron atoms, which become Fe^{2+}, with the other iron atoms remaining substantially unchanged.

27.6 **Explain the differences in the structures of the oxidized and reduced forms of the P-cluster in nitrogenase?** The structural changes accompanying oxidation of the P-cluster raise the possibility that the P-cluster may be involved in coupling electron and proton transfer in nitrogenase. The oxidized form of the P-cluster is accompanied by coordination of a serotonin and the amide nitrogen of a cystine residue to the clustered Fe atoms. Since both of these ligands will be protonated in their free states and may be deprotanated in their bond states, this raises the possibility that two-electron oxidation of the P-cluster simultaneously releases two protons. Transfer of electrons and protons to the FeMo-cofactor active site of nitrogenase needs to be synchronized, the change in structure suggest that the coupling of proton and electron transfer can also occur at the P-cluster, by controlling protonation of the exchangeable ligands.

27.7 **What metals are involved in the synthesis of acetyl groups?** Transfer of CH_3 is expected to involve Co(II), which is in the active site of cobalamin. Transfer and activation of CO is expected to involve an electron-rich metal such as the bioactive metals Fe(II) and Cu(I). Most likely the mechanism involves an insertion of a coordinated carbonyl into a metal alkyl bond; see Chapter 22, *d*-Metal organometallic chemistry.